MULTIVALUED LINEAR OPERATORS

PURE AND APPLIED MATHEMATICS

A Program of Monographs, Textbooks, and Lecture Notes

MONOGRAPHS AND TEXTBOOKS IN PURE AND APPLIED MATHEMATICS

1. *K. Yano,* Integral Formulas in Riemannian Geometry (1970)
2. *S. Kobayashi,* Hyperbolic Manifolds and Holomorphic Mappings (1970)
3. *V. S. Vladimirov,* Equations of Mathematical Physics (A. Jeffrey, ed.; A. Littlewood, trans.) (1970)
4. *B. N. Pshenichnyi,* Necessary Conditions for an Extremum (L. Neustadt, translation ed.; K. Makowski, trans.) (1971)
5. *L. Narici et al.,* Functional Analysis and Valuation Theory (1971)
6. *S. S. Passman,* Infinite Group Rings (1971)
7. *L. Dornhoff,* Group Representation Theory. Part A: Ordinary Representation Theory. Part B: Modular Representation Theory (1971, 1972)
8. *W. Boothby and G. L. Weiss, eds.,* Symmetric Spaces (1972)
9. *Y. Matsushima,* Differentiable Manifolds (E. T. Kobayashi, trans.) (1972)
10. *L. E. Ward, Jr.,* Topology (1972)
11. *A. Babakhanian,* Cohomological Methods in Group Theory (1972)
12. *R. Gilmer,* Multiplicative Ideal Theory (1972)
13. *J. Yeh,* Stochastic Processes and the Wiener Integral (1973)
14. *J. Barros-Neto,* Introduction to the Theory of Distributions (1973)
15. *R. Larsen,* Functional Analysis (1973)
16. *K. Yano and S. Ishihara,* Tangent and Cotangent Bundles (1973)
17. *C. Procesi,* Rings with Polynomial Identities (1973)
18. *R. Hermann,* Geometry, Physics, and Systems (1973)
19. *N. R. Wallach,* Harmonic Analysis on Homogeneous Spaces (1973)
20. *J. Dieudonné,* Introduction to the Theory of Formal Groups (1973)
21. *I. Vaisman,* Cohomology and Differential Forms (1973)
22. *B.-Y. Chen,* Geometry of Submanifolds (1973)
23. *M. Marcus,* Finite Dimensional Multilinear Algebra (in two parts) (1973, 1975)
24. *R. Larsen,* Banach Algebras (1973)
25. *R. O. Kujala and A. L. Vitter, eds.,* Value Distribution Theory: Part A; Part B: Deficit and Bezout Estimates by Wilhelm Stoll (1973)
26. *K. B. Stolarsky,* Algebraic Numbers and Diophantine Approximation (1974)
27. *A. R. Magid,* The Separable Galois Theory of Commutative Rings (1974)
28. *B. R. McDonald,* Finite Rings with Identity (1974)
29. *J. Satake,* Linear Algebra (S. Koh et al., trans.) (1975)
30. *J. S. Golan,* Localization of Noncommutative Rings (1975)
31. *G. Klambauer,* Mathematical Analysis (1975)
32. *M. K. Agoston,* Algebraic Topology (1976)
33. *K. R. Goodearl,* Ring Theory (1976)
34. *L. E. Mansfield,* Linear Algebra with Geometric Applications (1976)
35. *N. J. Pullman,* Matrix Theory and Its Applications (1976)
36. *B. R. McDonald,* Geometric Algebra Over Local Rings (1976)
37. *C. W. Groetsch,* Generalized Inverses of Linear Operators (1977)
38. *J. E. Kuczkowski and J. L. Gersting,* Abstract Algebra (1977)
39. *C. O. Christenson and W. L. Voxman,* Aspects of Topology (1977)
40. *M. Nagata,* Field Theory (1977)
41. *R. L. Long,* Algebraic Number Theory (1977)
42. *W. F. Pfeffer,* Integrals and Measures (1977)
43. *R. L. Wheeden and A. Zygmund,* Measure and Integral (1977)
44. *J. H. Curtiss,* Introduction to Functions of a Complex Variable (1978)
45. *K. Hrbacek and T. Jech,* Introduction to Set Theory (1978)
46. *W. S. Massey,* Homology and Cohomology Theory (1978)
47. *M. Marcus,* Introduction to Modern Algebra (1978)
48. *E. C. Young,* Vector and Tensor Analysis (1978)
49. *S. B. Nadler, Jr.,* Hyperspaces of Sets (1978)
50. *S. K. Segal,* Topics in Group Kings (1978)
51. *A. C. M. van Rooij,* Non-Archimedean Functional Analysis (1978)
52. *L. Corwin and R. Szczarba,* Calculus in Vector Spaces (1979)
53. *C. Sadosky,* Interpolation of Operators and Singular Integrals (1979)

54. *J. Cronin,* Differential Equations (1980)
55. *C. W. Groetsch,* Elements of Applicable Functional Analysis (1980)
56. *I. Vaisman,* Foundations of Three-Dimensional Euclidean Geometry (1980)
57. *H. I. Freedan,* Deterministic Mathematical Models in Population Ecology (1980)
58. *S. B. Chae,* Lebesgue Integration (1980)
59. *C. S. Rees et al.,* Theory and Applications of Fourier Analysis (1981)
60. *L. Nachbin,* Introduction to Functional Analysis (R. M. Aron, trans.) (1981)
61. *G. Orzech and M. Orzech,* Plane Algebraic Curves (1981)
62. *R. Johnsonbaugh and W. E. Pfaffenberger,* Foundations of Mathematical Analysis (1981)
63. *W. L. Voxman and R. H. Goetschel,* Advanced Calculus (1981)
64. *L. J. Corwin and R. H. Szczarba,* Multivariable Calculus (1982)
65. *V. I. Istrăţescu,* Introduction to Linear Operator Theory (1981)
66. *R. D. Järvinen,* Finite and Infinite Dimensional Linear Spaces (1981)
67. *J. K. Beem and P. E. Ehrlich,* Global Lorentzian Geometry (1981)
68. *D. L. Armacost,* The Structure of Locally Compact Abelian Groups (1981)
69. *J. W. Brewer and M. K. Smith, eds.,* Emmy Noether: A Tribute (1981)
70. *K. H. Kim,* Boolean Matrix Theory and Applications (1982)
71. *T. W. Wieting,* The Mathematical Theory of Chromatic Plane Ornaments (1982)
72. *D. B.Gauld,* Differential Topology (1982)
73. *R. L. Faber,* Foundations of Euclidean and Non-Euclidean Geometry (1983)
74. *M. Carmeli,* Statistical Theory and Random Matrices (1983)
75. *J. H. Carruth et al.,* The Theory of Topological Semigroups (1983)
76. *R. L. Faber,* Differential Geometry and Relativity Theory (1983)
77. *S. Barnett,* Polynomials and Linear Control Systems (1983)
78. *G. Karpilovsky,* Commutative Group Algebras (1983)
79. *F. Van Oystaeyen and A. Verschoren,* Relative Invariants of Rings (1983)
80. *I. Vaisman,* A First Course in Differential Geometry (1984)
81. *G. W. Swan,* Applications of Optimal Control Theory in Biomedicine (1984)
82. *T. Petrie and J. D. Randall,* Transformation Groups on Manifolds (1984)
83. *K. Goebel and S. Reich,* Uniform Convexity, Hyperbolic Geometry, and Nonexpansive Mappings (1984)
84. *T. Albu and C. Năstăsescu,* Relative Finiteness in Module Theory (1984)
85. *K. Hrbacek and T. Jech,* Introduction to Set Theory: Second Edition (1984)
86. *F. Van Oystaeyen and A. Verschoren,* Relative Invariants of Rings (1984)
87. *B. R. McDonald,* Linear Algebra Over Commutative Rings (1984)
88. *M. Namba,* Geometry of Projective Algebraic Curves (1984)
89. *G. F. Webb,* Theory of Nonlinear Age-Dependent Population Dynamics (1985)
90. *M. R. Bremner et al.,* Tables of Dominant Weight Multiplicities for Representations of Simple Lie Algebras (1985)
91. *A. E. Fekete,* Real Linear Algebra (1985)
92. *S. B. Chae,* Holomorphy and Calculus in Normed Spaces (1985)
93. *A. J. Jerri,* Introduction to Integral Equations with Applications (1985)
94. *G. Karpilovsky,* Projective Representations of Finite Groups (1985)
95. *L. Narici and E. Beckenstein,* Topological Vector Spaces (1985)
96. *J. Weeks,* The Shape of Space (1985)
97. *P. R. Gribik and K. O. Kortanek,* Extremal Methods of Operations Research (1985)
98. *J.-A. Chao and W. A. Woyczynski, eds.,* Probability Theory and Harmonic Analysis (1986)
99. *G. D. Crown et al.,* Abstract Algebra (1986)
100. *J. H. Carruth et al.,* The Theory of Topological Semigroups, Volume 2 (1986)
101. *R. S. Doran and V. A. Belfi,* Characterizations of C*-Algebras (1986)
102. *M. W. Jeter,* Mathematical Programming (1986)
103. *M. Altman,* A Unified Theory of Nonlinear Operator and Evolution Equations with Applications (1986)
104. *A. Verschoren,* Relative Invariants of Sheaves (1987)
105. *R. A. Usmani,* Applied Linear Algebra (1987)
106. *P. Blass and J. Lang,* Zariski Surfaces and Differential Equations in Characteristic $p > 0$ (1987)
107. *J. A. Reneke et al.,* Structured Hereditary Systems (1987)
108. *H. Busemann and B. B. Phadke,* Spaces with Distinguished Geodesics (1987)
109. *R. Harte,* Invertibility and Singularity for Bounded Linear Operators (1988)

110. *G. S. Ladde et al.*, Oscillation Theory of Differential Equations with Deviating Arguments (1987)
111. *L. Dudkin et al.*, Iterative Aggregation Theory (1987)
112. *T. Okubo*, Differential Geometry (1987)
113. *D. L. Stancl and M. L. Stancl*, Real Analysis with Point-Set Topology (1987)
114. *T. C. Gard*, Introduction to Stochastic Differential Equations (1988)
115. *S. S. Abhyankar*, Enumerative Combinatorics of Young Tableaux (1988)
116. *H. Strade and R. Farnsteiner*, Modular Lie Algebras and Their Representations (1988)
117. *J. A. Huckaba*, Commutative Rings with Zero Divisors (1988)
118. *W. D. Wallis*, Combinatorial Designs (1988)
119. *W. Więsław*, Topological Fields (1988)
120. *G. Karpilovsky*, Field Theory (1988)
121. *S. Caenepeel and F. Van Oystaeyen*, Brauer Groups and the Cohomology of Graded Rings (1989)
122. *W. Kozlowski*, Modular Function Spaces (1988)
123. *E. Lowen-Colebunders*, Function Classes of Cauchy Continuous Maps (1989)
124. *M. Pavel*, Fundamentals of Pattern Recognition (1989)
125. *V. Lakshmikantham et al.*, Stability Analysis of Nonlinear Systems (1989)
126. *R. Sivaramakrishnan*, The Classical Theory of Arithmetic Functions (1989)
127. *N. A. Watson*, Parabolic Equations on an Infinite Strip (1989)
128. *K. J. Hastings*, Introduction to the Mathematics of Operations Research (1989)
129. *B. Fine*, Algebraic Theory of the Bianchi Groups (1989)
130. *D. N. Dikranjan et al.*, Topological Groups (1989)
131. *J. C. Morgan II*, Point Set Theory (1990)
132. *P. Biler and A. Witkowski*, Problems in Mathematical Analysis (1990)
133. *H. J. Sussmann*, Nonlinear Controllability and Optimal Control (1990)
134. *J.-P. Florens et al.*, Elements of Bayesian Statistics (1990)
135. *N. Shell*, Topological Fields and Near Valuations (1990)
136. *B. F. Doolin and C. F. Martin*, Introduction to Differential Geometry for Engineers (1990)
137. *S. S. Holland, Jr.*, Applied Analysis by the Hilbert Space Method (1990)
138. *J. Okníński*, Semigroup Algebras (1990)
139. *K. Zhu*, Operator Theory in Function Spaces (1990)
140. *G. B. Price*, An Introduction to Multicomplex Spaces and Functions (1991)
141. *R. B. Darst*, Introduction to Linear Programming (1991)
142. *P. L. Sachdev*, Nonlinear Ordinary Differential Equations and Their Applications (1991)
143. *T. Husain*, Orthogonal Schauder Bases (1991)
144. *J. Foran*, Fundamentals of Real Analysis (1991)
145. *W. C. Brown*, Matrices and Vector Spaces (1991)
146. *M. M. Rao and Z. D. Ren*, Theory of Orlicz Spaces (1991)
147. *J. S. Golan and T. Head*, Modules and the Structures of Rings (1991)
148. *C. Small*, Arithmetic of Finite Fields (1991)
149. *K. Yang*, Complex Algebraic Geometry (1991)
150. *D. G. Hoffman et al.*, Coding Theory (1991)
151. *M. O. González*, Classical Complex Analysis (1992)
152. *M. O. González*, Complex Analysis (1992)
153. *L. W. Baggett*, Functional Analysis (1992)
154. *M. Sniedovich*, Dynamic Programming (1992)
155. *R. P. Agarwal*, Difference Equations and Inequalities (1992)
156. *C. Brezinski*, Biorthogonality and Its Applications to Numerical Analysis (1992)
157. *C. Swartz*, An Introduction to Functional Analysis (1992)
158. *S. B. Nadler, Jr.*, Continuum Theory (1992)
159. *M. A. Al-Gwaiz*, Theory of Distributions (1992)
160. *E. Perry*, Geometry: Axiomatic Developments with Problem Solving (1992)
161. *E. Castillo and M. R. Ruiz-Cobo*, Functional Equations and Modelling in Science and Engineering (1992)
162. *A. J. Jerri*, Integral and Discrete Transforms with Applications and Error Analysis (1992)
163. *A. Charlier et al.*, Tensors and the Clifford Algebra (1992)
164. *P. Biler and T. Nadzieja*, Problems and Examples in Differential Equations (1992)

165. *E. Hansen*, Global Optimization Using Interval Analysis (1992)
166. *S. Guerre-Delabrière*, Classical Sequences in Banach Spaces (1992)
167. *Y. C. Wong*, Introductory Theory of Topological Vector Spaces (1992)
168. *S. H. Kulkarni and B. V. Limaye*, Real Function Algebras (1992)
169. *W. C. Brown*, Matrices Over Commutative Rings (1993)
170. *J. Loustau and M. Dillon*, Linear Geometry with Computer Graphics (1993)
171. *W. V. Petryshyn*, Approximation-Solvability of Nonlinear Functional and Differential Equations (1993)
172. *E. C. Young*, Vector and Tensor Analysis: Second Edition (1993)
173. *T. A. Bick*, Elementary Boundary Value Problems (1993)
174. *M. Pavel*, Fundamentals of Pattern Recognition: Second Edition (1993)
175. *S. A. Albeverio et al.*, Noncommutative Distributions (1993)
176. *W. Fulks*, Complex Variables (1993)
177. *M. M. Rao*, Conditional Measures and Applications (1993)
178. *A. Janicki and A. Weron*, Simulation and Chaotic Behavior of α-Stable Stochastic Processes (1994)
179. *P. Neittaanmäki and D. Tiba*, Optimal Control of Nonlinear Parabolic Systems (1994)
180. *J. Cronin*, Differential Equations: Introduction and Qualitative Theory, Second Edition (1994)
181. *S. Heikkilä and V. Lakshmikantham*, Monotone Iterative Techniques for Discontinuous Nonlinear Differential Equations (1994)
182. *X. Mao*, Exponential Stability of Stochastic Differential Equations (1994)
183. *B. S. Thomson*, Symmetric Properties of Real Functions (1994)
184. *J. E. Rubio*, Optimization and Nonstandard Analysis (1994)
185. *J. L. Bueso et al.*, Compatibility, Stability, and Sheaves (1995)
186. *A. N. Michel and K. Wang*, Qualitative Theory of Dynamical Systems (1995)
187. *M. R. Darnel*, Theory of Lattice-Ordered Groups (1995)
188. *Z. Naniewicz and P. D. Panagiotopoulos*, Mathematical Theory of Hemivariational Inequalities and Applications (1995)
189. *L. J. Corwin and R. H. Szczarba*, Calculus in Vector Spaces: Second Edition (1995)
190. *L. H. Erbe et al.*, Oscillation Theory for Functional Differential Equations (1995)
191. *S. Agaian et al.*, Binary Polynomial Transforms and Nonlinear Digital Filters (1995)
192. *M. I. Gil'*, Norm Estimations for Operation-Valued Functions and Applications (1995)
193. *P. A. Grillet*, Semigroups: An Introduction to the Structure Theory (1995)
194. *S. Kichenassamy*, Nonlinear Wave Equations (1996)
195. *V. F. Krotov*, Global Methods in Optimal Control Theory (1996)
196. *K. I. Beidar et al.*, Rings with Generalized Identities (1996)
197. *V. I. Arnautov et al.*, Introduction to the Theory of Topological Rings and Modules (1996)
198. *G. Sierksma*, Linear and Integer Programming (1996)
199. *R. Lasser*, Introduction to Fourier Series (1996)
200. *V. Sima*, Algorithms for Linear-Quadratic Optimization (1996)
201. *D. Redmond*, Number Theory (1996)
202. *J. K. Beem et al.*, Global Lorentzian Geometry: Second Edition (1996)
203. *M. Fontana et al.*, Prüfer Domains (1997)
204. *H. Tanabe*, Functional Analytic Methods for Partial Differential Equations (1997)
205. *C. Q. Zhang*, Integer Flows and Cycle Covers of Graphs (1997)
206. *E. Spiegel and C. J. O'Donnell*, Incidence Algebras (1997)
207. *B. Jakubczyk and W. Respondek*, Geometry of Feedback and Optimal Control (1998)
208. *T. W. Haynes et al.*, Fundamentals of Domination in Graphs (1998)
209. *T. W. Haynes et al.*, Domination in Graphs: Advanced Topics (1998)
210. *L. A. D'Alotto et al.*, A Unified Signal Algebra Approach to Two-Dimensional Parallel Digital Signal Processing (1998)
211. *F. Halter-Koch*, Ideal Systems (1998)
212. *N. K. Govil et al.*, Approximation Theory (1998)
213. *R. Cross*, Multivalued Linear Operators (1998)

Additional Volumes in Preparation

MULTIVALUED LINEAR OPERATORS

Ronald Cross
University of Cape Town
Rondebosch, South Africa

MARCEL DEKKER, INC. NEW YORK • BASEL • HONG KONG

Library of Congress Cataloging-in-Publication Data

Cross, Ronald.
Multivalued linear operators / Ronald Cross.
p. cm.— (Monographs and textbooks in pure and applied mathematics; 213)
Includes bibliographical references and index.
ISBN 0-8247-0219-0
1. Linear operators. I. Title. II. Series.
QA329.2.C76 1998
515'.7246—dc21 98-24468
CIP

This book is printed on acid-free paper.

Headquarters
Marcel Dekker, Inc.
270 Madison Avenue, New York, NY 10016
tel: 212-696-9000; fax: 212-685-4540

Eastern Hemisphere Distribution
Marcel Dekker AG
Hutgasse 4, Postfach 812, CH-4001 Basel, Switzerland
tel: 44-61-261-8482; fax: 44-61-261-8896

World Wide Web
http://www.dekker.com

The publisher offers discounts on this book when ordered in bulk quantities. For more information, write to Special Sales/Professional Marketing at the headquarters address above.

Current printing (last digit)
10 9 8 7 6 5 4 3 2 1

PRINTED IN THE UNITED STATES OF AMERICA

Preface

The simplest naturally occurring example of a multivalued linear operator is the inverse T^{-1} of a linear map T from X to Y (where X and Y are vector spaces), defined by the set of solutions

$$T^{-1}y := \{x \in X : Tx = y\}$$

of the equation $Tx = y$.

A mapping T with domain a linear subspace $D(T)$ of X and range in $2^Y \backslash \emptyset$ (the set of nonempty subsets of Y) is called a multivalued linear operator, or linear relation, if it satisfies

$$T(\alpha x_1 + \beta x_2) = \alpha T x_1 + \beta T x_2$$

for all x_1, x_2 in $D(T)$ and nonzero scalars α, β.

Linear relations were introduced into functional analysis by J. von Neumann (see Bibliography), motivated by the need to consider adjoints of non-densely defined linear differential operators.

If the vector spaces are endowed with norms, one can define the inverse T^{-1}, closure $\overline{T}$, completion $\tilde{T}$ and adjoint T' for a linear relation T. Even if T is a linear operator (the single-valued case), its inverse, closure, completion and adjoint are in general multivalued. To compensate for this one has various properties such as $(T')^{-1} = (T^{-1})'$, $(\overline{T})' = (T')^{-}$, $(\overline{T}^{-1}) = (T^{-1})^{-}$, $R(\tilde{T}')^{\top} = N(\tilde{T})$, $\gamma(T^{-1}) = \|T\|^{-1}$ (see Notation index) and many others, which hold for arbitrary linear relations, but which are not meaningful in the single-valued context. As a result, the theory of linear relations would seem to have the potential for contributing to the enrichment and clarification of many aspects of operator theory, including those concerned with non-closable or non-densely-defined linear operators. Some elegant and transparent formulations and proofs have been achieved for theorems in the classical theory of linear operators in Banach spaces, such as the Closed Graph theorem and the Closed Range theorem (see III.4.2 and III.4.4).

There are some new propositions about linear operators which can be inferred as special cases of propositions derived in this book. Among the more succinct ones, we cite the characterisations of linear operators whose adjoints are compact or weakly compact (see Corollary V.5.15 and Theorem VIII.2.6; see also V.14.5), characterisations of operators whose adjoints are upper or lower semi-Fredholm operators (see Propositions V.1.9 and V.7.5), a characterisation of closed (in particular, bounded) upper semi-Fredholm operators in Banach spaces (see Theorem V.9.22 and its generalisation, Corollary V.12.5), a characterisation of continuity for closed operators in Banach spaces (Exercise V.10.6), and various equivalent formulations of the upper semi-Fredholm property from the list in Section V.18.

The book aims to set up a theoretical framework for the study of linear relations and to provide a set of basic concepts, rules, formulas, theorems and techniques which will constitute useful reference material for pursuing investigations either of a fundamental or of an applied nature. The chapters are all closely related and each is the subject of continuing research by the author, doctoral students and collaborators. Regrettably, owing to limitations of space and in order to keep the book reasonably self-contained, it has been necessary to omit some specialised topics which are currently the subject of vigorous research, particularly in areas relating to differential equations where much important work has been carried out in recent years (see the concluding remarks to Section III.10 and the references therein).

The book begins with a study of the algebra of linear relations in vector spaces (Chapter I), starting with relations on sets and going on to discuss the operational rules for linear relations, linear selections, dimensionality and the fundamental theorem of linear algebra, the notions of index and coindex, and a new version of the Fredholm product index theorem (Corollary I.6.12). From Chapter II onwards the underlying spaces are normed linear spaces. Chapter II presents the fundamental notions about linear relations which relate to normed vector spaces. The definitions and notations conform here (as in the rest of the book) to those current in the theory of linear operators. Continuity and openness are discussed and related to the norm and minimum modulus, and a new notion of a closable relation is introduced. The problem of best approximation with interpolative constraints is discussed as an illustrative example. Adjoint relations make their appearance in Chapter III. The topics include polar relationships, the gap between a pair of closed subspaces, a full discussion of the category theorems of functional analysis, state diagrams relating a given relation to its adjoint, and an illustrative example concerning a differential operator whose adjoint is multivalued. The first three chapters provide a self-contained foundation course on multivalued linear operators and conclude the introductory part of the book. Only the most elementary knowledge of the theory of normed linear spaces and their duals is required of the reader.

Chapter IV introduces certain functions defined on the class of all linear relations, called "quantities", which are used to study various special classes of linear relations in Chapter V. A new notion is introduced, that of density invariance, which is concerned with the behaviour, with respect to the quantities, of restrictions of linear relations to dense subspaces of their domains. Chapter V introduces the following classes of multivalued linear operators: compact, precompact, strictly singular, strictly cosingular, upper and lower semi-Fredholm, partially continuous (precompact or compact), and nowhere continuous relations. These are defined in terms of a canonically associated linear operator corresponding to any given linear relation. In the chapter the relationships of these classes to one another under inversion, adjoints, completion, and closure are fully exhibited. Topics presented include a detailed discussion of perturbation theory, including stability under complementation.

Chapter VI is intended to serve as an introduction to the spectral theory of linear relations and to provide the necessary background for Chapter VII. The latter chapter investigates relative boundedness and relative compactness of a relation, and the essential spectrum and its stability properties. Essential use is made of the perturbation theory developed in the earlier chapters.

Chapter VIII is devoted to a study of the second adjoint of a linear relation modelled on the role played by the second adjoint in the theory of weakly compact linear operators and Tauberian operators.

At the end of each chapter there is a Notes and Remarks section containing the historical background, remarks, acknowledgements and further references.

The interdependence of the chapters is as follows:

I → II → III → VIII
I → II → III → VI
I → II → III → IV → V → VI → VII

The Notation Index has been arranged lexicographically as far as it makes sense. The Index of Operational Quantities has been included to render explicit the role played by these functions and facilitate cross referencing. The Index is a combined subject and author index.

Every effort has been made to make the Bibliography comprehensive and relevant. It contains some in-press articles and preprints, and includes many items which have not received specific mention in the text. Its contents reflect the many diverse fields of application of multivalued linear operators.

My sincere thanks are due to colleagues, doctoral students and collaborators who have assisted me in this project, several of whom have checked the entire manuscript, or read substantial portions of it and offered suggestions. I wish to thank especially T. Alvarez, M. Gonzalez, J.J. Jaftha, P. Jeganathan, M.K. Kinyon, A.M. Krall, L.E. Labuschagne, S. Mabizela, R. Mennicken, M. Möller, M.Z. Nashed, P. Pillay and D. Wilcox. I am particularly indebted to Professor R. Mennicken for introducing me to the subject of linear relations. I am grateful to Professor M.K. Kinyon for his kind hospitality during my visit to South Bend in the summer of 1995, when 140 pages of the first draft were discussed in detail, resulting in many improvements in the first five chapters. Immediately prior to this period I had been fortunate in spending some days in valuable discussions with Professor M.Z. Nashed at Newark, Delaware. His published work with S.J. Lee has been an important source for the material of Chapter II. My thanks also go to Professor A.M. Krall for hospitality received at Pennsylvania State

University, State College, Pennsylvania, and for furnishing me with the particularly appropriate illustrative example of a differential operator with multivalued adjoint included in Chapter III, together with some pertinent background information. Dr. L.E. Labuschagne has read the final version in detail, and made numerous observations and suggestions which have substantially added to the quality and veracity of the final result. For financial support I am grateful to the Foundation for Research Development, Pretoria. Special thanks go to Wanda Fouquet for her generous efforts in the typing of the manuscript.

Ronald Cross

Contents

CHAPTER I

Linear Relations: Algebraic Properties

In this chapter the concept and basic properties of a multivalued linear operator defined on a vector space are developed. We begin by reviewing the elementary theory of relations on sets.

I.1 RELATIONS ON SETS

I.1.1 Preliminaries

Let $U, V, W, \ldots$ denote arbitrary nonempty sets. A **relation** T from U to V is any mapping having domain $D(T)$ a nonempty subset of U, and taking values in $2^V \backslash \emptyset$ (the collection of nonempty subsets of V). Such a mapping T is also known as a **set valued** (or **multivalued**) **mapping**. If T maps the points of its domain to singletons, then T is said to be a **single valued relation**, or **function**.

For $u \in U$, $u \notin D(T)$ we define $Tu = \emptyset$. With this convention, we have

(1) $$D(T) := \{u \in U : Tu \neq \emptyset\}.$$

The class of all relations from U to V will be denoted by $R(U, V)$. Examples of relations are functions, inverses of functions, adjoints of linear operators, partial order relations, equivalence relations, and convex processes.

If $T \in R(U, V)$, then the **graph** of T is the subset $G(T)$ of $U \times V$ defined by

(2) $$G(T) := \{(u, v) \in U \times V : u \in D(T), \ v \in T(u)\}.$$

A relation in $R(U, V)$ is uniquely determined by its graph, and conversely any nonempty subset of $U \times V$ uniquely determines a relation. (Some authors identify a relation with its graph, but we shall not do so here.)

Let $T \in LR(U, V)$. The **inverse** of T is the relation T^{-1} given by

(3) $$G(T^{-1}) := \{(v, u) \in V \times U : (u, v) \in G(T)\}.$$

Given a subset $M \subset U$, we write

(4) $$T(M) := \cup\{T(m) : m \in M \cap D(T)\}$$

called the **image** of M, with

$$R(T) := T(U) \qquad (= T(D(T)), \text{ by } (4))$$

called the **range** of T. If $R(T) = Y$, then T is called **surjective**. If T^{-1} is single valued, then T is called **injective**. If T is injective then it is easy to see that the following implication holds:

$$(5) \qquad T(u_1) = T(u_2) \implies u_1 = u_2 \qquad (u_1, u_2 \in D(T))$$

We shall denote $T(\{u\}) = T(u)$ by Tu; no distinction will be made between a single valued map and a map into V.

Let $\emptyset \neq N \subset V$. Then from (3) we have

$$(6) \qquad T^{-1}(N) = \{u \in D(T) : N \cap Tu \neq \emptyset\}$$

In particular, for $v \in R(T)$,

$$(7) \qquad T^{-1}v = \{u \in D(T) : v \in Tu\}.$$

It is clear that $D(T^{-1}) = R(T)$, and $R(T^{-1}) = D(T)$.

The **identity relation** defined on a nonempty subset E of U is denoted by I_E or simply I when E is understood; it is the relation in $R(U, U)$ whose graph is $G(I_E) = \{(e, e) : e \in E\}$

I.1.2 Exercise: Let T be an equivalence relation on U. Then $G(I_U) \subset G(T)$ and $T = T^{-1}$. Moreover, for $u_1, u_2 \in U$ we have the implication

$$(1) \qquad T(u_1) \cap T(u_2) \neq \emptyset \implies T(u_1) = T(u_2).$$

I.1.3 Composition of Relations

Let $T \in R(U, V)$ and $S \in R(V, W)$ where $R(T) \cap D(S) \neq \emptyset$. The **composition** (or **product**) ST in $R(U, W)$ is defined as follows:

$$(1) \qquad (ST)(u) := S(Tu) \qquad (u \in U)$$

The domain of ST is computed from (1) of I.1.1:

$$\begin{aligned} D(ST) &= \{u \in U : S(Tu) \neq \emptyset\} \\ &= \{u \in U : Tu \cap D(S) \neq \emptyset\}. \end{aligned}$$

Hence from (6) of I.1.1,

$$D(ST) = T^{-1}(D(S)) \tag{2}$$

From the definition of ST it is easily seen that

$$\begin{aligned} G(ST) = \{(u,w) \in U \times W : (u,v) \in G(T) \text{ and} \\ (v,w) \in G(S) \text{ for some } v \in V\} \end{aligned} \tag{3}$$

We list some elementary properties:

$$T = TI_U = I_V T \tag{4}$$

$$G(I_{D(T)}) \subset G(T^{-1}T) \tag{5}$$

$$(ST)^{-1} = T^{-1}S^{-1} \tag{6}$$

$$D(ST) = T^{-1}S^{-1}(R(S)) \tag{7}$$

Composition is associative, i.e. if $R \in R(W,Z)$ then

$$R(ST) = (RS)T \tag{8}$$

We also have

$$T \text{ injective} \iff T^{-1}T = I_{D(T)} \tag{9}$$

$$T \text{ single valued} \iff TT^{-1} = I_{R(T)} \tag{10}$$

I.1.4 Exercise Show that equivalence relations are idempotent (i.e. $TT = T$). Show that the converse is false by considering a linear projection defined on a vector space other than the identity.

I.1.5 Restrictions and Extensions of Relations

Let M be a subset of U such that $M \cap D(T) \neq \emptyset$. The **restriction** of T to M, denoted by $T|_M$, is defined by

$$\begin{aligned} T|_M &\in R(U,V), \\ D(T|_M) &= D(T) \cap M, \\ (T|_M)m &= Tm \text{ for } m \in M. \end{aligned}$$

Observe that

$$G(T|_M) = G(T) \cap (M \times V)$$
$$= \{(u, v) \in G(T) : u \in M\}$$

We have

(1) $$T|_M = T|_{M \cap D(T)}$$

Given two relations S and T in $R(U, V)$, we say that S is an **extension** of T if

(2) $$S|_{D(T)} = T$$

Clearly, if S is an extension of T, then $G(T) \subset G(S)$. (The converse is false however; see Corollary I.2.11(d)).

I.1.6 Example: The injection operator J_E^U.

Let E be a subset of U and let J_E^U (abbreviated J_E when U is understood) denote the natural injection map of E into U, i.e. $J_E \in R(E, U)$, $D(J_E) = E$ and $J_E(e) = e$ for all $e \in E$. Then we have

(1) $$T|_E = TJ_EJ_E^{-1}$$

Note that $I_E = I_U|_E$ (in particular, both sides are in $R(U, U)$). Thus from (1) we have

(2) $$I_E = J_E\, J_E^{-1}$$

Although I_E and J_E have identical graphs, I_E is not even an extension of J_E.

I.1.7 Exercise. Let $F \subset E \subset U$. Then

(1) $$J_E^U\, J_F^E = J_F^U.$$

I.2. LINEAR RELATIONS (MULTIVALUED LINEAR OPERATORS)

I.2.1 Definition and notation Let X and Y be vector spaces over the field $\mathbb{K} = \mathbb{R}$ or $\mathbb{C}$. A relation $T \in R(X, Y)$ is called a **linear relation** (or

multivalued linear operator) if for all $x, z \in D(T)$ and nonzero scalars α we have

$$Tx + Tz = T(x+z) \tag{1}$$

$$\alpha Tx = T(\alpha x) \tag{2}$$

Evidently the domain of a linear relation is a linear subspace.

The class of linear relations in $R(X,Y)$ will be denoted by $LR(X,Y)$. We write $LR(X,X) := LR(X)$.

It is clear that if M is a linear subspace of X and $T \in LR(X,Y)$, then

$$T|_M \in LR(X,Y)$$

and

$$TJ_M \in LR(M,Y)$$

I.2.2 Example Let $X = Y = \mathbb{R}^2$ and consider the matrix operator

$$A = \begin{pmatrix} 1 & 0 \\ 0 & 0 \end{pmatrix}.$$

Writing $x = \begin{pmatrix} x_1 \\ x_2 \end{pmatrix}$, $y = \begin{pmatrix} y_1 \\ y_2 \end{pmatrix}$ for arbitrary vectors in $\mathbb{R}^2$, we have

$$G(A) = \{\left(\begin{pmatrix} x_1 \\ x_2 \end{pmatrix}, \begin{pmatrix} y_1 \\ y_2 \end{pmatrix}\right) : \begin{pmatrix} x_1 \\ x_2 \end{pmatrix} \in \mathbb{R}^2, \begin{pmatrix} y_1 \\ y_2 \end{pmatrix} = \begin{pmatrix} x_1 \\ 0 \end{pmatrix}\}$$

and

$$A^{-1}\begin{pmatrix} x_1 \\ 0 \end{pmatrix} = \begin{pmatrix} x_1 \\ 0 \end{pmatrix} + A^{-1}(0) \tag{1}$$

where $A^{-1}(0) = \{\begin{pmatrix} 0 \\ x_2 \end{pmatrix} : x_2 \in \mathbb{R}\}$ is a linear subspace of $\mathbb{R}^2$.

It is simple to verify that A^{-1} is a linear relation with domain

$$D(A^{-1}) = \{\begin{pmatrix} x_1 \\ 0 \end{pmatrix} : x_1 \in \mathbb{R}\}.$$

Consider the relation $S \in LR(X)$ defined by

$$S\begin{pmatrix} x_1 \\ x_2 \end{pmatrix} = \begin{pmatrix} x_1 \\ 0 \end{pmatrix} + A^{-1}(0).$$

Then S is an extension of A^{-1}. Moreover, writing $J = J_{R(A)}$, we have according to (1) of Example I.1.6 that

$$SI_{R(A)} = SJJ^{-1} = A^{-1}$$

and hence $A = JJ^{-1}S^{-1}$; however, $A \neq S^{-1}$.

I.2.3 Proposition. *Let $T \in R(X,Y)$. The following properties are equivalent:*

(i) *T is a linear relation.*
(ii) *$G(T)$ is a linear subspace of $X \times Y$.*
(iii) *T^{-1} is a linear relation.*
(iv) *$G(T^{-1})$ is a linear subspace of $Y \times X$.*

PROOF. Let T be a linear relation and let (x_1, y_1), $(x_2, y_2) \in G(T)$, $0 \neq \alpha \in \mathbb{K}$. Then $y_1 \in Tx_1$ and $y_2 \in Tx_2$, and so

$$y_1 + y_2 \in Tx_1 + Tx_2 = T(x_1 + x_2).$$

Hence

$$(x_1, y_1) + (x_2, y_2) = (x_1 + x_2, y_1 + y_2) \in G(T) \tag{1}$$

Again, since $y_1 \in Tx_1$, so $\alpha y_1 \in \alpha Tx_1$ (as $\alpha \neq 0$) and hence

$$\alpha(x_1, y_1) = (\alpha x_1, \alpha y_1) \in G(T). \tag{2}$$

Finally, $(0,0) \in G(T)$ by (1), so (2) also holds for $\alpha = 0$. Therefore $G(T)$ is a subspace.

Conversely, suppose that $G(T)$ is a linear subspace of $X \times Y$. Then $D(T)$ is a linear subspace of X. Let $x_1, x_2 \in D(T)$. To show that

$$Tx_1 + Tx_2 \subset T(x_1 + x_2) \tag{3}$$

select $y_1 \in Tx_1$ and $y_2 \in Tx_2$. Then

$$(x_1 + x_2, y_1 + y_2) = (x_1, y_1) + (x_2, y_2) \in G(T).$$

Hence $y_1 + y_2 \in T(x_1 + x_2)$ and (3) follows.

Next, to show that

$$T(x_1 + x_2) \subset Tx_1 + Tx_2$$

let $y \in T(x_1 + x_2)$ and select $y_1 \in Tx_1$ arbitrarily. Then

$$(x_1 + x_2, y) = (x_1, y_1) + (x_2, y - y_1) \in G(T)$$

and hence $(x_2, y - y_1) \in G(T)$, whence $y - y_1 \in Tx_2$. Therefore $y \in y_1 + Tx_2 \subset Tx_1 + Tx_2$, as required. We have thus shown that $Tx_1 + Tx_2 = T(x_1 + x_2)$.

Now let $0 \neq \alpha \in \mathbb{K}$ and $x \in D(T)$. If $y \in Tx$, then $\alpha(x, y) = (\alpha x, \alpha y) \in G(T)$, so $\alpha y \in T(\alpha x)$. Hence $\alpha(Tx) \subset T(\alpha x)$. On the other hand if $y \in T\alpha x$, then $(x, \alpha^{-1} y) \in G(T)$, so $y \in \alpha Tx$. Thus $T(\alpha x) \subset \alpha(Tx)$. Hence $\alpha(Tx) = T(\alpha x)$. Therefore T is a linear relation.

The remaining equivalences now follow by symmetry. □

I.2.4 Corollary. *Let T be a linear relation. Then $T(0)$ and $T^{-1}(0)$ are linear subspaces.*

I.2.5 Corollary. *Let $T \in R(X,Y)$. Then T is a linear relation if and only if the equality*

$$\alpha T x_1 + \beta T x_2 = T(\alpha x_1 + \beta x_2) \tag{1}$$

holds for all $x_1, x_2 \in D(T)$ and nonzero $\alpha, \beta \in \mathbb{K}$.

In the sequel $X, Y, Z \ldots$ will denote vector spaces over $\mathbb{K} = \mathbb{R}$ or $\mathbb{C}$, and T will always denote an element of $LR(X,Y)$ except where stated otherwise. The term "operator" always refers to a single valued operator, while a multivalued linear operator will generally be referred to as a "linear relation".

I.2.6 Definition

The subspace $T^{-1}(0)$ is called the **null space** (or **kernel**) of T, and is denoted by $N(T)$. We shall be using both $N(T)$ and $T^{-1}(0)$ throughout the sequel.

I.2.7 Example Let P be the linear projection defined on $X = M + N$ with range M and null space N. Write $T = P^{-1}$. Then $T \in LR(X)$, $D(T) = M$, T is injective, and for each $x \in M$ we have

$$Tx = x + N.$$

I.2.8 Proposition.

(a) *For $x \in D(T)$ we have the following equivalence:*

$$y \in Tx \iff Tx = y + T(0) \tag{i}$$

In particular,

$$0 \in Tx \iff Tx = T(0) \tag{ii}$$

(b) *For $x_1, x_2 \in D(T)$ we have the equivalence:*

$$Tx_1 \cap Tx_2 \neq \emptyset \iff Tx_1 = Tx_2.$$

PROOF. (a) (i) Let $y \in Tx$. Then $y + T(0) \subset Tx + T(0) = Tx$. On the other hand for any $y_1 \in Tx$, we have

$$y_1 = y + (y_1 - y) \in y + Tx - Tx = y + T(0).$$

Hence $Tx = y + T(0)$. The converse is obvious.

(ii) We have $0 \in Tx$ if and only if $Tx = 0 + T(0)$ (by a(i)), if and only if $Tx = T(0)$ since $0 \in T(0)$ by Corollary I.2.4.

(b) Let $y \in Tx_1 \cap Tx_2$. Then by (a)(i) $Tx_1 = y + T(0) = Tx_2$. The converse is trivial. □

I.2.9 Corollary. *(a) T is single valued if and only if $T(0) = \{0\}$.*

(b) T is injective if and only if $N(T) = \{0\}$.

I.2.10 Corollary. *$TT^{-1}(0) = T(0)$ and $T^{-1}T(0) = T^{-1}(0)$.*

Proof. By Corollary I.2.4, $0 \in TT^{-1}(0)$. Hence $T(0) = TT^{-1}(0)$ by Proposition I.2.8. Substituting T^{-1} for T yields the second equality. □

I.2.11 Corollary. *We have*
(a) $TT^{-1}y = y + T(0)$ $(y \in R(T))$ *and* $T^{-1}Tx = x + T^{-1}(0)$ $(x \in D(T))$

(b) If $G(S) \subset G(T)$, then T is an extension of S if and only if $S(0) = T(0)$.

I.2.12 Exercise Let $T \in LR(X)$ be everywhere defined. Show that T is an equivalence relation if and only if T is reflexive.

I.2.13 Exercise. Show that T is injective if and only if the implication (5) of I.1.1 holds.

I.2.14 Exercise Let $S, T \in LR(X, Y)$.

(a) Suppose that $G(S) \subset G(T)$, $N(T) \subset N(S)$ and $R(T) \subset R(S)$. Show that $T = S$.

(b) Suppose that $D(T) = D(S)$ and $T(0) = S(0)$. Show that either $T = S$, or the graphs of T and S are incomparable.

I.2.15 Exercise Show that $TT^{-1}T = T$.

I.2.16 Proposition. *Let T be a relation from X to Y. Then T is a linear relation if and only if for all $x_1, x_2 \in D(T)$, and for all $\alpha, \beta \in \mathbb{K}$,*

$$\alpha T x_1 + \beta T x_2 \subset T(\alpha x_1 + \beta x_2). \tag{1}$$

PROOF. Let T be a linear relation. Let $\alpha, \beta \in \mathbb{K}$ and $x_1, x_2 \in D(T)$. Select $y_1 \in Tx_1$, $y_2 \in Tx_2$. Since $G(T)$ is a vector space, we have $(\alpha x_1 + \beta x_2, \alpha y_1 + \beta y_2) \in G(T)$ whence $\alpha y_1 + \beta y_2 \in T(\alpha x_1 + \beta x_2)$ and (1) follows.

Conversely, let T satisfy the condition (1) and let $(x_1, y_1), (x_2, y_2) \in G(T)$.

Then for $\alpha, \beta \in \mathbb{K}$ we have

$$\alpha y_1 + \beta y_2 \in \alpha T x_1 + \beta T x_2 \subset T(\alpha x_1 + \beta x_2)$$

showing that $\alpha(x_1, y_1) + \beta(x_2, y_2) \in G(T)$. Therefore $G(T)$ is a linear subspace of $X \times Y$. Consequently T is a linear relation by I.2.3. $\square$

I.3. OPERATIONAL RULES FOR A SINGLE LINEAR RELATION

The rules governing the operations of T on subsets are listed in the following proposition (where 0_X denotes the zero operator on X):

I.3.1 Proposition

(a) $T(\alpha M) = \alpha T(M) \quad (M \subset X, \alpha \in \mathbb{K}, \alpha \neq 0)$
(b) $T(M + N) \supset T(M) + T(N) \quad (M, N \subset X)$
(c) $T(M + N) = T(M) + T(N) \quad (M \subset X, N \subset D(T))$
(d) $TT^{-1}(M) = M \cap R(T) + T(0) \quad M \subset Y)$
(e) $T^{-1}T(M) = M \cap D(T) + T^{-1}(0) \quad (M \subset X)$
(f) $T^{-1}(0) \times \{0\} = G(T) \cap (X \times \{0\})$
$\qquad = G(T) \cap G(0_X)$
(g) $\{0\} \times T(0) = G(T) \cap (\{0\} \times Y)$
(h) $X \times R(T) = G(T) + (X \times \{0\})$
$\qquad = G(T) + G(0_X)$
(i) $D(T) \times Y = G(T) + (\{0\} \times Y)$.

PROOF.

(a) $$\begin{aligned} T(\alpha M) &= \cup\{T(\alpha m) : m \in M \cap D(T)\} \\ &= \cup\{\alpha(Tm) : m \in M \cap D(T)\} \\ &= \alpha \cup \{Tm : m \in M \cap D(T)\} \\ &= \alpha T(M) \end{aligned}$$

(b) Immediate.

(c) Let $m \in M$ and $n \in N \subset D(T)$. If $m+n \notin D(T)$, then $\emptyset = T(m+n) \subset Tm + Tn$ trivially (see I.1.1(3)). On the other hand if $m + n \in D(T)$, then $m \in D(T)$ and $T(m+n) = Tm + Tn \subset TM + TN$. Equality now follows from (b).

(d) and (e) Follow immediately from Corollary I.2.11.

Properties (f) –(i) are simple consequences of the definitions. □

Proposition I.3.1 yields a short proof of the linearity of the composition of two linear relations.

I.3.2 Corollary. *Let $T \in LR(X,Y)$ and $S \in LR(Y,Z)$. Then $ST \in LR(X,Z)$.*

PROOF. Let $x_1, x_2 \in D(ST)$ and let α, β be nonzero scalars. Then

$$\begin{aligned} \alpha(STx_1) + \beta(STx_2) &= \alpha S(Tx_1) + \beta S(Tx_2) \\ &= S(\alpha Tx_1) + S(\beta Tx_2) = S(T(\alpha x_1)) + S(T(\beta x_2)) \\ &= S(T(\alpha x_1) + T(\beta x_2)) = ST(\alpha x_1 + \beta x_2). \end{aligned}$$

Hence ST is linear, by Corollary I.2.5. □

I.4 THE ALGEBRA OF LINEAR RELATIONS

In this section we define the operations of addition and scalar multiplication in $LR(X,Y)$ and describe the laws governing these operations combined with the operations of composition and inversion.

I.4.1 Addition and scalar multiplication in $LR(X,Y)$

Let S and T be linear relations in $LR(X,Y)$ and let α be a scalar. The relations $S+T$ and αT are defined as follows:

(1) $(S+T)x := Sx + Tx \quad (x \in X)$.
(2) $(\alpha T)x := \alpha(Tx) \quad (x \in X, \alpha \in \mathbb{K})$.

It follows that

(3) $D(S+T) = D(S) \cap D(T)$

(4) $D(\alpha T) = D(T)$.

To verify (3), let $x \in D(S) \cap D(T)$. Then $Sx + Tx = (S+T)x \neq \emptyset$ by I.1.1(1) and so $x \in D(S+T)$. Conversely, if $x \in D(S+T)$, then since $(S+T)x = Sx + Tx \neq \emptyset$ we have $Sx \neq \emptyset$ and $Tx \neq \emptyset$. Therefore $x \in D(S) \cap D(T)$.

It is evident from the above that αT and $S+T$ are linear relations. Moveover

(5) $G(S+T) = \{(x,y) \in X \times Y : y = s+t$, where $(x,s) \in G(S)$ and $(x,t) \in G(T)\}$

(6) $G(\alpha T) = \{(x, \alpha y) \in X \times Y : (x,y) \in G(T)\}$.

We furthermore have

(7) $\alpha(\beta T) = (\alpha\beta)T \quad (\alpha, \beta \in \mathbb{K})$.

(8) $S + T = T + S$.

and if R, S, T are in $LR(X,Y)$ then

(9) $R + (S+T) = (R+S) + T$.

The rules needed for combining the various operations, in particular the right and left distributive laws, are listed in the next proposition.

Observe that for linear relations $T \in LR(X,Y)$ and $S \in LR(W,Z)$, the composition ST is defined if and only if $W = Y$.

I.4.2 Proposition. *Let R, S, T be linear relations.*

(a) $TT^{-1} = I + (TT^{-1} - TT^{-1})$

(b) $\lambda(ST) = (\lambda S)T = S(\lambda T) \quad (0 \neq \lambda \in \mathbb{K})$.

(c) *If $G(S) \subset G(T)$ then, $G(SR) \subset G(TR)$.*

(d) $(R+S)Tx \subset (RT + ST)x \qquad (x \in X)$
and if T is single valued, then

$$(R+S)T = RT + ST.$$

(e) *$T(R+S)$ is an extension of $TR + TS$. Furthermore, if $D(T)$ contains the ranges of both R and S (in particular, if $D(T)$ is the whole space) then*

$$T(R+S) = TR + TS.$$

Proof. (a) For $y \in R(T)$ we have by Proposition I.2.8 that $TT^{-1}(y) = y + T(0)$ and $TT^{-1}(y) - TT^{-1}(y) = T(0)$, whence $TT^{-1} = I_{R(T)} + TT^{-1} - TT^{-1} = I + TT^{-1} - TT^{-1}$. (Here the subscript $R(T)$ has been dropped on account of I.4.1(3)).

(b) Let $\lambda \neq 0$. We have

$$\begin{aligned}
(x,z) \in G((\lambda S)T) &\iff (\exists y)(x,y) \in G(T), (y,z) \in G(\lambda S)\\
&\iff (\exists y)(x,y) \in G(T), (y,\lambda^{-1}z) \in G(S) \quad \text{(I.4.1(2))}\\
\text{(1)} \qquad &\iff (x,z) \in G(\lambda(ST)) \quad \text{(I.4.1(2))}\\
&\iff (x,\lambda^{-1}z) \in G(ST) \quad \text{(I.4.1(2))}\\
&\iff (\exists y)(x,\lambda^{-1}y) \in G(T), (\lambda^{-1}y,\lambda^{-1}z) \in G(S)\\
&\iff (\exists y)(x,y) \in G(\lambda T), (y,z) \in G(S)\\
\text{(2)} \qquad &\iff (x,z) \in G(S(\lambda T))
\end{aligned}$$

Hence from (1) and (2), $(\lambda S)T = \lambda(ST) = S(\lambda T)$.

(c) Let $(z,y) \in G(SR)$. Then $(z,x) \in G(R)$ and $(x,y) \in G(S) \subset G(T)$ for some $x \in D(S)$, whence $(z,x) \in G(TR)$.

(d) Let $R, S \in LR(Y,Z)$, $T \in LR(X,Y)$. We have

$$\begin{aligned}
G((R+S)T) &= \{(x,z) : (x,y) \in G(T) \text{ and } (y,z) \in G(R+S)\}\\
&= \{(x,z) : (x,y) \in G(T), (y,z_1) \in G(R), (y,z_2) \in G(S), z_1 + z_2 = z\}\\
&\subset \{(x,z) : (x,z_1) \in G(RT), (x,z_2) \in G(ST), z_1 + z_2 = z\}\\
&= G(RT + ST).
\end{aligned}$$

It is obvious that $RT + ST = (R+S)T$ if T is single valued.

(e) Let $T \in LR(X,Y)$ and $R, S \in LR(Z,X)$. We shall verify that

$$\text{(1)} \qquad G(TR + TS) \subset G(T(R+S))$$

We have

$$\begin{aligned}
G(TR+TS) &= \{(z,y) : y = r + s \text{ where } (z,r) \in G(TR) \text{ and } (z,s) \in G(TS)\}\\
&= \{(z,y) : y = r + s, (z,x_1) \in G(R), (x_1,r) \in G(T),\\
&\qquad (z,x_2) \in G(S), (x_2,s) \in G(T)\}\\
&\subset \{(z,y) : y = r + s, (z,x_1+x_2) \in G(R+S), (x_1+x_2, r+s) \in G(T)\}\\
&= \{(z,y) : (z,x) \in G(R+S), (x,y) \in G(T)\}\\
&= G(T(R+S)).
\end{aligned}$$

Now let $z \in D(TR+TS)$. Then by (1), $z \in D(T(R+S))$. Let $y \in T(R+S)z$. Then $y \in Tx$, where $x \in (R+S)z$. Write $x = x_1 + x_2$, where $x_1 \in Rz$, $x_2 \in Sz$. Then

$$y \in T(x_1 + x_2) = Tx_1 + Tx_2 \subset TRz + TSz.$$

Therefore $T(R+S)z \subset TRz+TSz$. Hence from (1), $T(R+S)z = TRz+TSz$. If the ranges of R and S are contained in $D(T)$, then $D(T(R+S)) = D(R+S) = D(R) \cap D(S) = D(TR) \cap D(TS) = D(TR+TS)$, and the stated equality now follows. □

The examples below show that additional conditions are needed to ensure left and right distributivity in (d) and (e) of the above proposition.

I.4.3 Examples

(a) Put $R = -S$ in (d) of Proposition I.4.2, let S be single valued and nonzero, and let $T(0) = Y \neq \{0\}$. Then

$$(R+S)T(0) = (S-S)T(0) = 0$$

and

$$(RT+ST)(0) = (ST-ST)(0) = ST(0) = S(Y) \neq 0.$$

This shows that $RT+ST$ *is not in general an extension of* $(R+S)T$ (see Corollary I.2.11(b)).

(b) *We construct an example making* $T(R+S)$ *in Proposition* I.4.3(e) *a proper extension of* $TR+TS$. Let $R, S, T \in LR(\mathbb{R}^2)$ be defined by

$$R = I$$

$$G(S) = \{\left(\begin{pmatrix} x \\ 0 \end{pmatrix}, \begin{pmatrix} w \\ x \end{pmatrix}\right) : x, w \in \mathbb{R}\}$$

$$T = I|_M \text{ where } M = \{\begin{pmatrix} 0 \\ y \end{pmatrix} : y \in \mathbb{R}\}$$

Then $D(TR) = D(T)$

$D(TS) = \{\binom{x}{0} : x \in \mathbb{R}\}$

Hence $D(TR+TS) = \{0\}$

Now $G(R+S) = \{(\binom{x}{0}, \binom{z}{x}) : x, z \in \mathbb{R}\}$

Hence $D(T(R+S)) = (R+S)^{-1}(D(T)) = \{\binom{x}{0} : x \in \mathbb{R}\}$

Therefore $TR+TS \neq T(R+S)$.

In this example S is the inverse of the single valued map $\begin{pmatrix} 0 & 1 \\ 0 & 0 \end{pmatrix}$.

(c) Let $R, S, T \in LR(X)$. If $RT = TR$ and $ST = TS$ then

(1) $$G((R+S)T) \subset G(T(R+S))$$

(d) *Equality fails in* (1) *even if* R, S *and* T *are all single valued*, as is seen by considering the following example: $S = -R = I$, and T any linear operator with $D(T) \neq X$.

I.4.4 Exercise. Let $\lambda \neq 0$. Show that $\lambda T = (\lambda I_Y)T = T(\lambda I_X)$, and $(\lambda T)^{-1} = \lambda^{-1}T^{-1}$.

I.5 LINEAR SELECTIONS

I.5.1 Definition

A linear operator A is called a **selection** (or **single valued part**) of T if

$$T = A + T - T \quad \text{and} \quad D(A) = D(T). \tag{1}$$

If A is a selection of T then we have for all $x \in D(T)$

$$Tx = Ax + T(0). \tag{2}$$

The relation $T - T$ given by $G(T - T) = D(T) \times T(0)$ is known as the **singular part**, or the **indeterminacy**, of T and is denoted by some authors by T_∞.

From (2) we have $R(T) = R(A) + T(0)$, but this sum may or may not be direct. A method of constructing selections is provided by the following proposition.

I.5.2 Proposition. *If P is a linear projection with domain $R(T)$ and kernel $T(0)$ then PT is a selection of T. Conversely, if A is a selection of T, and if $R(A) \cap T(0) = \{0\}$, then the linear projection defined on $R(T)$, with range $R(A)$ and kernel $T(0)$, satisfies $A = PT$.*

PROOF. Let P be given. Then $PT(0) = 0$ i.e. PT is single valued. Then for $y \in Tx$ we have $Tx = y + T(0) = Py + (I - P)y + T(0) = PTx + T(0)$.

Conversely, let A be selection of T and let P be the linear projection defined on $R(T)$ with range $R(A)$ and kernel $T(0)$. For $x \in D(T)$ we have

$$Tx = Ax + T(0).$$

Hence

$$PTx = PAx = Ax - (I - P)Ax = Ax. \quad \square$$

Proposition I.5.2 shows that the class of selections of A with the property $R(A) \cap T(0) = \{0\}$, is in one–to–one correspondence with the class of linear projections with domain $R(T)$ and kernel $T(0)$.

All the selections of T can be obtained from the next proposition.

I.5.3 Proposition. *Let A be a selection of T. Then*

$$G(A) \cap (\{0\} \times T(0)) = \{0\} \times \{0\}, \tag{1}$$

and

$$G(T) = G(A) + (\{0\} \times T(0)). \tag{2}$$

The linear projection P defined on $G(T)$ with range $G(A)$ and kernel $\{0\} \times T(0)$ has the property that for $(x, y) \in G(T)$

$$P(x, y) = (x, Ax). \tag{3}$$

Conversely, any linear projection P defined on $G(T)$ with kernel $\{0\} \times T(0)$ determines a selection A whose graph is given by

$$G(A) = PG(T). \tag{4}$$

PROOF. Let A be a selection of T. Then clearly (1) holds. The equality (2) is a consequence of I.5.1(2). Let P be a linear projection defined on $G(T)$ with range $G(A)$ and kernel $\{0\} \times T(0)$. Then for $x \in D(T)$ we have from I.5.1(2),

$$\begin{aligned} P\{(x, y) : y \in Tx\} &= P\{(x, Ax) + \{0\} \times T(0)\} \\ &= (x, Ax) \end{aligned}$$

from which (3) follows.

Conversely, let P be any linear projection defined on $G(T)$ with kernel $\{0\} \times T(0)$. Write

$$P(x, y) := (Cx, Dy).$$

Then C and D are single valued and idempotent. We shall verify that $PG(T)$ is the graph of a single valued relation. Suppose that

$$(Cx, Dy_1), (Cx, Dy_2) \in PG(T).$$

Then

$$(0, D(y_1 - y_2) \in PG(T).$$

Hence

$$\begin{aligned} (0, D(y_1 - y_2) &= P(x, y) \text{ for some } y \in Tx \\ &= (Cx, Dy). \end{aligned}$$

Then $Cx = 0$ and so $P(x, y) = (0, Dy) \in G(T)$ where $(0, Dy) \in N(P)$. Therefore

$$P(0, Dy) = (C(0), D^2 y) = (0, Dy) = (0, 0).$$

Hence $D(y) = 0$, i.e. $Dy_1 = Dy_2$, proving single valuedness.

We conclude that $P(G(T))$ is of the form $\{(x, Ax) : x \in D(T)\}$ where A is a linear operator. Now we have

$$\begin{aligned} G(T) &= PG(T) + (I - P)G(T) \\ &= G(A) + \{0\} \times T(0) \end{aligned}$$

where A is a selection of T whose graph is $PG(T)$. □

I.5.4 Example Let P be a linear projection defined on X, and let $T = P^{-1}$. Then since

$$Tx = x + T(0) \qquad (x \in R(P))$$

the map $I|_{R(P)}$ is a selection of T. In this case $R(I|_{R(P)}) \cap N(P) = R(P) \cap N(P) = \{0\}$, so another selection is the projection P. Thus

$$P^{-1}x = Px + P^{-1}(0) \qquad (x \in R(P)).$$

The corresponding decompositions of P^{-1} into single valued and singular parts are

$$P^{-1} = I|_{R(P)} + (P^{-1} - P^{-1})$$

and

$$P^{-1} = P + (P^{-1} - P^{-1}).$$

I.5.5 Example Let X denote the vector space $C[a, b]$ of all continuous real valued functions defined on some given compact interval $[a, b]$ in $\mathbb{R}$. Define $T \in LR(X)$ by $D(T) = X$ and

$$Tf = \int f(t)dt \tag{1}$$

Clearly, the function $Af = \int_a^x f(t)dt$ is a selection of T.

All the selections of T are determined by the linear projections P defined on $G(T)$ with kernel $\{0\} \times T(0)$, i.e. the subspace $\{(0, \lambda) : \lambda \in \mathbb{R}\}$.

Here $T(0)$ is the one–dimensional subspace consisting of the constant functions on $[a, b]$. It is clear that $R(A) \cap T(0) = \{0\}$.

I.5.6 Example Let E be a non zero subspace of X and define T by $Tx = x + E \quad (x \in X)$. Then T is linear, I is a selection of T, and $R(I) \cap T(0) = E \neq \{0\}$.

I.6 DIMENSION OF DOMAIN AND RANGE

An important property of linear relations is the invariance of finite codimensionality, described in the following proposition.

I.6.1 Proposition. *Let $T \in LR(X,Y)$, and let $M \subset X$. Then*

$$\dim R(T)/T(M) \leq \dim D(T)/D(T) \cap M \leq \dim X/M.$$

In particular, if M is a finite codimensional subspace of $D(T)$, then $T(M)$ is a finite codimensional subspace of $R(T)$.

PROOF. If $M \subset X$, we may select $E, F \subset X$ such that $F \subset D(T)$ with $M \cap D(T) + F = D(T)$, $M \cap D(T) \cap F = \{0\}$, and $E \cap D(T) = \{0\}$ with $X = M + (E + F)$ where $M \cap (E+F) = \{0\}$. Then

$$\dim R(T)/T(M) \leq \dim D(T)/D(T) \cap M = \dim F \leq \dim(E+F) = \dim X/M. \quad \square$$

I.6.2 Notation For a given linear relation $T \in LR(X,Y)$, let q_T (or q, if T is understood) denote the natural quotient map of Y onto $Y/T(0)$.

I.6.3 Proposition. *The linear relation qT is single valued.*

PROOF. Let $qy_1, qy_2 \in qTx$. Then $qy_1 - qy_2 \in qTx - qTx = qT(0) = 0$. $\square$

The single valued version of the next proposition is known as "*the fundamental theorem of linear algebra*".

I.6.4 Proposition. $\dim D(T) + \dim T(0) = \dim D(T^{-1}) + \dim T^{-1}(0)$.

PROOF. We have

$$(1) \qquad \dim R(T) = \dim R(T)/T(0) + \dim T(0).$$

Clearly,

$$(2) \qquad \dim D(qT) = \dim N(qT) + \dim R(qT).$$

Now $D(T) = D(qT)$ and $N(qT) = \{x \in D(T) : 0 \in qTx\} = \{x \in D(T) : Tx = T(0)\} = N(T)$ (see Proposition I.2.8). Also $R(qT) = R(T)/T(0)$.

Hence from (1) and (2),

$$\dim D(T) + \dim T(0) = \dim N(T) + \dim R(T). \quad \square$$

I.6.5 Exercise *Let $M \subset D(T)$. Show that*

$$\dim R(T)/T(M) = \dim D(T)/M$$

if and only if $N(T) \subset M$.

I.6.6 Definitions For a linear relation $T \in LR(X,Y)$ we define

$$\alpha(T) = \dim N(T)$$
$$\beta(T) = \dim Y/R(T)$$

called the **nullity** and **deficiency** of T respectively.

The **index** $\kappa(T)$ of T is defined as

$$\kappa(T) = \alpha(T) - \beta(T)$$

provided $\alpha(T)$ and $\beta(T)$ are not both infinite. If $\alpha(T)$ and $\beta(T)$ are both infinite, then T is said to have no index.

The index of T^{-1} is called the **coindex** of T and is denoted by $c\kappa(T)$. Thus

$$c\kappa(T) = \dim T(0) - \operatorname{codim} D(T).$$

All single valued, or everywhere defined, linear relations have a coindex. In case T is single valued and everywhere defined, then $c\kappa(T) = 0$.

I.6.7 Exercise Let $T \in LR(X,Y)$, where X and Y are finite dimensional. Show that

$$c\kappa(T) + \dim X = \kappa(T) + \dim Y.$$

In particular, if X and Y have the same dimension, then $c\kappa(T) = \kappa(T)$.

I.6.8 Lemma. *Let $X = M + N$, $M \cap N = \{0\}$ and suppose that $D(T) = X$, $N(T) \subset N$. Then $R(T) = T(M) + T(N)$, and $T(M) \cap T(N) = T(0)$.*

PROOF. Since T is everywhere defined, it is clear that $R(T) = T(M) + T(N)$ (see I.3.1(c)).

Now let $y \in T(M) \cap T(N)$. We have $y \in Tm$ and $y \in Tn$ where $m \in M$, $n \in N$. Then $0 \in Tm - Tn = T(m-n)$, and so $m - n \in N(T) \subset N$, by Proposition I.2.8. Therefore $m \in N$, whence $m = n = 0$ since $M \cap N = \{0\}$. Hence $T(M) \cap T(N) = T(0)$. $\square$

I.6.9 Lemma. *Let $S \in LR(Y, Z)$. Then*

$$\beta(q_S ST) = \beta(ST)$$

and

$$\alpha(q_S ST) = \alpha(ST).$$

PROOF. Let $R(ST) + E = Z$ where $E \cap R(ST) = \{0\}$. Write $q := q_S$. Then since $N(q) = S(0) \subset R(ST)$, Lemma I.6.8 yields immediately

(1) $$R(qST) + qE = qZ, \quad R(qST) \cap qE = \{0\}.$$

Furthermore since $q|_E$ is injective we have $\dim qE = \dim E$, i.e. $\beta(qST) = \beta(ST)$.

Since $N(q_{ST} ST) = N(ST)$, the second equality follows from the inclusions

$$N(ST) \subset N(q_S ST) \subset N(q_{ST} ST). \quad \square$$

I.6.10 Corollary. *We have*

$$\alpha(qT) = \alpha(T)$$

and

$$\beta(qT) = \beta(T).$$

PROOF. Since $N(qT) = N(T)$, we have $\alpha(qT) = \alpha(T)$. The second equality follows immediately from Lemma I.6.9. $\square$

I.6.11 Theorem. *Let $T \in LR(X, Y)$ and $S \in LR(Y, Z)$, and suppose that $D(S) = Y$. Then*

(1) $$\alpha(ST) + \beta(T) + \beta(S) + \dim T(0) \cap N(S) = \beta(ST) + \alpha(T) + \alpha(S)$$

PROOF. Consider first the case when S is single valued. Let ψ be the linear relation from $N(ST)/N(T)$ to $R(T) \cap N(S)$ defined by

(2) $$\psi[x] := Tx.$$

Put

$$A := R(T) \cap N(S).$$

The relation ψ is easily seen to be injective and surjective. Hence by Proposition I.6.4,

$$\dim N(ST)/N(T) + \dim \psi[0] = \dim A \tag{3}$$

Noting that $\psi[0] = T(0) \cap N(S)$, we have

$$\alpha(ST) + \dim T(0) \cap N(S) = \alpha(T) + \dim A \tag{4}$$

Choose a subspace B of $N(S)$ such that

$$N(S) = A + B, \quad A \cap B = \{0\} \tag{5}$$

Then

$$\alpha(S) = \dim A + \dim B. \tag{6}$$

Now for $y \in Tx \cap B$, we have $y \in A \cap B = \{0\}$. Hence

$$R(T) \cap B = \{0\} \tag{7}$$

Choose a subspace C of Y such that

$$Y = R(T) + B + C, \quad (R(T) + B) \cap C = \{0\} \tag{8}$$

Then

$$\beta(T) = \dim B + \dim C \tag{9}$$

Now by (5), $N(S) = A + B \subset R(T) + B$. Hence from (8), S is injective on C and therefore by Lemma I.6.8 and Proposition I.3.1(c),

$$R(S) = R(ST) + S(C), \; R(ST) \cap S(C) = S(0) \tag{10}$$

and

$$\dim C = \dim S(C). \tag{11}$$

Consequently by (10) and (11),

$$\beta(S) + \dim C = \beta(ST). \tag{12}$$

Adding (4), (9) and (12) gives

$$(13)\qquad \begin{aligned}\alpha(ST) + \beta(T) + \beta(S) + \dim T(0) \cap N(S) + \dim C \\ = \alpha(T) + \dim A + \dim B + \dim C + \beta(ST) \\ = \alpha(T) + \alpha(S) + \dim C + \beta(ST).\end{aligned}$$

If $\dim C = \infty$, then $\beta(T) = \infty$ (by (8)) and $\beta(ST) = \infty$ (by (11) and (12)). Consequently (1) is satisfied, since both sides are infinite.

On the other hand if $\dim C < \infty$, then (1) follows from (13) after cancelling $\dim C$. This completes the proof for the case when $S(0) = \{0\}$.

For the general case, writing $q := q_S$, we now have

$$\alpha(qST) + \beta(T) + \beta(qS) + \dim T(0) \cap N(S) = \beta(qST) + \alpha(T) + \alpha(qS).$$

The theorem now follows from Lemma I.6.9 and Corollary I.6.10. □

The special case of Theorem I.6.9 when $\alpha(T), \alpha(S), \beta(T)$ and $\beta(S)$ are all finite is a generalisation of the algebraic version of the *Index Theorem* for the product of Fredholm operators.

I.6.12 Corollary (Index Theorem for the product).

Let $T \in LR(X,Y)$, $S \in LR(Y,Z)$, $D(S) = Y$, and suppose that S and T have finite indices. Then ST has a finite index and

$$(1)\qquad \kappa(ST) = \kappa(S) + \kappa(T) - \dim T(0) \cap N(S).$$

I.6.13 Examples (a) *Corollary I.6.12 fails without the assumption $D(S) = Y$, even if $R(T) \subset D(S)$.*

Let $X = Y = Z = \mathbb{R}^3$. Define T to be the identity restricted to the subspace $\{(x,0,0) : x \in \mathbb{R}\}$ and S the identity restricted to $\{(x,y,0) : x,y \in \mathbb{R}\}$. Then $\kappa(ST) = -2$, $\kappa(S) = -1$, and $\kappa(T) = -2$.

(b) Let $X = Y = Z = \mathbb{R}^2$. Define

$$S = \begin{pmatrix} 0 & 0 \\ 0 & 1 \end{pmatrix}$$

and let $T = S^{-1}$. We have

$$\begin{aligned} N(T) &= \{0\} \times \{0\}, \quad \alpha(T) = 0 \\ R(T) &= \mathbb{R} \times \mathbb{R}, \quad \beta(T) = 0 \\ N(S) &= \mathbb{R} \times \{0\}, \quad \alpha(S) = 1 \\ R(S) &= \{0\} \times \mathbb{R}, \quad \beta(S) = 1 \\ N(ST) &= \{0\} \times \{0\}, \quad \alpha(ST) = 0 \\ R(ST) &= \{0\} \times \mathbb{R}, \quad \beta(ST) = 1 \\ T(0) \cap N(S) &= \mathbb{R} \times \{0\}. \end{aligned}$$

Then $\kappa(ST) = -1$, $\kappa(S) = 0$, $\kappa(T) = 0$ and $\dim T(0) \cap N(S) = 1$.

I.6.12 Exercise Let X, Y, Z be finite dimensional, $T \in LR(X, Y)$, $S \in LR(Y, Z)$ and $D(S) = Y$. Show that

$$\dim ST(0) + \dim T(0) \cap N(S) = \dim S(0) + \dim T(0). \tag{1}$$

I.7 ADDITIONAL EXAMPLES

1. Let Ω be a directed set, and X the vector space of all convergent nets in a topological vector space Y defined on Ω. Then the convergence

$$\lim_{\Omega} = \{(x, y) \in X \times Y : x \to y\}$$

is a linear relation from X into Y.

2. Let X and Y be topological vector spaces such that Y is complete, and let f be a continuous linear function from a dense subspace M of X into Y. Then the relation T defined by

$$G(T) = \{(x, y) \in X \times Y : y \in \lim_{t \to x} f(t)\}$$

is a linear relation from X into Y. Notice that $\overline{G(f)} = G(T)$.

3. Let Σ be a measure space and Y a topological vector space. Denote by X the vector space of all functions from Σ into Y for which there exists $y \in Y$ such that

$$f(y) = \int_{\Sigma} f \circ x \text{ for all } f \in Y' ,$$

where Y' is the dual space of Y. Then the weak integral

$$\int_{\Sigma} = \{(x, y) \in X \times Y : f(y) = \int_{\Sigma} f \circ x \text{ for all } f \in Y'.\}$$

is a linear relation from X into Y.

4. If T is an everywhere defined linear operator from X into X different from the identity, then T^{-1} is a non reflexive linear relation on X.

5. Let $D(T) = X$. Then $T^{-1}T$ is reflexive. Moreover, if A is any selection of T then

$$T^{-1}T = T^{-1}A = A^{-1}T.$$

6. Let (X, X^*) and (Y, Y^*) be dual pairs (i.e. X^* and Y^* are linear subspaces of the vector space of all linear functionals defined on X and Y respectively). Let $A : X^* \to Y^*$ be an everywhere defined linear operator. Define the relation $T \in R(X, Y)$ by the equation

$$(Ax^*)(y) = x^*(x)$$

for all $x \in X$, $x^* \in X^*$ and $(x, y) \in G(T)$. Verify that T is a linear relation.

I.8 NOTES AND REMARKS

Some authors define a relation to be a non empty subset of $U \times V$ (see e.g. Cech [1, p.25]). Property (ii) of Proposition I.2.3, namely "$G(T)$ is a subspace", is often used as the definition of a linear relation (e.g. R. Arens [1], and De Wilde [1]). Books and articles which are concerned chiefly with convex processes, such as Aubin and Frankowska [1], usually define a linear relation by the inclusion property of Proposition I.2.16.

Other types of restriction besides the one introduced in Subsection I.1.5, sometimes referred to as the "domain restriction", can be considered; see for example Cech [1, p.27].

Linear relations made their first appearance in functional analysis in J. von Neumann [1]. Convex processes were introduced by R.T. Rockafellar [1,2].

Let us briefly review the relationship between linear relations and certain other types of set valued maps between (real) vector spaces X and Y. Let $F \in R(X, Y)$. Then F is called

(a) **affine** if for all $\alpha \in \mathbb{R}$ and $x_1, x_2 \in X$,

$$\alpha F x_1 + (1 - \alpha) F x_2 \subset F(\alpha x_1 + (1 - \alpha) x_2) \tag{1}$$

(b) **convex** if for $\alpha \in [0, 1]$ and $x_1, x_2 \in X$, the inclusion (1) holds

(c) a **convex process** if for all $\alpha, \beta \geq 0$ and $x_1, x_2 \in X$,

$$\alpha F x_1 + \beta F x_2 \subset F(\alpha x_1 + \beta x_2) \tag{2}$$

(d) **linear** if for all $\alpha, \beta \in \mathbb{R}$ and $x_1, x_2 \in X$, the inclusion (2) holds. (Definition (d) coincides with the characterisation I.2.16 – see also R. Arens [1]).

Evidently we have the inclusions:

linear $\subset$ affine $\subset$ convex

and

$$\text{linear} \subset \text{convex process} \subset \text{convex}.$$

S.M. Robinson [1] gives the following example: Let $T \in R(\mathbb{R}, \mathbb{R}^2)$ be defined by

$$D(T) = \{x : x \geq 0\}$$
$$Tx = \{(y, z) : y^2 \leq zx \text{ and } z \geq 0\} \qquad (x \in D(T)).$$

Then T is a convex process. However, T is not a linear relation.

A theorem of Rockafellar [2, p.414] states: *If $F \in R(X,Y)$ is a convex process with $D(F) = X$ and $F(0) = \{0\}$, then F is single valued and linear.* This theorem is contained in the following more recent result of F. Deutsch and I. Singer [1] characterizing single valued affine maps: *If $F \in R(X,Y)$, $D(F) = X$ and $F(x_0)$ is a singleton for some $x_0 \in X$, then F is convex if and only if F is single valued and affine.*

Affine relations can be reduced to linear relations (Labuschagne [9]).

The result I.2.12 (Exercise) and Examples 1-3 of Section I.7 are due to A. Száz and G. Száz [1].

Proposition I.3.1 is from R. Mennicken and B. Sagraloff [1].

Subsections I.4.1 and I.4.2 are due to R. Arens [1] apart from some modifications and extensions.

Proposition I.5.3 is due to S.J. Lee and M.Z. Nashed [5].

If $S \in LR(X,Y)$ and $T \in LR(X,Y)$ are everywhere defined linear operators D. Hadwin [1] has shown that

$$\alpha(ST) + \beta(S) + \beta(T) = \beta(ST) + \alpha(S) + \alpha(T). \tag{3}$$

D. Sarason [1] gave an algebraic proof of the Fredholm index theorem for the product, which is equivalent to (3) when the indices of S and T are both finite. Corollary I.6.10 is an extension of this for the multivalued case. Hadwin's result is covered by Theorem I.6.11 which is due to the author (Cross [19]).

CHAPTER II

Normed Linear Relations

In this chapter we present the fundamental notions about linear relations which relate to normed vector spaces. The definitions and notations conform to those in current use in the theory of linear operators.

II.1 THE NORM OF A LINEAR RELATION

Throughout the sequel $X, Y, Z, \ldots$ will denote normed vector spaces, and T a linear relation in $LR(X,Y)$, unless otherwise specified.

We write $B_X := \{x \in X : \|x\| \leq 1\}$.

II.1.1 Notation: The operator Q_E^X. For a given closed linear subspace E of X let Q_E^X (or simply, Q_E) denote the natural quotient map with domain X and null space E.

We shall denote $Q^Y_{\overline{T(0)}}$ by Q_T, or simply by Q when T is understood.

II.1.2 Proposition. *QT is single valued.*

PROOF. Let $x \in D(T)$ and let $z_1, z_2 \in QTx$. Then $z_1 - z_2 \in QTx - QTx = QT(0) \subset Q\overline{T(0)} = 0$. Hence $z_1 = z_2$. □

II.1.3 Definitions We define

$$\|Tx\| := \|QTx\| \qquad (x \in D(T)) \tag{1}$$

$$\|T\| := \|QT\| \tag{2}$$

called the **norm** of Tx and T respectively. If $D(T) = X$ and if $\|T\| < \infty$, then we shall say that T is **bounded**.

Let U and V be nonempty subsets of a normed space. We define the **distance** between U and V by the formula

$$d(U,V) := \inf\{\|u - v\| : u \in U, v \in V\} \tag{3}$$

We shall also write $d(x,V)$, or $d(V,x)$, for the distance between $\{x\}$ and V.

The main justification for this terminology comes from Propositions II.1.5, II.1.6 and II.1.7, despite the fact that $LR(X,Y)$ is not a vector space, and nonzero relations can have zero norm.

II.1.4 Proposition.

(a) *If* $y_1, y_2 \in Tx$, *then* $d(y_1, T(0)) = d(y_2, T(0))$.

(b) *We have for* $x \in D(T)$,

$$\inf_{y \in Tx} d(y, T(0)) = \sup_{y \in Tx} d(y, T(0)).$$

(c) $\|Tx\| = d(y, T(0))$ *for any* $y \in T(x)$.

(d) $\|Tx\| = d(Tx, T(0)) = d(Tx, 0) \quad (x \in D(T))$.

PROOF. (a) Since $y_1 - y_2 \in T(0)$ we have

$$\begin{aligned} d(y_1, T(0)) &= \inf_{z \in T(0)} \|y_1 - z\| \\ &= \inf_{z \in T(0)} \|y_2 + (y_1 - y_2 - z\| \\ &= \inf_{z \in T(0)} \|y_2 - z\| = d(y_2, T(0)). \end{aligned}$$

(b) Let $z \in Tx$. By (a), we have

$$\begin{aligned} d(z, T(0)) &= \sup_{y \in Tx} d(y, T(0)) \\ &= \inf_{y \in Tx} d(y, T(0)) \end{aligned}$$

(c) For any $y \in Tx$, we have $Tx = y + T(0)$ by Proposition I.2.8 and hence $\|Tx\| = \|QTx\| = \|Qy\| = d(y, \overline{T(0)}) = d(y, T(0))$.

(d) Fix $z \in Tx$. Then

$$\begin{aligned} d(Tx, 0) &= \inf_{y \in Tx} \|y\| \\ &= \inf_{y \in Tx} \|z - (z - y)\| \\ &= \inf_{k \in T(0)} \|z - k\| = d(z, T(0)) = \|Qz\| = \|QTx\| = \|Tx\|. \quad \square \end{aligned}$$

II.1.5 Proposition.

(a) *For* $S, T \in LR(X, Y)$ *we have*

$$\|Sx + Tx\| \leq \|Sx\| + \|Tx\| \quad (x \in D(T + S)).$$

(b) *For all* $\alpha \in \mathbb{K}$ *and* $x \in D(T)$ *we have*

$$\|\alpha Tx\| = |\alpha| \, \|Tx\|.$$

PROOF. (a) Let $s \in Sx$, $t \in Tx$. Then by Proposition II.1.4, $s + t \in (S + T)x = Sx + Tx$, and hence

$$\begin{aligned}\|Sx + Tx\| &= d(s + t, (S + T)(0)) \\ &\leq d(s, S(0) + T(0)) + d(t, S(0) + T(0)) \\ &\leq d(s, S(0)) + d(t, T(0)) = \|Sx\| + \|Tx\|\,.\end{aligned}$$

(b) We have

$$\begin{aligned}\|\alpha Tx\| &= \|Q(\alpha T)(x)\| \qquad (\text{since } \alpha T(0) = T(0) \text{ for } \alpha \neq 0) \\ &= \|\alpha QTx\| = |\alpha|\,\|QTx\| = |\alpha|\,\|Tx\|\,. \quad \square\end{aligned}$$

II.1.6 Proposition. $\|T\| = \sup_{x \in B_X} \|Tx\|$.

PROOF. $\|T\| = \|QT\| = \sup_{x \in B_X} \|QTx\| = \sup_{x \in B_X} \|Tx\|$. $\square$

II.1.7 Proposition.

(a) *For $S, T \in LR(X, Y)$ we have*

$$\|S + T\| \leq \|S\| + \|T\|\,.$$

(b) $\|\alpha T\| = |\alpha|\,\|T\| \quad (\alpha \in \mathbb{K})$.

PROOF. From Propositions II.1.5(a) and II.1.6

$$\begin{aligned}\|S + T\| &= \sup\{\|Sx + Tx\| : x \in B_X \cap D(S) \cap D(T)\} \\ &\leq \sup\{\|Sx\| + \|Tx\| : x \in B_X \cap D(S) \cap D(T)\} \\ &\leq \sup\{\|Sx\| : x \in B_X \cap D(S) \cap D(T)\} \\ &\qquad + \sup\{\|Tx\| : x \in B_X \cap D(S) \cap D(T)\} \\ &\leq \sup\{\|Sx\| : x \in B_{D(S)}\} \\ &\qquad + \sup\{\|Tx\| : x \in B_{D(T)}\} = \|S\| + \|T\|\,.\end{aligned}$$

(b) Combine Propositions II.1.5(b) and II.1.6. $\square$

We warn that it is not in general true that $\|S\| - \|T\| \leq \|S - T\|$ (see Exercise II.1.12).

II.1.8 Example *Computing the norm of Q_E^{-1}.*

Let E be a proper closed subspace of X. Then for $y \in D(Q_E^{-1})$ we have

(1) $$\left\|Q_E^{-1}y\right\| = \|y\|$$

Indeed writing $Q = Q_E$, if $y = Qx$, then $\|Q^{-1}Qx\| = d(Q^{-1}Qx, 0)$ (Proposition II.1.4) $= \inf_{e \in E} \|x+e\| = \|Qx\| = \|y\|$. Thus $\left\|Q_E^{-1}\right\| = 1$ $(E \neq X)$.

II.1.9 Example *Inverse of a projection map.*

Let P be a linear projection (not necessarily bounded) with domain X. We have for $y \in D(P^{-1}) = P(X)$, that $y = Py$, and hence $P^{-1}y = y + P^{-1}(0)$. Therefore $\left\|P^{-1}y\right\| = d(y + P^{-1}(0), P^{-1}(0)) = d(y, P^{-1}(0)) = \|Q_{P^{-1}}y\|$ by Proposition II.1.4(c). Consequently by Proposition II.1.6.

(1) $$\|P^{-1}\| = \begin{cases} 1 & \text{if } \overline{N(P)} \not\supset P(X) \\ 0 & \text{otherwise.} \end{cases}$$

The geometric characterisation of the norm is the following:

II.1.10 Proposition.

(a) $\|T\| < \infty$ *if and only if there exists a number* $\lambda > 0$ *for which*

(1) $$TB_{D(T)} \subset \lambda B_{R(T)} + T(0)$$

(b) *If* $\|T\| < \infty$, *then*

(2) $$\|T\| = \inf_{\lambda > 0} \{\lambda : TB_{D(T)} \subset \lambda B_{R(T)} + T(0)\}.$$

PROOF. (a) Suppose that $\|T\| < \infty$. Then by Proposition II.1.4(d)

$$\sup_{\|x\| \leq 1} \sup_{y \in Tx} d(y, T(0)) = \|T\| < \infty.$$

Hence if $x \in B_{D(T)}$ and $y \in Tx$, then there exists $k \in T(0)$ such that, given $\epsilon > 0$, we have

$$\|y - k\| < \|T\| + \epsilon,$$

i.e. $y - k \in (\|T\| + \epsilon)B_{R(T)}$. Therefore

(3) $$y \in (\|T\| + \epsilon)B_{R(T)} + T(0) \text{ as required.}$$

Conversely, suppose that (1) holds. Let $x \in B_{D(T)}$ and consider $y \in Tx$. Then $y = \lambda y_1 + k$ where $\|y_1\| \leq 1$ and $k \in T(0)$. Hence $\|y - k\| \leq \lambda$. Therefore $d(y, T(0)) \leq \lambda$. Consequently by Proposition II.1.4, $\|T\| \leq \lambda < \infty$.

(b) Let $\|T\| < \infty$. If $\|T\| = 0$, then $TB_{D(T)} \subset \overline{T(0)}$ and so (2) follows trivially. Now let $\|T\| > 0$. By (3)

$$\inf_{\lambda>0}\{\lambda : TB_{D(T)} \subset \lambda B_{R(T)} + T(0)\} \leq \|T\|\,.$$

Now suppose that α is any number satisfying $0 < \alpha < \|T\|$. Choose $x \in B_{D(T)}$ and $y \in Tx$ so that

$$\alpha < d(y, T(0)) \tag{4}$$

Assume that $TB_{D(T)} \subset \alpha B_{R(T)} + T(0)$. Then there exists $y_1 \in B_{R(T)}$, $k \in T(0)$ such that $y = \alpha y_1 + k$. Then

$$\|y - k\| \leq \alpha < d(y, T(0))$$

by (4), a contradiction. We conclude that

$$\alpha < \inf\{\lambda > 0 : TB_{D(Y)} \subset \lambda B_{R(T)} + T(0)\}$$

and the result follows. □

II.1.11 Exercise Use Proposition II.1.10 to verify (1) of II.1.9.

II.1.12 Exercise Let $X \neq \{0\}$, $S = I_X$, $G(T) = X \times X$. Show that

(1) $\|Sx - Tx\| \ngeq \|Sx\| - \|Tx\| \quad (x \neq 0)$
(2) $\|S - T\| \ngeq \|S\| - \|T\|$.

II.1.13 Exercise Let $S, T \in LR(X, Y)$. Then for $x \in N(S)$ we have

$$\|(S + T)x\| \leq \|Tx\|\,. \tag{1}$$

Letting T be nonzero and single valued, and $G(S) = D(T) \times R(T)$ shows that strict inequality may hold in (1).

II.1.14 Exercise Show that Proposition II.1.10 remains valid if closed balls are replaced by open balls.

II.1.5 Exercise Use Exercise II.1.14 to verify that $\|Q_N\| = 1$ whenever N is a proper closed subspace of X.

II.2 THE MINIMUM MODULUS

II.2.1 Definition The **minimum modulus** of T is the quantity

$$\gamma(T) := \sup\{\lambda : \|Tx\| \geq \lambda d(x, N(T)) \text{ for } x \in D(T)\}$$

A useful equivalent formula is (II.2.2) below.

II.2.2 Proposition. *We have*

$$(1) \qquad \gamma(T) = \begin{cases} \infty, & \text{if } D(T) \subset \overline{N(T)} \\ \inf & \{\frac{\|Tx\|}{d(x,N(T))} : x \in D(T), x \notin \overline{N(T)}\} \text{ otherwise.} \end{cases}$$

PROOF. Write $\gamma_1(T)$ for the expression in (1). Let $\lambda \geq 0$ satisfy

$$(2) \qquad \|Tx\| \geq \lambda d(x, N(T))$$

for all $x \in D(T)$.

If $D(T) \subset \overline{N(T)}$, then $\lambda < \gamma_1(T)$. Otherwise, we have $D(T) \not\subset \overline{N(T)}$ and then for $x \in D(T) \backslash \overline{N(T)}$, $\dfrac{\|Tx\|}{d(x, N(T))} \geq \lambda$ by (2). Hence $\lambda \leq \gamma_1(T)$. Taking the supremum over λ in (2) gives $\gamma(T) \leq \gamma_1(T)$.

Now suppose that $\gamma(T) < \gamma_1(T)$. Then $\gamma(T) < \infty$ and so $D(T) \not\subset \overline{N(T)}$. Furthermore, there exists $x \in D(T)$ such that $\|Tx\| < \gamma_1(T)d(x, N(T))$, which contradicts (1). Hence $\gamma(T) = \gamma_1(T)$. □

II.2.3 Example Let f be a discontinuous linear functional. Then $N(f)$ is dense in the whole space and hence $\gamma(f) = \infty$.

A geometric characterisation of $\gamma(T)$ is the following.

II.2.4 Proposition. *We have*

$$(1) \qquad \gamma(T) = \sup\{\lambda : TB_{D(T)} \supset \lambda B_{R(T)}\}.$$

PROOF. Let γ_1 denote the expression on the right hand side of (1). Suppose $\gamma_1 \neq \infty$ and that $\gamma_1 > \gamma(T)$. Then there exists $(x, y) \in G(T)$ such that

$$(2) \qquad 1 = \|y\| < \gamma_1 d(x, N(T)) \cdot (1 + 2\epsilon)^{-1}$$

where $\epsilon > 0$ is so small that $\gamma_1 > (1 + 2\epsilon)\gamma(T)$. By (1), $y \in \gamma_1^{-1}(1 + \epsilon)TB_{D(T)}$, so $x_1 \in \gamma_1^{-1}(1 + \epsilon)B_{D(T)}$ may be chosen so that $y \in Tx_1$. Since $x_1 - x \in N(T)$, we have from (2) that $1+2\epsilon < \gamma_1 d(x, N(T)) = \gamma_1 d(x_1, N(T)) \leq \gamma_1 \|x_1\| \leq 1+\epsilon$, a contradiction. Therefore $\gamma_1 \leq \gamma(T)$.

To establish the opposite inequality, it is sufficient to consider the case when $\gamma_1 < \infty$. In that case, suppose that $\gamma_1 < \alpha < \gamma(T)$. Then $TB_{D(T)} \not\supset \alpha B_{R(T)}$ and there exists $y \in R(T)$ with $\|y\| = 1$ for which $y \notin \alpha^{-1}TB_{D(T)}$. Thus

$$(3) \qquad \|x\| \leq \alpha^{-1} \implies y \notin Tx.$$

Let $y \in Tx_0$. Fix β such that $\alpha < \beta < \gamma(T)$. Then

$$1 = \|y\| \geq \|Tx_0\| \geq \beta d(x_0, N(T)) \quad (\text{since } \gamma(T) > \beta).$$

Choose $x_1 \in N(T)$ so that $\|Tx_0\| \geq \alpha \|x_0 + x_1\|$. Writing $x = x_0 + x_1$, we have $Tx_0 = Tx$, and so $y \in Tx$ where $1 = \|y\| \geq \|Tx\| \geq \alpha \|x\|$, contradicting (3). Therefore $\gamma_1 \geq \gamma(T)$.

It only remains to consider the case where $\gamma_1 = \infty$. In this case we have

$$TB_{D(T)} \supset \bigcup_{x \in \mathbb{N}} \{nB_{R(T)}\} = R(T)$$

whence by Corollary I.2.11

$$B_{D(T)} + T^{-1}(0) \supset D(T).$$

Hence for each $\epsilon > 0$,

$$\epsilon B_{D(T)} + T^{-1}(0) \supset D(T)$$

and so taking the intersection over $\epsilon > 0$,

$$\overline{N(T)} \supset D(T)$$

and hence $\gamma(T) = \infty$ by Proposition II.2.2. □

The next formula shows that the norm and minimum modulus are dual constructs.

II.2.5 Theorem. *We have the formula*

(1) $$\gamma(T) = \left\|T^{-1}\right\|^{-1}$$

PROOF. Let $\lambda \geq 0$. We have the following chain of implications:

$$\begin{aligned} T^{-1}B_{R(T)} \supset \lambda B_{D(T)} &\implies B_{R(T)} + T(0) \supset \lambda TB_{D(T)} \\ &\implies T^{-1}(B_{R(T)} + T(0)) \supset \lambda B_{D(T)} + T^{-1}(0) \\ &\implies T^{-1}B_{R(T)} + T^{-1}(0) \supset \lambda B_{D(T)} + T^{-1}(0) \\ &\implies T^{-1}B_{R(T)} \supset \lambda B_{D(T)} \end{aligned}$$

Hence

(2) $$T^{-1}B_{R(T)} \supset \lambda B_{D(T)} \iff \lambda TB_{D(T)} \subset B_{R(T)} + T(0).$$

These are two cases to consider.

Case 1: $\|T\| < \infty$. We have from Proposition II.1.10

$$\begin{aligned}\|T\| &= \inf\{\lambda > 0 : \lambda^{-1}TB_{D(T)} \subset B_{R(T)} + T(0)\} \\ &= (\sup\{\lambda > 0 : \lambda TB_{D(T)} \subset B_{R(T)} + T(0)\})^{-1} \\ &= (\sup\{\lambda > 0 : T^{-1}B_{R(T)} \supset \lambda B_{D(T)}\})^{-1} \text{ by (2)} \\ &= \gamma(T^{-1})^{-1} \text{ by Proposition II.2.4.}\end{aligned}$$

Case 2: $\|T\| = \infty$. By Proposition II.1.10 we have:

$$TB_{D(T)} \not\subset \lambda B_{R(T)} + T(0) \text{ for all } \lambda > 0.$$

Hence $$\lambda TB_{D(T)} \not\subset B_{R(T)} + T(0) \text{ for all } \lambda > 0.$$

Therefore $$T^{-1}B_{R(T)} \not\supset \lambda B_{D(T)} \text{ for all } \lambda > 0, \text{ by (2).}$$

Hence by Proposition II.2.4, $\gamma(T^{-1}) = 0$ and so $\|T\|^{-1} = \gamma(T^{-1})$. Hence in either case, $\|T\| = \gamma(T^{-1})^{-1}$.

The formula (1) now follows on replacing T by T^{-1}. □

II.3 CONTINUITY AND OPENNESS

II.3.1 Definitions

Let S be an arbitrary relation from one topological space to another. Then S is said to be **continuous** if for each neighbourhood V in $R(S)$, the inverse image $S^{-1}(V)$ is a neighbourhood in $D(S)$. The relation S is called **open** if whenever U is a neighbourhood in $D(S)$, the image $S(U)$ is a neighbourhood in $R(S)$.

We note that

(1) *$S$is continuous if and only if S^{-1} is open.*

It is simple to verify that the composition of continuous single valued relations is continuous. Conditions will be derived under which the composition of linear relations is continuous or open (see Theorem II.3.11 and Corollary II.3.13).

Let U_X denote the open unit ball of X, i.e.

(2) $$U_X := \{x \in X : \|x\| < 1\}.$$

Let E be a closed subspace of X. A simple verification shows that

(3) $$Q_E U_X = U_{Q_E(X)}$$

(4) $$Q_E B_X = B_{Q_E(X)}.$$

We note the following simple topological criterion for continuity:

(5) *T is continuous if and only if $T\overline{\Omega} \subset \overline{T\Omega}$ for every subset Ω of $D(T)$.* (See Exercise II.3.21).

II.3.2 Proposition. *Let $T \in LR(X, Y)$. Then*

(a) *T is continuous if and only if $\|T\| < \infty$.*
(b) *T is open if and only if $\gamma(T) > 0$.*

In particular:

(c) *If $\dim D(T) < \infty$, then T is continuous.*
(d) *If $\dim R(T) < \infty$, then T is open.*

PROOF. (a) Let T be continuous. Then there exists an open neighbourhood U in $D(T)$ such that $T^{-1}U_{R(T)} \supset U$. Hence

$$T^{-1}U_{R(T)} - T^{-1}U_{R(T)} \supset U - U,$$

an open neighbourhood of $\{0\}$ in $D(T)$. Also $2T^{-1}U_{R(T)} = T^{-1}(U_{R(T)} - U_{R(T)}) = T^{-1}U_{R(T)} - T^{-1}U_{R(T)}$. Consequently there exists $\lambda > 0$ such that

$$T^{-1}(U_{R(T)}) \supset \lambda U_{D(T)}$$

and hence

$$T^{-1}(B_{R(T)}) \supset \frac{1}{2}\lambda B_{D(T)}.$$

Therefore

$$TT^{-1}(B_{R(T)}) = B_{R(T)} + T(0) \supset \frac{1}{2}\lambda T B_{D(T)}$$

and hence $\|T\| < \infty$ by Proposition II.1.10.

Conversely, assume that $\|T\| < \infty$. Let V be an open ball neighbourhood in $R(T)$, with center y. Then

$$V_0 = V - \{y\} = \alpha U_{R(T)} \quad \text{for some } \alpha > 0.$$

By Proposition II.1.10, we have for some $\lambda > 0$,

$$TU_{D(T)} \subset \lambda U_{R(T)} + T(0).$$

Hence

$$U_{D(T)} + T^{-1}(0) \subset \lambda T^{-1} U_{R(T)} = \alpha^{-1} \lambda T^{-1} V_0.$$

Thus

$$T^{-1}(V_0) = T^{-1}(V - y) \supset \lambda^{-1} \alpha U_{D(T)} + T^{-1}(0).$$

Hence

$$T^{-1}V - T^{-1}y + T^{-1}y \supset \lambda^{-1}\alpha U_{D(T)} + T^{-1}y$$
$$\text{i.e.} \qquad T^{-1}V \supset \lambda^{-1}\alpha U_{D(T)} + T^{-1}y$$

which is a neighbourhood in $D(T)$.

Therefore T is continuous.

(b) Follows immediately from (a), using Theorem II.2.5.

(c) Follows from the continuity of the linear operator QT, since $\|T\| = \|QT\|$.

(d) Substitute T^{-1} for T in (c). □

II.3.3 Example. Let P be a linear projection operator. If $\overline{N(P)} \not\supset R(P)$, then since $\|P^{-1}\| = 1$ (see Example II.1.9), P^{-1} is continuous (Proposition II.3.2), and $\gamma(P) = 1$ (Theorem II.2.5); otherwise, if $\overline{N(P)} \supset R(P)$, then $\|P^{-1}\| = 0$ and so $\gamma(P) = \infty$ (see Example II.1.9). In either case, P is open.

II.3.4 Proposition. *We have*

(a) $N(T) \subset N(QT)$ *with equality if* $T(0)$ *is relatively closed in* $R(T)$.
(b) $\gamma(T) \leq \gamma(QT)$ *with equality if* $T(0)$ *is relatively closed in* $R(T)$.

PROOF. (a)

$$\begin{aligned} N(T) &= \{x : (x, 0) \in G(T)\} \\ &= \{x : Tx \cap T(0) \neq \emptyset\} \text{ by Proposition I.2.8} \\ &\subset \{x : Tx \cap \overline{T(0)} \neq \emptyset\} \\ &= N(QT) \end{aligned}$$

with equality whenever $T(0)$ is closed in $R(T)$.

(b) $\gamma(T) = \sup\{\lambda : \|Tx\| \geq \lambda d(x, N(T)) \text{ for all } x \in D(T)\} \leq \sup\{\lambda : \|Tx\| \geq \lambda d(x, N(QT)) \text{ for } x \in D(T)\}$, with equality when $T(0)$ is closed in $R(T)$, by part (a). The result now follows from the equality $\|Tx\| = \|QTx\|$. □

The next example shows that equality can fail to hold in (a) and (b) of Proposition II.3.4.

II.3.5 Example Let f be a discontinuous linear functional defined everywhere on an infinite dimensional normed space X and let $T = f^{-1}$. Then $T(0)$ is a dense hyperplane in X and so $Q_T = O_X$. Hence

$$N(T) = O \neq \mathbb{K} = N(QT)$$

and we have

$$\gamma(T) = O, \gamma(QT) = \infty.$$

II.3.6 Remark From Proposition II.3.2 and II.3.4 we see that

$$T \textit{ is open } \Longrightarrow QT \textit{ is open.}$$

The converse is false, as shown by Example II.3.5.

II.3.7 Proposition. *Let T be open. Then*

$$R(T) \cap \overline{T(0)} \subset T\overline{T^{-1}(0)}$$

with equality if $T^{-1}(0)$ is relatively closed in $D(T)$ (for example, if T is injective).

PROOF. Let T be open and let $\{y_n\}$ be a sequence in $T(0)$ such that $\lim y_n = y \in R(T)$. Assume first that T is injective. By the continuity of T^{-1} we have

$$\lim \left\|T^{-1}(y_n - y)\right\| \leq \lim \left\|T^{-1}\right\| \|y_n - y\| = 0.$$

Hence $\left\|T^{-1}y\right\| = 0$, whence $y \in N(T^{-1}) = T(0)$ since T^{-1} is single valued.

Turning to the general case, let $Q := Q_{T^{-1}}$. From II.3.1(3) we see that TQ^{-1} is open. Hence from what has been proved we have

$$D(QT^{-1}) \cap \overline{N(QT^{-1})} = N(QT^{-1}).$$

Combining with Proposition II.3.4 we obtain

$$\begin{aligned} R(T) \cap \overline{T(0)} = D(T^{-1}) \cap \overline{N(T^{-1})} &\subset D(QT^{-1}) \cap N\overline{(QT^{-1})} \\ &= N(QT^{-1}) = T\overline{(T^{-1}(0))}. \quad \square \end{aligned}$$

II.3.8 Example The following example shows that the inclusion in Proposition II.3.7 may be proper: Let P be a linear projection whose null space is a dense proper subspace of the domain of P. Then P is open and $R(P) \cap \overline{P(O)} = \{0\} \neq P\overline{N(P)} = R(P)$.

II.3.9 Corollary. *If T is open and $N(T)$ is closed, then $N(T) = N(QT)$ and $\gamma(T) = \gamma(QT)$.*

PROOF. Combine II.3.4 with II.3.7. □

II.3.10 Proposition. *Let M be a nonempty subset of $R(T)$ and let $\gamma(T) < \infty$. Then for $N \subset D(T)$ we have*

$$d(TN, M) \geq \gamma(T) d(N, T^{-1}M)$$

PROOF. Let $\epsilon > 0$ and choose $m \in M$ and $x \in N$ such that

$$d(TN, M) > d(Tx - m, 0) - \epsilon \tag{1}$$

Now $Tx - m = T(x - T^{-1}m)$. Hence

$$\begin{aligned}
d(Tx - m, 0) &= d(Tx - m - T(0), 0) = d(Tx, m + T(0)) \\
&= d(Tx, TT^{-1}m) = \inf_{h \in T^{-1}m} d(Tx, Th) \\
&= \inf_{h \in T^{-1}m} d(T(x-h), 0) = \inf_{h \in T^{-1}m} \|T(x-h)\| \quad \text{(see II.1.4)} \\
&\geq \gamma(T) \inf_{h \in T^{-1}m} d(x - h, T^{-1}(0)) \\
&= \gamma(T) \inf_{h \in T^{-1}m} d(x, h + T^{-1}(0)) \\
&= \gamma(T) d(x, T^{-1}m) \geq \gamma(T) d(x, T^{-1}M) \geq \gamma(T) d(N, T^{-1}M).
\end{aligned}$$

Hence from (1)

$$d(TN, M) > \gamma(T) d(N, T^{-1}M) - \epsilon. \quad \square$$

II.3.11 Theorem. *Let $S \in LR(Y, Z)$, $T \in LR(X, Y)$. Then*

$$\gamma(ST) \geq \gamma(S|_{R(T)})\gamma(T) \quad (\infty.0 \textit{ excluded}) \tag{1}$$

with $\gamma(ST) = \infty$ whenever $\gamma(T) = \infty$ (even if $\gamma(S|_{R(T)}) = 0$). Moreover we have the implication

$$S^{-1}(0) \subset R(T) \implies \gamma(ST) \geq \gamma(S)\gamma(T). \tag{2}$$

PROOF. Let $x \in D(ST)$. Since $ST = (S|R(T))T$ (see I.1.3(2)), we shall assume without loss of generality that $S = S|R(T)$. First assume $\gamma(S), \gamma(T) < \infty$. We have from Proposition II.3.10,

$$\begin{aligned}\|STx\| = d(STx, ST(0)) &\geq \gamma(S)d(Tx, S^{-1}ST(0)) \\ &\geq \gamma(S)\gamma(T)d(x, T^{-1}S^{-1}ST(0))\end{aligned}$$

(since $S^{-1}ST(0) = T(0) + S^{-1}(0) \subset R(T)$)

$$= \gamma(S)\gamma(T)d(x, T^{-1}S^{-1}(0)).$$

Hence the inequality (1) follows.

Now assume that $S^{-1}(0) \subset R(T)$ (here we no longer suppose that $S = S|_{R(T)}$). In that case $S^{-1}ST(0) = T(0) \cap D(S) + S^{-1}(0) \subset R(T)$ and therefore

$$d(Tx, S^{-1}ST(0)) \geq \gamma(T)d(x, T^{-1}S^{-1}(0))$$

and hence, as before, we obtain

$$\|STx\| \geq \gamma(S)\gamma(T)d(x, S^{-1}T^{-1}(0))$$

from which the implication (2) follows.

Now let $\gamma(T) = \infty$. Then $T^{-1}(0)$ is dense in $D(T)$ (see II.2.2). Hence $d(x, T^{-1}(0)) = 0$ for all x in $D(ST)$. But then $N(ST)$ is dense in $D(ST)$ which means that $\gamma(ST) = \infty$.

Finally, suppose that $\gamma(S) = \infty$, $\gamma(T) < \infty$. Since now $d(Tx, S^{-1}(0)) = 0$, hence $\gamma(T)d(x, T^{-1}S^{-1}(0)) = 0$ (II.3.10). So $\gamma(ST) = \infty$ if $\gamma(T) \neq 0$.

The proof of the implication (2) follows similarly. □

II.3.12 Remarks

(1) Theorem II.3.11 already shows that the inequality

$$\|STx\| \geq \gamma(S)d(Tx, T(0) \cap D(S) + S^{-1}(0))$$

holds for all $x \in D(ST)$.

(2) If $0.\infty$ occurs on the right hand side of II.3.11(1), then $\gamma(ST) = \infty$. However, the case $\infty.0$ does not work. Consider $f = S$, $f^{-1} = T$, where $f : X \to \mathbb{C}$ is an unbounded linear functional. Then $\gamma(ff^{-1}) = \gamma(I) = 1$, $\gamma(f^{-1}) = \|f\|^{-1} = 0$ and $\gamma(f) = \infty$ (since $\overline{N(f)} = X$).

The following corollary to Theorem II.3.10 gives a sufficient condition for the composition of two continuous linear relations to be continuous.

II.3.13 Corollary. *Let $T \in LR(X,Y)$ and $S \in LR(Y,Z)$. Then*

$$\|ST\| \leq \|S\| \, \|I_{D(S)}T\| \quad (\infty.0 \text{ excluded}) \tag{1}$$

with $\|ST\| = 0$ whenever $\|S\| = 0$ even if $\|I_{D(S)}T\| = \infty$. Moreover, we have the implication

$$T(0) \subset D(S) \implies \|ST\| \leq \|S\| \, \|T\| \tag{2}$$

PROOF. The inequality (1) follows at once from (1) of Theorem II.3.11, using II.2.5(1).

Now suppose that $T(0) \subset D(S)$. Then for $x \in D(ST)$ and $y \in Tx$ we have, since $x \in T^{-1}(D(S))$, that $Tx \subset D(S) + T(0) = D(S)$, and so $I_{D(S)}Tx = Tx \cap D(S) = Tx$. Therefore

$$\|ST\| \leq \|S\| \, \|T|_{D(ST)}\| \tag{3}$$

and (2) follows. □

II.3.14 Remarks

(1) There is no conclusion if the right hand side of II.3.13(1) is of the form $\infty.0$ (see II.3.12(2)).

(2) We shall later give an example of a pair of open linear operators whose composition is not open (see Example V.I.13). Hence continuous linear relations exist whose composition is defined but not continuous.

(3) By considering the restriction of T to a one dimensional subspace of $D(ST)$, it follows from (2) of Corollary II.3.13 that

$$T(0) \subset D(S) \implies \|STx\| \leq \|S\| \, \|Tx\|$$

holds for each $x \in D(ST)$.

II.3.15 Theorem. *(Uniform Boundedness Principle).*

Let $\{Y_\alpha : \alpha \in \Lambda\}$ be a family of normed spaces, X a fixed Banach space, and $\{T_\alpha : \alpha \in \Lambda\}$ a family of bounded linear relations such that $T_\alpha \in LR(X, Y_\alpha)$ for each $\alpha \in \Lambda$. Suppose that

$$\sup\{\|T_\alpha x\| : \alpha \in \Lambda\} < \infty \quad \textit{for each } x \in X.$$

Then

$$\sup\{\|T_\alpha\| : \alpha \in \Lambda\} < \infty.$$

PROOF. Consider the sequence of closed sets $A_n = \{x \in X : \|T_\alpha x\| \leq n$ for $\alpha \in \Lambda\}$. Since $X = \cup_{n=1}^{\infty} A_n$, by Baire's Category theorem, A_n contains a non empty open set for some $n \in \mathbb{N}$. Let B be a ball, with centre z and radius r, contained in A_n. For $x \in rB_X$ we have $x + z \in B$ and

$$\|T_\alpha x\| = \|T_\alpha(x + z - z)\| \leq \|T_\alpha(x+z)\| + \|T_\alpha z\| \leq 2n.$$

Hence $\|T_\alpha\| \leq 2n/r$ for all $\alpha \in \Lambda$. □

II.3.16 Corollary. *Let (T_n) be a sequence of bounded linear relations in $LR(X,Y)$, and let X be complete. If T is an everywhere defined relation in $LR(X,Y)$ such that $\overline{T(0)} \supset T_n(0)$ $(n \in \mathbb{N})$ and $\lim_n \|T_n x - Tx\| = 0$ for all $x \in X$, then T is bounded.*

II.3.17 Corollary. *Let X be a normed space and let W be a subset of X such that*

$$\sup_{w \in W} |x'w| < \infty \qquad (x' \in X').$$

Then W is bounded.

PROOF. Consider W as a subset of X'' and apply Theorem II.3.15. □

II.3.18 Proposition. *If $X = M + N$ where $M \cap N = \{0\}$ and M is closed and finite codimensional, then the projection defined on X with range M and kernel N is bounded.*

PROOF. Let $\{x_1, \ldots, x_n\}$ be a basis for N and let $N_i := sp\{x_1, \ldots, x_{i-1}, x_{i+1}, \ldots, x_n\}$. Since each $M + N_i$ is closed (see e.g. Goldberg [3; p.16]) there exists by the Hahn Banach theorem an $f_i \in X'$ such that $f_i(x_i) = 1$ and $f_i(x) = 0$ for $x \in M + N_i$. Define $Qx = \sum_{i=1}^{n} f_i(x)x_i$ $(x \in X)$. Then Q is a bounded projection with range N and null space M, and $P = I - Q$ is the required projection. □

II.3.19 Corollary. *If M is a finite codimensional subspace of $D(T)$, closed in $D(T)$, and if $T|_M$ is continuous, then T is continuous.*

PROOF. Let P be as in Proposition II.3.18. We have

$$\begin{aligned}\|Tx\| &\le \|TPx\| + \|T(I-P)x\| \\ &\le (\|T|_M\|\,\|P\| + \|T|_N\|\,\|I-P\|)\,\|x\|\end{aligned}$$

where $\|T|_N\| < \infty$ by Proposition II.3.2(c). □

II.3.20 Proposition. *Let T be open and injective and let $S \in LR(X,Y)$ be a linear operator satisfying $\|S\| < \gamma(T)$. Then $T+S$ is open and injective.*

PROOF. We have for $x \in D(T+S)$,

$$\begin{aligned} \|(T+S)x\| &\geq \|Tx\| - \|Sx\| \quad (\text{since } S(0) = \{0\}) \\ &\geq \gamma(T)\,\|x\| - \|S\|\,\|x\| = (\gamma(T) - \|S\|)\,\|x\|. \quad \square \end{aligned}$$

II.3.21 Exercise Use Proposition II.3.2 to verify II.3.1(5).

II.3.22 Exercise Prove Corollary II.3.16. Observe that is is enough to consider the case when T is single valued.

II.4 CONTINUOUS SELECTIONS

In this section we explore the possiblity of finding a continuous selection for a given linear relation T.

II.4.1 Proposition. *If T has a continuous selection A, then T is continuous. Furthermore*

(1) $$\|T\| \leq \|A\|.$$

PROOF. Let A be a continuous selection of T. We have for $x \in D(T)$,

(2) $$Tx = Ax + T(0)$$

and hence

$$\|Tx\| = \inf\{\|y\| : y \in Tx\} \leq \|Ax\| \leq \|A\|\,\|x\|.$$

Therefore T is continuous. $\square$

A sufficient condition for the existence of a continuous selection is provided by the next proposition.

II.4.2 Proposition. *If T is continuous, and if $T(0)$ is the kernel of a continuous projection P defined on $R(T)$, then PT is a continuous selection of T.*

PROOF. Since P and T are continuous and since P is everywhere defined on $R(T)$, the continuity of PT follows from Corollary II.3.13. □

II.4.3 Terminology

Let M be a linear subspace of X. Any subspace N satisfying

$$M \cap N = \{0\} \quad \text{and} \quad M + N = X \tag{1}$$

is called a **complement** of M.

Every pair of subspaces M, N satisfying (1) determines a linear projection P by means of the equalities

$$M = R(P), \; N = N(P). \tag{2}$$

If the projection P is bounded, then M is said to be **topologically complemented** in X, and $N = N(P)$ is called a **topological complement** of M.

It is a consequence of the Closed Graph theorem that if a closed subspace M of a Banach space has a closed complementary subspace N, then M is topologically complemented by N.

Every closed subspace of a Hilbert space is topologically complemented by its orthogonal complement. Furthermore, a Banach space X has the property that all its closed subspaces are complemented if and only if X is isomorphic to a Hilbert space (see Lindenstrauss and Tzafriri [1]).

The finite dimensional subspaces, and the closed finite codimensional subspaces of a normed space are topologically complemented. (No other topologically complemented subspaces may exist in some spaces – see Gowers and Maurey [1]).

On the product space $X \times Y$ we shall define the norm

$$\|(x, y)\| = \|x\| + \|y\| . \tag{3}$$

Associated with T we define the linear relation T_G by

$$\begin{cases} T_G \in LR(X, X \times Y) \\ D(T_G) = D(T) \\ T_G x = \{(x, y) \in X \times Y : y \in Tx\} \end{cases} \tag{4}$$

We have the following analogue of Proposition II.4.2.

II.4.4 Proposition. *Let T be continuous, and let $\{0\} \times T(0)$ be topologically complemented in $G(T)$. Then T has a continuous selection, namely, the relation whose graph is $PG(T)$, where P is any bounded projection defined on $G(T)$ with kernel $\{0\} \times T(0)$.*

PROOF. Since T is continuous, therefore T_G is continuous; indeed,

$$\|T_G x\| = \inf\{\|x\|+\|y\| : y \in Tx\} = \|x\|+\|Tx\| \leq \|x\|+\|T\|\,\|x\| = (1+\|T\|)\,\|x\|\,.$$

Then PT_G is single valued, and continuous (since $D(P) = R(T_G)$). It follows that

$$PT_G x = (x, Ax) \qquad (x \in D(T))$$

where A is linear and single valued. Now

$$\|(x, Ax)\| = \|x\| + \|Ax\| = \|PT_G x\| \leq \|P\|\,\|T_G x\| = \|P\|\,(\|x\| + \|Tx\|)$$

Hence

$$\|Ax\| \leq (\|P\| - 1 + \|P\|\,\|T\|)\,\|x\|$$

showing that A is a continuous selection of T. □

II.4.5 Corollary. *Let $\{0\} \times T(0)$ be topologically complemented in $G(T)$, or in $X \times Y$. Then T has a continuous selection if and only if T is continuous.*

PROOF. If $\{0\} \times T(0)$ is topologically complemented in $X \times Y$, then it is already topologically complemented in $G(T)$. Therefore T has a continuous selection by Proposition II.4.4. The converse is obvious (see II.4.1(1)). □

II.4.6 Corollary. *Let H be a Hilbert space, let $T \in LR(X, H)$ and let $T(0)$ be closed. Then T has a continuous selection if and only if T is continuous.*

A significant special case of Proposition II.4.4 follows from II.3.18:

II.4.7 Proposition. *Let $T(0)$ be finite dimensional. Then T has a continuous selection if and only if T is continuous.*

II.4.8 Definition If the range of the linear relation T is finite dimensional, then T is said to be of **finite rank**.

II.4.9 Exercise Let T have a continuous selection, and suppose that $F \in LR(X, Y)$ is continuous and of finite rank. Then $T + F$ has a continuous selection.

II.5 CLOSED AND CLOSABLE LINEAR RELATIONS

II.5.1 Definitions

The closure of a linear relation $T \in LR(X,Y)$ is the relation $\overline{T}$ defined by

(1) $$G(\overline{T}) := \overline{G(T)}.$$

It is clear that $\overline{T} \in LR(X,Y)$.

The relation T is called **closed** if its graph $G(T)$ is closed in $X \times Y$, or equivalently, if $T = \overline{T}$.

We list some simple properties:

(2) *T is closed if and only if T^{-1} is closed*
(3) $(\overline{T})^{-1} = (\overline{T^{-1}})$
(4) $G(\overline{T}J_{D(T)}) = \overline{G(TJ_{D(T)})}$
(5) *if T is closed, then $T(0)$ is closed*
(6) *if T is continuous and $D(T)$ and $T(0)$ are closed, then T is closed.*

We should note that even though $G(T) \subset G(\overline{T})$, $\overline{T}$ need not be an extension of T; we only have $Tx \subset \overline{T}x$ for all $x \in D(T)$.

II.5.2 Proposition. $Q\overline{T} = \overline{QT}$ *(where $Q = Q_T$).*

PROOF. Let $(x, Qy) \in G(\overline{QT}) = \overline{G(QT)}$. Choose a sequence (x_n, QTx_n) in $G(QT)$ converging to (x, Qy). Then there exists a sequence (k_n) in $T(0)$ and a sequence $\{y_n\}$, with $y_n \in Tx_n$, such that $y_n + k_n \to y$. Then $(x_n, y_n + k_n)$ is a sequence in $G(T)$ converging to $(x, y) \in G(\overline{T})$. Thus $(x, Qy) \in G(Q\overline{T})$. Hence $G(\overline{QT}) \subset G(Q\overline{T})$.

For the reverse inclusion, let $(x, Qy) \in G(Q\overline{T})$. Then $(x, y_1) \in G(\overline{T})$ where $Qy = Qy_1$. Choose a sequence (x_n, y_n) in $G(T)$ converging to (x, y_1). Then

$$(x_n, Qy_n) = (x_n, QTx_n) \to (x, Qy_1) = (x, Qy)$$

showing that $(x, Qy) \in \overline{G(QT)}$, as required. □

II.5.3 Proposition. *The following properties are equivalent:*

(i) *T is closed.*

(ii) *QT is closed and $T(0)$ is closed.*

PROOF. If T is closed, then by Proposition II.5.2, QT is closed; and $T(0)$ is closed by II.5.1 (5). Conversely, suppose that (ii) holds. Then by Proposition II.5.2, $D(T) = D(QT) = D(Q\overline{T}) = D(\overline{T})$, and for $x \in D(T)$ we have $Q\overline{T}x = QTx$, so $\overline{T}x = \overline{T}x + T(0) = Tx + T(0) = Tx$. Hence $T = \overline{T}$, i.e. T is closed. □

II.5.4 Definition

The linear relation T is said to be **closable** if $\overline{T}$ is an extension of T; i.e., if

(1) $$Tx = \overline{T}x \text{ for all } x \in D(T).$$

II.5.5 Proposition. *The following properties are equivalent:*

(i) *T is closable*

(ii) *$T(0) = \overline{T}(0)$*

(iii) *$TJ_{D(T)}$ is closed.*

PROOF. The equivalence of (i) and (iii) is a consequence of II.5.1(4), while the equivalence of (i) and (ii) follows from I.2.11 (b). □

II.5.6 Corollary. *T^{-1} is closable if and only if $N(T) = N(\overline{T})$.*

II.5.7 Proposition. *The following properties are equivalent:*

(i) *T is closable*

(ii) *QT is closable and $T(0)$ is closed.*

In particular, if T is continuous and $T(0)$ is closed, then T is closable.

PROOF. If T is closable then $T(0) = \overline{T}(0)$ by Proposition II.5.5. Hence $T(0)$ is closed. Since $\overline{T}x = Tx$ for $x \in D(T)$, we have $Q\overline{T}J_{D(T)} = QTJ_{D(T)}$ and hence from Proposition II.5.2, $\overline{QT}J_{D(T)} = QTJ_{D(T)}$. Therefore QT is closable.

Conversely, let QT be closable and $T(0)$ closed. Then for $x \in D(T)$ we have

$$\begin{aligned} \overline{QT}x &= QTx \\ &= Q\overline{T}x \qquad \text{by Proposition II.5.2} \end{aligned}$$

Hence $Tx = \overline{T}x$, since $\overline{T(0)} = T(0)$, i.e., T is closable. □

II.5.8 Proposition. *We have*

$$\|\overline{T}\| \leq \|T\|$$

with equality holding if $\overline{T}(0) = \overline{T(0)}$; *for example, if* T *is closable.*

PROOF. Suppose that $\|\overline{T}\| > \|T\|$ and choose $\epsilon > 0$ and $(x_0, y_0) \in G(\overline{T})$ such that $\|x_0\| \leq 1$ and

$$(1) \qquad d(y_0, \overline{T}(0)) > \|T\| + 2\epsilon.$$

Next choose $(x, y) \in G(T)$ such that $\|x - x_0\| + \|y - y_0\| < \epsilon$, and $\|x\| \leq 1$. Then since

$$|d(y_0, T(0)) - d(y, T(0))| \leq \|y - y_0\| < \epsilon$$

we obtain from (1) that

$$\|T\| + 2\epsilon < d(y_0, \overline{T}(0)) \leq d(y_0, T(0)) \leq d(y, T(0)) + \epsilon \leq \|T\| + \epsilon$$

which is a contradiction. Therefore $\|\overline{T}\| \leq \|T\|$.

Now let $\overline{T}(0) = \overline{T(0)}$. We have for $x \in D(T)$, and $y \in Tx \subset \overline{T}x$, $\|Tx\| = d(Tx, \overline{T}(0)) = d(\overline{T}x, \overline{T}(0)) = \|\overline{T}x\|$. Hence $\|T\| \leq \|\overline{T}\|$ and equality follows. □

II.5.9 Corollary. *We have*

$$\gamma(\overline{T}) \geq \gamma(T)$$

with equality holding if $\overline{N(T)} = N(\overline{T})$.

II.5.10 Example *A continuous linear operator* T *from* $\ell_1(S)$ *into* ℓ_2 *such that* $N(T) = 0$ *and* $N(\overline{T})$ *is infinite dimensional.* Let $\{x_\alpha : \alpha \in S\}$ be a normalised Hamel basis for ℓ_2 and let Z_1 be ℓ_2 renormed by $\left\|\sum_{\alpha \in S} c_\alpha x_\alpha\right\| := \sum_{\alpha \in S} |c_\alpha|$. Let Z be the normed space $\{f \in \mathbb{K}^S : f(\alpha) \neq 0$ for at most finitely many $\alpha, \|f\| = \sum_{\alpha \in S} |f(\alpha)|\}$. Then Z_1 is isometric to Z under the correspondence $\{c_\alpha\} \leftrightarrow \Sigma c_\alpha x_\alpha$ and Z is a linear subspace of $\ell_1(S)$. It is easy to see that Z is dense in $\ell_1(S)$. Indeed if $f \in \ell_1(S)$ then $\{\alpha : f(\alpha) \neq 0\}$ is countable; letting $\{\alpha_1, \alpha_2 \ldots\}$ be the set of α's for which $f(\alpha) \neq 0$, then the sequence (f_n) in $\ell_1(S)$ defined by

$$\begin{aligned} f_n(\alpha_k) &:= f(\alpha_k), \qquad 1 \leq k \leq n \\ &:= 0, \qquad k > n \\ f_n(\alpha) &:= f(\alpha) = 0 \quad \text{otherwise} \end{aligned}$$

satisfies $\|f_n - f\| = \sum_{k=n+1}^{\infty} |f(\alpha_k)| \to 0 \quad (n \to \infty)$. Let $T \in LR(\ell_1(S), \ell_2)$ be defined by $T\{c_\alpha\} = \Sigma c_\alpha x_\alpha$. Then $\overline{T}$ maps $\ell_1(S)$ onto ℓ_2. Hence $\ell_1(S)/N(\overline{T})$ is isomorphic to ℓ_2, which implies that $N(\overline{T})$ cannot be finite dimensional since no closed finite codimensional subspace of $\ell_1(S)$ is separable. Evidently $N(T) = 0$.

II.5.11 Example Let M be a nonclosed subspace of X and let $T \in LR(X)$ be the relation whose graph is $X \times M$. Then $G(\overline{T}) = \overline{G(T)} = X \times \overline{M}$. Since $\overline{T}$ is not an extension of T, T is not closable. Here we have $T(0) = M$ and $\overline{T}(0) = \overline{M}$.

II.5.12 Example Let T be any linear relation with $T(0) \neq \overline{T(0)}$. Let F be any finite dimensional subspace of $D(T)$. Then $T|_F$ is continuous (II.3.2(c)). However, $\overline{(T|_F)}(0) = \overline{T(0)} \neq T(0)$. Therefore $T|_F$ is not closable (by II.5.5). Hence

(1) *T continuous $\not\Rightarrow$ $T(0)$ closed*

(2) *T continuous $\not\Rightarrow$ T closable.*

II.5.13 Proposition. *Let T be closed and let $F \subset Y$ be finite dimensional. Then $Q_F T$ is closed.*

PROOF. First assume that T is single valued. For simplicity of notation, write $Q := Q_F^Y$. Let (x_n, Qy_n) (where $y_n \in Tx_n$) be a sequence in $G(QT)$ converging to $(x, Qy) \in X \times QY$. Then there is a sequence $\{f_n\} \subset F$ such that $y_n - y + f_n \to 0$.

Case 1: Suppose that $\{f_n\}$ is unbounded. Then for some subsequence $\{f_{n'}\}$ we have $\|f_{n'}\| \to \infty$ and $\|y_{n'} + f_{n'}\| / \|f_{n'}\| \to 0$. Since $f_{n'}/\|f_{n'}\|$ is bounded in the finite dimensional space F, there exists a subsequence $\{f_{n''}\}$ of $\{f_{n'}\}$ and $f \in F$ such that $f_{n''}/\|f_{n''}\| \to f$. But then $y_{n''}/\|f_{n''}\| \to -f$ and $x_{n''}/\|f_{n''}\| \to 0$. Since T is closed, this implies that $(0, -f) \in G(T)$. Hence $f = 0$, contradicting $\|f\| = 1$. Therefore $\{f_n\}$ is bounded.

Case 2: $\{f_n\}$ is bounded. Then $\{f_n\}$ has a convergent subsequence in the finite dimensional space F. Passing to a subsequence if necessary, we may suppose that $\{f_n\}$ is convergent. Let $f = \lim_n f_n$. Then $y_n \to y - f$. Since T is closed, we now have $(x, y - f \in G(T)$, where $f \in F$. Therefore $(x, Qy) \in G(QT)$.

It remains to show that T can be assumed to be single valued. For this note that since $Q_F T(0) = T(0) + F$ is closed by the finite dimensionality of F, it is clear from Proposition II.5.3 that we need only show that $Q_{Q_F T} Q_F T$ is closed. But $Q_{Q_F T}(Q_F T) = Q_{Q_T F}(Q_T T)$, where $Q_T T$ is closed and single valued, and

$Q_T F$ is a finite dimensional subspace of $Y/T(0)$. The required property therefore follows from the first part applied to Q_T. □

II.5.14 Exercise A linear projection defined on a normed space is closed if and only if its kernel and range are closed (see II.5.1(6)).

II.5.15 Exercise Let X be complete and T closed. Show that if T is open, then $R(T)$ is closed.

II.5.16 Exercise Let T be closed and let $S \in LR(X,Y)$ be single valued and continuous with $D(S) \supset \overline{D(T)}$. Show that $T + S$ is closed.

II.5.17 Proposition. *Let $S \in LR(Z,X)$ be closed and let the relation T have closed range and satisfy $\alpha(T) < \infty$ and $\gamma(T) > 0$. Then TS is closed.*

PROOF. We shall first assume that S is single valued.

Let $(z_n, y_n) \to (z, y)$ where $y_n \in TSz_n$. Then $y \in Tx$ for some $x \in D(T)$, since $R(T)$ is closed. By the continuity of T^{-1} we have

$$\left\|T^{-1}T(Sz_n - x)\right\| = d(Sz_n - x, N(T)) \to 0.$$

Select a sequence $\{w_n\}$ in $N(T)$ such that $Sz_n + w_n - x \to 0$. We shall verify that $\{w_n\}$ is bounded. Suppose $\{w_n\}$ is unbounded and select a subsequence w'_n such that $\|w'_n\| \to \infty$. Then $Sz'_n/\|w'_n\| + w'_n/\|w'_n\| \to 0$. Since $N(T)$ is finite dimensional, the bounded sequence $\{w'_n/\|w'_n\|\}$ has a convergent subsequence $\{w''_n/\|w''_n\|\}$. Therefore $S(z''_n/\|w''_n\|)$ converges, where $z''_n/\|w''_n\| \to 0$. Since S is closed, we have $Sz''_n/\|w''_n\| \to 0$. Hence $1 = \|w''_n\|/\|w''_n\| \to 0$, a contradiction. It follows that (w_n) is bounded and therefore has a subsequence (w_{k_n}) converging to some element $w \in N(T)$. Hence $Sz_{k_n} \to x - w$. Since S is closed, we have $z \in D(S)$ and $Sz = x - w$. Then $Sz \in D(T)$, and $TSz = Tx$, showing that TS is closed.

For the general case, we note that since $S(0)$ is closed,

$$TS = TQ_S^{-1}Q_SS. \tag{1}$$

Hence it remains only to verify that TQ_S^{-1} satisfies the appropriate conditions. We have

$$R(TQ_S^{-1}) = T(Q_S^{-1}Q_SX) = T(X) \tag{2}$$

$$\gamma(TQ_S^{-1}) \geq \gamma(T)\gamma(Q_S^{-1}) > 0 \tag{3}$$

by Theorem II.3.11.

Finally, $\dim(TQ_S^{-1})^{-1}(0) = \dim Q_S(N(T)) \leq \dim N(T)$ and hence

$$\alpha(TQ_S^{-1}) < \infty. \tag{4}$$

Hence the result follows. □

II.5.18 Exercise Let $S \in LR(Y, Z)$ be closed and T single valued and bounded. Then ST is closed.

II.5.19 Exercise Show that $\overline{T(0)} \subset \overline{T}(0)$.

II.6 THE MINIMUM MODULUS UNDER RESTRICTION

II.6.1 Proposition. *Let M be a linear subspace of $D(T)$. Then*

$$\gamma(T|_{M+N(T)}) \geq \gamma(T).$$

PROOF. Since the relations T and $T|_{M+N(T)}$ have the same kernel, the inequality is clear. □

II.6.2 Corollary. $\gamma(T^{-1}|_{T(M)}) \geq \gamma(T^{-1})$ $(M \subset D(T))$.

PROOF. Replace T by T^{-1} in Proposition II.6.1, noting that $T(M)+N(T^{-1}) = T(M)$. □

II.6.3 Proposition. *Let $N(T)$ be topologically complemented in $D(T)$ (or in X). Then*

$$0 \leq \|P\|^{-1}\, \gamma(T) \leq \gamma(T|_{R(P)}) \leq \gamma(T)$$

where P is any continuous projection defined on $D(T)$ (or on X) with kernel $N(T)$.

PROOF. Note that $N(T)$ is relatively closed in $D(T)$. Suppose that P is defined on $D(T)$, with $N(P) = N(T)$. Let $\epsilon > 0$ be arbitrary. Let $z \in D(T)$ and choose $n \in N(T)$ so that $\|Pz + n\| \leq (1+\epsilon)d(Pz, N(T))$. We have

$$\begin{aligned} \frac{\|TPz\|}{\|Pz\|} &= \frac{\|TPz\|}{d(Pz, N(T))} \frac{d(Pz, N(T))}{\|Pz\|} \geq \frac{\gamma(T)}{1+\epsilon} \frac{\|Pz+n\|}{\|Pz\|} \\ &\geq \frac{\gamma(T)}{1+\epsilon} \inf_{x \in D(T)} \frac{\|x\|}{\|Px\|} \\ &\geq \frac{\gamma(T)}{1+\epsilon} \left(\sup_{x \in D(T)} \frac{\|Px\|}{\|x\|} \right)^{-1} \\ &= \frac{\gamma(T)}{1+\epsilon} \|P\|^{-1}. \end{aligned}$$

Since $T|_{R(P)}$ is injective, it follows that

$$\left\|\gamma(T|_{R(P)})\right\| \geq \gamma(T)/\left\|P\right\|.$$

For the second inequality, choose $x \in D(T)$ such that

$$\|Tx\| < (\gamma(T)+\epsilon)d(x, N(T)).$$

Hence

$$\begin{aligned}\gamma(T|_{R(P)}) &\leq \frac{\|TPx\|}{\|Px\|} = \frac{\|Tx\|}{d(x,N(T))}\,\frac{d(x,N(T))}{\|Px\|}\\ &< (\gamma(T)+\epsilon)\frac{d(Px,N(T))}{\|Px\|}\\ &\leq \gamma(T)+\epsilon.\end{aligned}$$

For the second case, when $N(T)$ is topologically complemented in X, note that $N(T)$ is then topologically complemented in $D(T)$. Hence, for the given projection P defined on X, it is sufficient to observe that

$$R(P) \cap D(T) = R(P|_{D(T)}). \quad \square$$

II.6.4 Corollary. *Let M be a closed finite codimensional subspace of $D(T)$ such that $M + N(T) = D(T)$ and $M \cap N(T) = \{0\}$. Then there exists $c > 0$ such that*

$$c\gamma(T) \leq \gamma(T|_M) \leq \gamma(T).$$

II.6.5 Example Let M be a proper closed finite codimensional subspace of X, and let P be a linear projection defined on X with kernel M. Then

$$\gamma(P) = 1$$

(see II.1.9 and II.2.5). However

$$\gamma(P|_M) = \infty.$$

Hence the restriction $M \cap N(T) = \{0\}$ in Corollary II.6.4 is necessary.

II.6.6 Exercise Let $N(T)$ be finite dimensional, and let M be a linear subspace of $D(T)$ such that $\dim M > \dim N(T)$. Show that

$$\gamma(T) \leq \|T|_M\|.$$

(See Lemma V.5.3).

II.6.7 Exercise Let P be any linear projection defined on $D(T)$ such that $N(P) \subset N(T)$. Show that $\gamma(T|_{R(P)}) \leq \gamma(T)$.

II.7 EXAMPLE FROM APPROXIMATION THEORY

Set valued mappings, and in particular linear relations, arise naturally in Approximation Theory. In this section we shall give a brief introduction to a problem of best approximation with interpolating constraints. The results obtained are illustrative applications of Sections II.1 and II.4 of this chapter.

II.7.1 Definitions Let M be an n-dimensional subspace of X and let $\{\phi_1, \phi_2, \ldots, \phi_k\}$ be a set of $k \leq n$ continuous linear functionals defined on X such that the set of restrictions $\{\phi_1|_M, \phi_2|_M, \ldots, \phi_k|_M\}$ is linearly independent.

Define the relation $T \in R(X, X)$ by

$$(1) \qquad Tx := \{m \in M : \phi_i(m) = \phi_i(x), \quad 1 \leq i \leq k\} \qquad (x \in X)$$

The elements of Tx are said to **interpolate** x relative to the set $\{\phi_1, \phi_2, \ldots, \phi_k\}$.

Let $\{m_1, m_2, \ldots, m_n\}$ be the vector basis of M orthogonal to $\{\phi_i : 1 \leq i \leq k\}$, i.e.,

$$(2) \qquad \phi_i(m_j) = \delta_{ij} \qquad (1 \leq i \leq k, 1 \leq j \leq n).$$

Define the linear operator $L : X \to X$ by

$$(3) \qquad Lx := \sum_{i=1}^{k} \phi_i(x) m_i \qquad (x \in X)$$

II.7.2 Proposition. *We have*

(a) *T is an everywhere defined linear relation*

(b) $T(0) = \bigcap_{i=1}^{k} N(\phi_i|_M)$. *Hence* $\dim T(0) = n - k$, *and* $T(0) = \{0\}$ *if and only if* $n = k$.

(c) *L is a bounded linear projection with range* $\text{sp}\{m_i : 1 \leq i \leq k\}$ *and kernel* $\bigcap_{i=1}^{k} N(\phi_i)$

(d) *L is a continuous selection of T. Thus*

$$T = L + T - T.$$

Furthermore, $L = LT$.

PROOF. (a) Let $x \in X$ and consider $m \in M$ defined by

$$m := \sum_{i=1}^{k} \phi_i(x) m_i.$$

Then $\phi_i(x) = \phi_i(m)$ for $i = 1, 2, \ldots, k$ and therefore $x \in D(T)$. Hence $D(T) = X$.

To verify that T is linear, let (x_0, m_0), $(x, m) \in G(T)$. Then $\phi_i(x_0) = \phi_i(m_0)$ and $\phi_i(x) = \phi_i(m)$ for $i = 1, 2, \ldots, k$. Hence

$$\phi_i(\alpha x_0 + \beta x_0) = \phi_i(\alpha m_0 + \beta m) \qquad (\alpha, \beta \in \mathbb{R})$$

showing that $\alpha(x_0, m_0) + \beta(x, y) \in G(T)$. Therefore $T \in LR(X)$.

(b) Follows immediately from (1) of Definition II.7.1.

(c) Evidently L is bounded, since

$$\|Lx\| = \left\| \sum_{i=1}^{k} \phi_i(x) m_i \right\| \leq (\sum_{i=1}^{k} \|\phi_i\| \, \|m_i\|) \, \|x\| \qquad (x \in X).$$

We have for $x \in X$,

$$\begin{aligned} L^2 x &= L(\sum_{i=1}^{k} \phi_i(x) m_i) = \sum_{j=1}^{k} \phi_j(\sum_{i=1}^{k} \phi_i(x) m_i) m_j \\ &= \sum_{j=1}^{k} \phi_j(x) m_j = Lx, \qquad \text{by II.7.1(2).} \end{aligned}$$

Hence $L^2 = L$. Thus L is a linear projection defined on X. Next, since $Lm_j = \sum_{i=1}^{k} \phi_i(m_j) m_i = m_j \;\; (1 \leq j \leq k)$, we have $L(X) = \mathrm{sp}\{m_i : i \leq i \leq k\}$. Finally, $Lx = 0$ if and only if $\phi_i(x) = 0$ for $i = 1, 2, \ldots, k$. Hence $N(L) = \bigcap_{i=1}^{k} N(\phi_i)$.

(d) Let $x \in X$. We verify by direct computation that $Lx + T(0) = Tx$. We have for $m \in T(0)$,

$$\phi_j(Lx + m) = \phi_j(\sum_{i=1}^{k} \phi_i(x) m_i + m) = \phi_j(x) + \phi_j(0) = \phi_j(x).$$

Therefore $Lx + m \in Tx$, and hence $Lx + T(0) = Tx$.

Since L is bounded (by (c) above) it follows that L is a continuous selection of T.

Finally, we have for $x \in X$,

$$LTx = L(Lx + T(0)) = Lx + LT(0) = Lx. \quad \square$$

II.7.3 Definitions Let K be a convex subset of X. For a given $x \in X$, the set $P_K(x)$ of **best approximations** from x to K is defined by

$$P_K(x) := \{z \in K : \|x - z\| = d(x, K)\}. \tag{1}$$

The set K is called **proximinal** if $P_K(x)$ is non empty for each $x \in X$, and **Chebyshev** if $P_K(x)$ is a singleton for every $x \in X$.

An element $x_0 \in Tx$ will be a best approximation to x from Tx if

$$\|x - x_0\| = d(x, Tx)$$

and the set of best approximations to x from Tx is the set $P_{Tx}(x)$. The map

$$x \to P_{Tx}(x)$$

is called the **parameter map** for the relation T.

It is obvious that $P_{Tx}(x) = \{x\}$ whenever $x \in Tx$; in that case x is its own best approximation.

II.7.4 Proposition. *We have*

$$P_{Tx}(x) = P_{T(0)}(x - Lx) + Lx. \tag{1}$$

In particular, if $x \in N(L)$, then

$$P_{Tx}(x) = P_{T(0)}x. \tag{2}$$

If T is single valued (i.e., if $\dim M = k$) then

$$P_{Tx}(x) = Lx. \tag{3}$$

PROOF. Let $z \in P_{T(0)}(x - Lx)$. Then

$$\|x - Lx - z\| = d(x - Lx, T(0))$$

i.e.

$$\|x - (z + Lx)\| = d(x, Lx + T(0)) \\ = d(x, Tx) \qquad \text{by II.7.2(d).}$$

Therefore $z + Lx \in P_{Tx}(x)$. Hence

$$P_{T(0)}(x - Lx) + Lx \subset P_{Tx}(x).$$

For the reverse inclusion, let $z \in P_{Tx}(x)$. Then

$$\|z - x\| = d(x, Tx) = d(x - Lx, Tx - Lx) \\ = d(x - Lx, T(0)) = \|z - Lx - (x - Lx)\| .$$

Thus $z - Lx \in P_{T(0)}(x - Lx)$, as required.

Now suppose that $x \in N(L)$. Then $Lx = 0$ and (2) follows.

Finally, if T is single valued, then $P_{T(0)}(x - Lx) = \{0\}$ and (3) follows. □

II.7.5 Remark Proposition II.7.4 shows that the problem of parametric approximation with interpolatory contraints can be reduced to an ordinary best approximation problem from the fixed $n - k$ dimensional subspace $T(0)$ to M.

II.8 NOTES AND REMARKS

S.M. Robinson [1] defines the norm of a convex process T by

(1) $$\|T\| := \sup\{d(Tx, 0) : x \in B_{D(T)}\}$$

(see also R.T. Rockafellar [1]). This definition is adopted for linear relations in normed spaces by S.J. Lee and M.Z. Nashed [6]. Propositions II.1.4 and II.1.6 show that the definition (1) is equivalent to our definition II.1.3.

The norm $\|T\|$ is not a true "norm" (except in the single valued case) since a nonzero relation may have zero norm; this is why the reverse triangle inequality fails (see Exercise II.1.12). The terminology has been retained because of its operator theoretic connotation.

The paper Robinson [1] generalises several familiar results for continuous linear operators in Banach spaces to convex processes (hence in particular, to linear relations). There are important differences however as the following example from that paper shows. Define the convex process T from $\mathbb{R}^2$ to $[0, \infty) \subset \mathbb{R}$ by

$$T(y, z) := \begin{cases} \{x : xz \geq y^2\} & \text{for } z > 0 \text{ and all } y \\ [0, \infty) & \text{for } z = 0,\ y = 0 \\ \emptyset & \text{otherwise} \end{cases}$$

Then since $T(1, 1/n) = \{x : x \geq n\}$, we have $\|T\| = \infty$, while $D(T)$ is finite dimensional. For a further discussion, see Robinson [1] and Lee and Nashed [6].

The definition of minimum modulus II.2.1 is formally identical with T. Kato's "reduced minimum modulus" of a linear operator (Kato [2; p.231]. The minimum modulus of a linear relation in the more general setting of locally convex spaces is used in several papers of R. Mennicken and B. Sagraloff (see Mennicken and Sagraloff [1,2,3]). The formula of Theorem II.2.5 relating $\|T\|$ and $\gamma(T)$ does not appear to be well-known. Also not very well-known it seems, are the inequalities for $\gamma(ST)$ and $\|ST\|$ (Theorem II.3.11 and Corollary II.3.13). For a discussion of the minimum modulus of some differential operators see Goldberg [2; VI.3].

The notion of continuity as defined in III.3.1 conforms to one of the standard topological definitions of continuity for a function. Since the linear relation T is continuous if and only if $\|T\| < \infty$ (Proposition II.3.2), this notion of continuity can be seen to coincide with that of *upper semi-continuity* for set valued maps (see e.g. J.-P Aubin and H. Frankowska [1; p.39]). In common with some other authors (see e.g. Mennicken and Sagraloff [4], Möller [2]) we have termed certain relations "open" which are otherwise known as "relatively open".

Linear selections have been investigated in Lee and Nashed [5, 6] and have found several applications. (See Lee and Nashed [6]).

Our definition of closable linear relation (Definition II.5.4) is a natural generalization of the notion of closable linear operator. One may observe, for example, that Proposition 1.3 of Gohberg, Goldberg and Kaashoek [1; p.289] characterizing closable linear operators is the single valued version of Proposition II.5.5.

The example from Approximation Theory (Section II.7) is from F. Deutsch and S. Mabizela [1] where the reader is referred for further details. Other related papers are D.A. Ault, F.R. Deutsch and P.D. Morris [1], F. Deutsch [1, 2], F. Deutsch, W. Li and S.-H. Park [1], F. Deutsch, W. Pollul and I. Singer [1], S. Mabizela [1], S. Mabizela and J. Zhong [1]. For the Approximation Theory background, see E.W. Cheney [1], P.J. Davis [1], and I. Singer [1]. (Further references may be found in the Bibliography.)

CHAPTER III

Adjoints of Linear Relations

III.1 THE ADJOINT

If X is a normed linear space, then X' will denote the norm dual of X; i.e. the space of all continuous linear functionals x' defined on X, with the norm $\|x'\| = \inf\{\lambda : |x'x| \leq \lambda \|x\| \text{ for all } x \in X\}$.

We shall adopt the following notation: If $K \subset X$ and $L \subset X'$, then

$$K^{\perp} := \{x' \in X' : x'x = 0 \text{ for all } x \in K\}$$
$$L^{\top} := \{x \in X : x'x = 0 \text{ for all } x' \in L\}.$$

Clearly, $K^{\perp}$ and $L^{\top}$ are closed linear subspaces of X' and X respectively. Moreover, if M is a linear subspace of X, then $M^{\perp\top} = \overline{M}$. If N is a linear subspace of X', then $N^{\top\perp}$ is the closure of N is the $\sigma(X', X)$–topology (or weak*–topology) of X'.

III.1.1 Definition The adjoint T' of T is defined by

(1) $$G(T') = G(-T^{-1})^{\perp} \subset Y' \times X'$$

where

(2) $$\langle (y, x), (y', x') \rangle := \langle x, x' \rangle + \langle y, y' \rangle .$$

This means that

(3) $(y', x') \in G(T')$ *if and only if* $y'y - x'x = 0$ *for all* $(x, y) \in G(T)$.

In (3) we have $y'y = x'x$ for all $y \in Tx$, $x \in D(T)$. Hence

(4) $$x' \in T'y' \iff y'Tx = x'x \text{ for all } x \in D(T).$$

Therefore x' is an extension of $y'T$. We can thus characterize the adjoint as follows:

(5) $$G(T') = \{(y', x') \in Y' \times X' : x' \text{ is an extension of } y'T\}.$$

From (5) we see that if T is densely defined, then T' will be single valued (since in that case, $y'T$ will have a unique extension to all of X).

III.1.2 Proposition. *T' is a closed linear relation in $LR(Y', X')$, and we have:*

$$D(T') = \{y' \in Y' : y'T \text{ is continuous and single valued}\}.$$

Furthermore, given $y' \in D(T')$, $x \in D(T)$, we have $T'y'x = y'Tx \in \mathbb{K}$.

PROOF. By (5) of III.1.1, $y'T = x'|_{D(T)}$ is continuous on $D(T)$, and $y'T(0) = x'(0) = 0$. □

III.1.3 Proposition.

(a) $(\overline{T})' = T'$
(b) $(T')^{-1} = (T^{-1})'$
(c) $(\lambda T)' = \lambda T'$ $(\lambda \neq 0)$.

PROOF. (a) and (b) are immediate consequences of the definition.

(c) For $\lambda \neq 0$ we have:

$$\begin{aligned} G((\lambda T)') &= \{(y', x') : y'y = x'x \text{ for } (x,y) \in G(\lambda T)\} \\ &= \{(y', \lambda x') : y'(\lambda y) = (\lambda x')x \text{ for } (x,y) \in G(T)\} \\ &= G(\lambda T'). \quad \square \end{aligned}$$

III.1.4 Proposition

(a) $N(T') = R(T)^{\perp}$
(b) $T'(0) = D(T)^{\perp}$
(c) $N(\overline{T}) = R(T')^{\top}$
(d) $\overline{T}(0) = D(T')^{\top}$

PROOF.

(a)
$$\begin{aligned} y' \in (T')^{-1}(0) &\iff (0, y') \in G((T')^{-1}) \\ &\iff (y', 0) \in G(T') \\ &\iff y'y = 0 \text{ for all } y \in R(T) \\ &\iff y' \in R(T)^{\perp}. \end{aligned}$$

(b) Follows from (a) upon substituting T^{-1} for T, using III.1.3 (b).

(c) We have

$$\begin{aligned} x \in N(\overline{T}) &\iff (x,0) \in G(\overline{T}) = G(-(T')^{-1})^{\top} \\ &\iff x'x = y'(0) \text{ for all } (y',x') \in G(T') \\ &\iff x'x = 0 \text{ for all } (y',x') \in G(T') \\ &\iff x'x = 0 \text{ for all } x' \in R(T') \\ &\iff x \in R(T')^{\top} \end{aligned}$$

(f) Substitute T^{-1} for T in (c), using Proposition III.1.3.

III.1.5 Proposition. *Let $S, T \in LR(X,Y)$.*

(a) $G(S' + T') \subset G((S+T)')$.

(b) *If $D(T) \subset D(S)$, and if S is continuous (for example, if $D(S') = Y'$), then*

$$S' + T' = (S+T)'.$$

(c) *$(S+T)'$ is an extension of $S' + T'$ if and only if $(D(S) \cap D(T))^{\perp} = D(S)^{\perp} + D(T)^{\perp}$.*

PROOF. (a) The proof proceeds by direct computation. Let $(y', x') \in G(S' + T')$. Then $(y', x_1') \in G(S')$ and $(y', x_2') \in G(T')$ where $x' = x_1' + x_2'$. Let $(x, s+t) \in G(S+T)$ where $(x,s) \in G(S)$ and $(x,t) \in G(T)$. Then

$$y'(s+t) - x'x = y'(s) + y'(t) - x_1'x - x_2'x = 0.$$

Hence $(y', x') \in G((S+T)')$.

(b) By part (a), $D(S'+T') \subset D((S+T)')$. If now $y' \in D((S+T)')$, then $0 = y'(S+T)(0) = y'S(0) + y'T(0)$, which can only be the case if $0 = y'S(0) = y'T(0)$, i.e. $y'S$ and $y'T$ are single valued. Since S is continuous so is $y'S$ and hence also $y'T$, as $y'Tx = y'(S+T)x - y'Sx$ for $x \in D(T) = D(S+T) \subset D(S)$. Hence by III.1.2, $y' \in D(T') \cap D(S') = D(S' + T')$. Therefore $D(S' + T') = D((S+T)')$. Hence by III.1.1(4), $S' + T' = (S+T)'$.

(c) By I.2.11 (b), $(S+T)'$ is an extension of $S'+T'$ if and only if $(S+T)'(0) = S'(0) + T'(0)$. The desired conclusion now follows from III.1.4 (b). □

Sufficient conditions for $(ST)' = T'S'$ to hold will now be given. Part (a) of the theorem below is due to Kascic [1].

III.1.6 Theorem. *(a) Let $T \in LR(X,Y)$ and $S \in LR(Y,Z)$. Then $G(T'S') \subset G((ST)')$. Furthermore, if either*

(i) $R(T') = X'$ *and* $D(S) \subset R(T)$, *or*

(ii) $D(S') = Z'$ *and* $R(T) \subset D(S)$,

then

$$(ST)' = T'S'. \tag{1}$$

(b) Let M be a finite codimensional subspace of $D(T)$, closed in $D(T)$. Then

$$(TJ_M^X)' = (J_M^X)'T'. \tag{2}$$

PROOF. (a) If $(z',x') \in G(T'S')$ then $(z',y') \in G(S')$ and $(y',x') \in G(T')$ for some $y' \in Y'$. Then $z'z = y'y = x'x$ for all $(z,x) \in G(ST)$. Hence $(z',x') \in G((ST)')$. We have shown that

$$G(T'S') \subset G((ST)'). \tag{3}$$

Suppose that (i) holds. Let $(z',x') \in G((ST)')$. Since $x' \in X' = R(T')$, there exists $y' \in Y'$ for which $(y',x') \in G(T')$. Now let $(y,z) \in G(S)$. Then $y \in D(S) \subset R(T)$, and there exists $x \in D(T)$ such that $(x,y) \in G(T)$. Therefore $x'x = y'y$. Also $(x,z) \in G(ST)$. Hence $x'x = z'z$, and so $y'y = z'z$. Consequently $(z',y') \in G(S')$. This ensures that $(z',x') \in G(T'S')$. Therefore $G((ST)') \subset G(T'S')$. The equality (1) now follows from (3).

Now suppose that (ii) holds. Since $(S^{-1})' = (S')^{-1}$ by III.1.3 (b), we have $R((S^{-1})') = Z'$, and $D(T^{-1}) \subset R(S^{-1})$. Hence, by the first part, we have

$$(T^{-1}S^{-1})' = (S^{-1})'(T^{-1})'$$

which is the same as (1) by III.1.3 (b).

(b) Write $J := J_M^X$. It is easy to see that $D(T') \subset D((TJ)')$. Also $D(J'T') = D(T')$ as J' is everywhere defined. Furthermore $(TJ)'$ is single valued (see III.1.4(b)). We shall now verify that $D((TJ)') \subset D(T')$. Let $y' \in D((TJ)')$. Then $y'T$ is continuous and single valued on M (see III.1.2) and hence on $D(T)$, by Proposition II.3.19. Therefore $y' \in D(T')$. Hence $(TJ)'$ and $J'T'$ have the same domain.

In view of (3), it only remains to show that $G(J'T') \supset G((TJ)')$. Let $(y',m') \in G((TJ)')$. Since $x' := y'T$ is a continuous single valued extension of $m' = y'TJ$, we now have $(y',x') \in G(T')$ and $(x',m') \in G(J')$. Consequently $(y',m') \in G(J'T')$, as required. $\square$

III.1.7 Example We exhibit a pair S, T for which $D(S')$ is the whole space; and yet $(ST)' \neq T'S'$.

Let X be a Hilbert space, M an infinite dimensional closed subspace of X of infinite codimension. Let P be the orthogonal projection defined on X with range M and let $T_1 = P^{-1}P$. Denote the orthogonal complement of M by N. Let S_1 be an injective and continuous linear operator with domain N and non closed range in X (for example, an injective compact operator).

We shall first verify that $S_1 T_1$ is not closed. Let $z \in \overline{R(S_1)} \backslash R(S_1)$ and (z_n) a sequence in $R(S_1)$ converging to z. Let $y_n = S_1^{-1}(z_n)$. Then (y_n) does not converge (since $z \notin R(S_1)$). However, $y_n \in D(S_1) \subset N(T_1)$. Since $T_1 = T_1^{-1}$, $(0, y_n) \in G(T_1)$. Since $(y_n, z_n) \in G(S_1)$, we have $(0, z_n) \in G(S_1 T_1)$. Now $z \notin R(S_1 T_1) \subset R(S_1)$, and hence $(0, z) \notin G(S_1 T_1)$. Since $(0, z_n) \to (0, z)$, this shows that $S_1 T_1$ is not closed.

Put $S = T_1'$ and $T = S_1'$. Since T_1 and S_1 are closed, we have $T_1 = T_1''$ and $S_1 = S_1''$ (by the reflexivity of Hilbert space). Hence $T'S' = S_1'' T_1'' = S_1 T_1$, which we have shown to be not closed, while $(ST)'$ is closed. Therefore $(ST)' \neq T'S'$.

The next proposition is a standard result in the theory of normed spaces. A proof can be found in Goldberg [3; p.25].

III.1.8 Proposition. *Let M be a linear subspace of X.*

(a) *$X'/M^\perp$ is isomorphic to M' under the norm preserving isomorphism U defined by*

$$U(Q_{M^\perp} x') = x'|_M \qquad (x' \in X').$$

(b) *If M is closed, then $(X/M)'$ is isomorphic to $M^\perp$ under the norm preserving isomorphism V defined by*

$$(Vz')x = z'(Q_M x) \qquad (z' \in (X/M)').$$

Whenever we write $X'/M^\perp = M'$ and $(X/M)' = M^\perp$, these equalities are to be interpreted in terms of the natural isometries U and V described in III.1.8.

III.1.9 Proposition. *Let E be a subspace of X. Then*

$$(1) \qquad (Q_E^X)' = J_{E^\perp}^{X'} \qquad \textit{(if } E \textit{ is closed)}$$

$$(2) \qquad (J_E^X)' = Q_{E^\perp}^{X'}.$$

PROOF. Using the canonical isometry which identifies $(X/\overline{E})'$ with $E^\perp$ (Proposition III.1.8), we have (assuming E is closed)

$$(Q_E^X)' : E^\perp \to X', \quad \text{and}$$

$$(Q_E^X)'x'(x) = x'(Q_E^X x) = x'(x) \tag{3}$$

for $x \in X$, $x' \in E^\perp$.
Also, $J_{E^\perp}^{X'} : E^\perp \to X'$, and

$$(J_{E^\perp}^{X'} x')(x) = x'(x) \tag{4}$$

for $x \in X$, $x' \in E^\perp$.
Comparing (3) and (4) establishes (1).

By a similar application of Proposition III.1.8, we have

$$Q_{E^\perp}^{X'} : X' \to E' \quad \text{and}$$

$$Q_{E^\perp}^{X'} x'(e) = x'e \tag{5}$$

for $x' \in X'$, $e \in E$.
Also, $(J_E^X)' : X' \to E'$, and

$$(J_E^X)'x'(e) = x'(J_E^X e) = x'e \tag{6}$$

for $x' \in X'$, $e \in E$.
The equality (2) now follows from (5) and (6). □

III.1.10 Theorem. *We have*

(a) $(QT)' = T' J_{T(0)^\perp}$
(b) $(TJ_{D(T)})' = Q_{T'}T'$
(c) $(QTJ_{D(T)})' = Q_{T'}T'J_{T(0)^\perp}$.

PROOF. Combine Theorem III.1.6 and Proposition III.1.9. □

III.1.11 Corollary. *We have*

$$D(T') = D((QT)') \tag{1}$$

and

$$T'y' = (QT)'y' \qquad (y' \in D(T')). \tag{2}$$

PROOF. Combine Theorem III.1.10 with Proposition III.1.4 (d). □

III.1.12 Remark Corollary III.1.11 shows that the adjoints of the relations T and QT have the same graph. In the case when T is closable and single valued, then $D(T')^{\top\perp} = Y'$ and the two adjoints are identical (see III.3.8).

III.1.13 Corollary. $\|T'\| \leq \|T\|$.

PROOF. Assume without loss of generality that $\|T\| < \infty$. We have from Theorem III.1.10 that

$$(QTJ)' = Q_{T'}T'J_{T(0)^\perp} \qquad (\text{where } J = J_{D(T)}) \tag{1}$$

which, by Proposition III.1.4, has domain $D(T')$ and moreover

$$(QTJ)'(y') = Q_{T'}T'J_{T(0)^\perp}(y') \qquad (y' \in D(T')). \tag{2}$$

Since QTJ is everywhere defined, $(QTJ)'$ is single valued. Hence

$$\begin{aligned}\|T'\| = \left\|Q_{T'}T'J_{T(0)^\perp}\right\| &= \sup\{\|(QTJ)'y'\| : y' \in B_{D(T')}\} \\ &= \sup\{|y'(QTJx)| : y' \in B_{D(T')}, x \in B_{D(T)}\} \\ &\qquad\qquad (\text{since } D(T') \subset T(0)^\perp) \\ &\leq \sup\{\|y'\|\,\|QTJ\|\,\|x\| : y' \in B_{D(T')}, x \in B_{D(T)}\} \\ &\leq \|QTJ\| \\ &= \|T\|. \quad \square\end{aligned}$$

III.1.14 Proposition. *Let P be a bounded projection defined on X. Then*

$$N(P)^\perp + R(P)^\perp = X' \quad \textit{and} \quad N(P)^\perp \cap R(P)^\perp = \{0\}.$$

PROOF. It is simple to verify that P' is a bounded projection defined on X'. Now by III.1.4(a)

$$\begin{aligned}R(P)^\perp &= N(P') \\ N(P)^\perp &= R(I-P)^\perp = N((I-P)') = R(P')\end{aligned}$$

Hence $X' = N(P') + R(P') = N(P)^\perp + R(P)^\perp$, and $N(P') \cap R(P') = \{0\} = N(P)^\perp \cap R(P)^\perp$. $\square$

III.1.15 Proposition. *If* $\overline{R(T)} + M = Y$ *where* $\dim M < \infty$, *then*

$$R((Q^Y_M T)') = R(T').$$

PROOF. Write $Q = Q^Y_M$. Since $\dim M < \infty$ there exists a bounded projection P of Y onto $\overline{R(T)}$ with kernel M (by Proposition II.4.7). By Proposition III.1.14 we have

$$R(T)^\perp + M^\perp = Y', \quad R(T)^\perp \cap M^\perp = \{0\}.$$

By Theorem III.1.6(a),

$$(QT)' = T'Q' = T' J^{Y'}_{M^\perp}.$$

Hence

$$R((QT)') = T'(M^\perp) = T'(M^\perp + N(T')) = T'(M^\perp + R(T)^\perp) = R(T'). \quad \square$$

III.2 POLAR RELATIONSHIPS

III.2.1 Definitions Let X be a normed space. For a given nonempty subset A of X, the **polar** of A is defined to be the set

(1) $$A^0 = \{x' \in X' : |x'x| \leq 1 \text{ for all } x \in A\}.$$

Note that A^0 is weak*–closed, and also absolutely convex (i.e., $|\alpha| + |\beta| \leq 1$ and $y, z \in A$ imply $\alpha y + \beta z \in A$). Furthermore, if $|\text{co}|\, A$ denotes the absolute convex hull of A, then

(2) $$(|\text{co}|\, A)^0 = A^0.$$

Let C be a nonempty subset of X'. The **prepolar** of C is defined to be the set

(3) $$C_0 = \{x \in X : |x'x| \leq 1 \text{ for all } x' \in C\}.$$

Obviously C_0 is absolutely convex and closed, and

(4) $$(|\text{co}|\, C)_0 = C_0.$$

From the Bipolar Theorem (see Robertson and Robertson [1], p.35) we also have

(5) $$(A^0)_0 = (|\text{co}|\, A)^- \qquad \text{(norm closure)}$$

(6) $$(C_0)^0 = \text{ weak*–closure of } |co|\, C\ .$$

If A, C are linear subspaces, then $A^0 = A^\perp$ and $C_0 = C^\top$.

Evidently, $(B_X)^0 = B_{X'}$ and $(B_{X'})_0 = B_X$.

III.2.2 Theorem. *Let $A \subset D(T)$. Then*

$$(TA)^0 \cap D(T') = (T')^{-1}(A^0). \tag{1}$$

PROOF. First suppose that $A \subset D(T)$. We have

$$\begin{aligned}(TA)^0 \cap D(T') &= \{y' \in D(T') : |y'y| \le 1 \text{ whenever } y \in TA\} \\ (T')^{-1}(A^0) &= \{y' \in D(T') : y' \in (T')^{-1}x' \text{ for some } x' \in A^0\} \\ &= \{y' \in D(T') : (y', x') \in G(T') \text{ for some} \\ &\qquad x' \text{ satisfying } |x'x| \le 1 \text{ for all } x \in A\}. \end{aligned} \tag{2}$$

Suppose that $y' \in (T')^{-1}(A^0)$. Then $y' \in D(T')$ and we have:

$$\begin{aligned} & y' \in (T')^{-1}x' \text{ for some } x' \in A^0 \\ \Longrightarrow\ & x' \in T'y' \quad (\text{where } x' \in A^0) \\ \Longrightarrow\ & x'|_{D(T)} = y'T \quad (\text{see III.1.2}) \\ \Longrightarrow\ & |y'Tx| \le |x'x| \le 1 \text{ for all } x \in A \subset D(T). \end{aligned}$$

Hence for all $y \in TA$, we have $|y'y| \le 1$, whence $y' \in (TA)^0$.

For the reverse inclusion, let $y' \in (TA)^0 \cap D(T')$. For any $y \in TA$, we have $|y'y| \le 1$. Hence if $x \in A$ and $y \in Tx$, then $|y'y| = |x'x| \le 1$ for $x' \in T'y'$. Therefore, by (2), $y' \in (T')^{-1}(A^0)$. □

We remark that $D(T')$ cannot be omitted from (1). For example, let $S : \ell_2 \to \ell_2$ be an injective compact operator. Put $T = S^{-1}$, and $A = \{0\}$. Then $(TA)^0 = \ell_2$, while $(T^{-1})'(A^0) = R(S') \ne \ell_2$.

III.2.3 Corollary.

(a) *If $C \subset D(T')$ then*

$$(T'C)_0 \cap D(T) = T^{-1}(C_0)$$

(b) *If $C \subset R(T)$, then*

$$T'(C^0) = R(T') \cap (T^{-1}C)^0.$$

PROOF. The equality (a) is obtained by interchanging the roles of T and T', while (b) follows by substituting T^{-1} for T, in Theorem III.2.2. □

III.2.4 Remarks Various orthogonality relationships follow from III.2.2 and III.2.3. For example, putting $C = D(T')$ in III.2.3 (a) gives

(1) $$R(T')^{\top} \cap D(T) = T^{-1}(D(T')^{\top}).$$

Hence

(2) $$D(T')^{\top} \cap R(T) = T(R(T')^{\top}).$$

Combining with III.1.4 gives

(3) $$T(N(\overline{T})) = \overline{T}(0) \cap R(T).$$

Similarly

(4) $$T^{-1}\overline{T}(0) = N(\overline{T}) \cap D(T)$$

III.2.5 Exercise Verify:

$$\overline{T}x \cap R(T) = Tx \cap T(N(\overline{T})) \qquad (x \in D(T))$$

III.3 THE MINIMAL GAP BETWEEN A PAIR OF SUBSPACES

III.3.1 Definitions Let X be a Banach space, and let M and N be closed subspaces of X. The quantity

(1) $$\gamma(M,N) = \sup\{\lambda \in [0,1] : \lambda B_X \cap (M+N) \subset (M \cap B_X) + N\}$$

is called the **minimal gap between M and N.**

The ordered pair (M, N) is said to be **open** if $\gamma(M,N) > 0$.

The above definitions are due to Kato (see [2]). Observe that

(2) $$\gamma(M,N) \leq 2\gamma(N,M)$$

and so (M, N) is open if and only if (N, M) is open.

Conditions equivalent to $\gamma(M,N) > 0$ are listed in Theorem III.3.9 below, due to Mennicken and Sagraloff [1,4]. By substituting the graphs of suitable operators for the subspaces M and N, the Open Mapping, Closed Graph and Closed Range theorems for linear relations in Banach spaces will be obtained as corollaries in Section III.4.

To prove the lemmas required for Theorem III.3.9 we require, in addition to $\gamma(M,N)$, the two related quantities defined as follows:

(3) $$\alpha(M,N) := \max\{\lambda \in [0,1] : \lambda B_X \cap \overline{M+N} \subset \overline{M \cap B_X + N}\}$$

(4) $$\beta(M,N) := \sup\{\lambda \in [0,1] : \lambda B_X \cap \overline{M+N} \subset M \cap B_X + N\}$$

We shall write

$$B := B_X.$$

III.3.2 Lemma. $\alpha(M,N) = \beta(M,N) = \gamma(M,N)$.

PROOF. We first show that

$$\alpha(M,N) \le \beta(M,N). \tag{1}$$

We may clearly assume that $\alpha := \alpha(M,N) > 0$. It is sufficient to show that for each $q \in (0,1)$,

$$\alpha B \cap \overline{M+N} \subset (1-q)^{-1}(M \cap B) + N. \tag{2}$$

Let $0 \ne y_0 \in \alpha B \cap \overline{M+N}$. From (2) we have

$$\alpha B \cap \overline{M+N} \subset \overline{(M \cap B) + N}. \tag{3}$$

Since $B \cap (M+N)^-$ is a neighbourhood of zero in $(M+N)^-$, we obtain

$$\overline{(M \cap B) + N} \subset (M \cap B) + N + \alpha q B \cap \overline{M+N}. \tag{4}$$

Using (3) and (4) we construct by induction two sequences $\{x_i\}$ and $\{y_i\}$ such that for each $i \in \mathbb{N}$

$$\begin{cases} y_i = q^i x_i + y_{i+1}, & x_i \in (M \cap B) + N \\ y_i \in q^i \alpha B \cap \overline{M+N} \subset q^i \overline{(M \cap B) + N} \end{cases} \tag{5}$$

We clearly have $\lim_{i\to\infty} y_i = 0$. Therefore

$$y_0 = \lim_{i\to\infty}(y_0 - y_i) = \sum_{i=0}^{\infty} q^i x_i.$$

Now choose $u_i \in M \cap B$ and $v_i \in N$ such that $x_i = u_i + v_i$. Since $\|u_i\| \le 1$, we have from (5) that

$$\|x_i\| = q^{-i}\,\|y_i - y_{i+1}\| \le \alpha(1+q).$$

Consequently the sequence $\{x_i\}$ is bounded. It follows that the sequence $\{v_i\}$ is bounded and hence that the series $\sum_{i=0}^{\infty} q^i v_i$ is convergent. Therefore

$$y_0 = \sum_{i=0}^{\infty} q^i u_i + \sum_{i=0}^{\infty} q^i v_i \in (1-q)^{-1}(B \cap M) + N,$$

since $M \cap B$ and N are closed. This proves (2). Hence (1) is established.

Next, the inequality $\beta(M,N) \leq \gamma(M,N)$ is trivial. Finally to prove that $\gamma(M,N) \leq \alpha(M,N)$ it is enough to show that

$$\overline{B \cap (M+N)} = B \cap \overline{M+N}. \tag{6}$$

Let $x \in B \cap (\overline{M+N})$. Let $\{x_n\}$ be any sequence in $M+N$ such that $x_n \to x$.

Set

$$y_n \begin{cases} := \|x_n\|^{-1} \|x\| \, x_n \text{ if } x_n \neq 0 \\ := 0 \quad \text{otherwise.} \end{cases}$$

Then $y_n \in B \cap M+N$ and $y_n \to x$, which together imply that $x \in \overline{B \cap (M+N)}$. Hence

$$B \cap \overline{M+N} \subset \overline{B \cap (M+N)}.$$

Since the reverse inclusion is trivial, the equality (6) follows. The proof of the lemma is now complete. □

III.3.3 Lemma. *(M,N) is open if and only if $M+N$ is closed.*

PROOF. Assume that $M+N$ is closed. Since

$$M+N = \bigcup_{n \in \mathbb{N}} n(B \cap M+N)$$

therefore by Baire's Category Theorem, there exists $n_0 \in \mathbb{N}$, $\lambda > 0$ and $x_0 \in M+N$ such that

$$U := x_0 + \lambda(B \cap (M+N)) \subset n_0(\overline{B \cap M+N}).$$

Hence

$$U - U = 2\lambda(B \cap (M+N)) \subset 2n_0(\overline{B \cap M+N}),$$

showing that $\alpha(M,N) > 0$. Therefore (M,N) is open, by Lemma III.3.2.

Conversely, if $\gamma(M,N) > 0$ and thus $\beta := \beta(M,N) > 0$, it follows that

$$\overline{M+N} = \bigcup_{n \in \mathbb{N}} (nB \cap \overline{M+N}) \subset M+N.$$

Hence $M+N$ is closed. □

III.3.4 Lemma. *Let p be a continuous seminorm on X and let $K_p := \{x \in X : p(x) \leq 1\}$. Then*

$$(M \cap K_p)^0 = M^\perp + K_p^0.$$

PROOF. Clearly $M^\perp + K_p^0 \subset (M \cap K_p)^0$. For the reverse inclusion, let $x_0' \in (M \cap K_p)^0$. For $m \in M$ we have $|x_0'(m)| \leq p(m)$ (as $m/p(m) \in K_p$ for $p(m) \neq 0$, while by linearity $x_0'(m) = 0$ for $p(m) = 0$). By the Hahn-Banach Theorem there is an extension x' of $x_0'|_M$ such that $|x'(x)| \leq p(x)$ for $x \in X$. Then $|x'(x)| \leq 1$ for $x \in K_p$ and we have

$$x_0' = x' + (x_0' - x') \in K_p^0 + M^\perp.$$

Therefore $(M \cap K_p)^0 \subset M^\perp + K_p^0$ as required. $\square$

The next lemma is straightforward and is left as an exercise.

III.3.5 Lemma. *If A is a subset of X,*

$$(A + M)^0 = A^0 \cap M^\perp.$$

III.3.6 Lemma. *Let $\lambda > 0$. We have the equivalence*

$$\lambda B \cap (M + N) \subset (B \cap M) + N \iff \lambda(B + N) \cap M \subset B + (M \cap N).$$

PROOF. Assume $\lambda B \cap (M + N) \subset (B \cap M) + N$. Let $x \in \lambda(B + N) \cap M$. Write $x := \lambda(b + n)$ where $b \in B$, $n \in N$, and $\lambda(b + n) := m \in M$. Then $\lambda b = m - \lambda n \in \lambda B \cap (M + N) \subset (B \cap M) + N$. Consequently $\lambda b = a + n_1$ where $a \in B \cap M$ and $n_1 \in N$, and so $x = \lambda(b + n) = a + n_1 + \lambda n = m$, whence $n_1 + \lambda n \in M \cap N$. Therefore $x \in B + (M \cap N)$, showing that $\lambda(B + N) \cap M \subset (B \cap M) + N$.

Conversely, assume that $\lambda(B + N) \cap M \subset B + (M \cap N)$, Let $x \in \lambda B \cap (M + N)$. Write $x := \lambda b$ where $b \in B$. Then $\lambda b = m + n$ where $m \in M$, $n \in N$. We have $\lambda b - n = m \in \lambda(B + N) \cap M \subset B + (M \cap N)$ and so $\lambda b - n = m = b_1 + c$ where $b_1 \in B$ and $c \in M \cap N$. Then $b_1 = m - c \in B \cap M$ and so $x = \lambda b = b_1 + c + n \in B \cap M + N$, as required. $\square$

III.3.7 Proposition $\gamma(M, N) = \gamma(N^\perp, M^\perp)$.

PROOF. By Lemma III.3.2 it is enough to show that $\gamma(N^\perp, M^\perp) = \alpha(M, N)$.

We shall first verify that $\gamma(N^\perp, M^\perp) \leq \alpha(M,N)$. We may clearly suppose that $\gamma(N^\perp, M^\perp) > 0$. Now choose $0 < \lambda < \gamma(N^\perp, M^\perp)$ arbitrarily. Noting that $B^0 = B_{X'}$, we have the inclusion

$$\lambda B^0 \cap (M^\perp + N^\perp) \subset (B^0 \cap N^\perp) + M^\perp \tag{1}$$

which according to Lemma III.3.6 is equivalent to

$$\lambda(B^0 + M^\perp) \cap N^\perp \subset B^0 + (M^\perp \cap N^\perp). \tag{2}$$

Hence by Lemmas III.3.4 and III.3.5,

$$\lambda(B \cap M)^0 \cap N^\perp \subset B^0 + (M+N)^\perp \ , \qquad \text{i.e.,}$$
$$\lambda((B \cap M) + N)^0 \subset (B \cap (M+N))^0.$$

Hence by the bipolar formula

$$\lambda^{-1}\overline{((B \cap M) + N)} \supset \overline{B \cap (M+N)}.$$

Consequently we have from (6) of Lemma III.3.2,

$$\lambda B \cap (\overline{M+N}) \subset \overline{(B \cap M) + N}$$

showing that $\gamma(N^\perp, M^\perp) \leq \alpha(M,N)$.

It only remains to verify the reverse inequality $\alpha(M,N) \leq \gamma(N^\perp, M^\perp)$. Assume without loss of generality that $\alpha(M,N) > 0$ and choose λ arbitrarily in the interval $0 < \lambda < \alpha(M,N)$. Then $\lambda B \cap \overline{M+N} \subset \overline{(M \cap B) + N}$. Hence

$$\lambda^{-1}(B \cap \overline{M+N})^0 \supset (M \cap B + N)^0$$

which by Lemmas III.3.4 and III.3.5 is equivalent to

$$\lambda^{-1}(B^0 + (M+N)^\perp) \supset (M \cap B)^0 \cap N^\perp.$$

Hence by a second application of III.3.4 and III.3.5,

$$\lambda^{-1}(B^0 + (M^\perp \cap N^\perp)) \supset (M^\perp + B^0) \cap N^\perp$$

which is equivalent to (2) above. From the equivalence of (1) and (2), it follows that $\lambda \leq \gamma(N^\perp, M^\perp)$. Therefore $\alpha(M,N) \leq \gamma(N^\perp, M^\perp)$. $\square$

III.3.8 Lemma. *We have the implication*

$$\gamma(M,N) > 0 \implies M^{\perp} + N^{\perp} = (M \cap N)^{\perp}.$$

PROOF. Obviously $(M\cap N)^{\perp} \supset M^{\perp}+N^{\perp}$. For the reverse inclusion, suppose that $0 < \lambda < \gamma(M,N)$. By Lemma III.3.6,

$$\lambda(B+N) \cap M \subset B + (M \cap N). \tag{1}$$

Now since

$$\lambda(\overline{B+N}) = \overline{\lambda B + N} \subset (\lambda+\epsilon)B + N$$

and since (1) holds arbitrarily for λ in the given interval, we have for sufficiently small $\epsilon > 0$,

$$\lambda(\overline{B+N}) \cap M \subset B + (M \cap N). \tag{2}$$

Since $\overline{B+N}$ is a closed absolutely convex neighbourhood of zero, the Minkowski functional $p(x) = \inf\{\lambda \geq 0 : x \in \lambda(\overline{B+N})\}$ is well defined for each $x \in X$ since there always exists $\lambda > 0$ such that $x \in \lambda(B+N)$. Now p is a seminorm on X (see e.g., Köthe [2; p.159]), which is continuous since $B \subset B+N$, and $B_p = \overline{B+N}$. From Lemma III.3.5 we have $B_p^0 = B^0 \cap N^{\perp}$. Together with Lemmas III.3.4 and III.3.5 we obtain, upon taking the polars of both sides of (2),

$$\lambda^{-1}(B_p \cap M)^0 = \lambda^{-1}(B_p^0 + M^{\perp}) = \lambda^{-1}(B^0 \cap N^{\perp} + M^{\perp}) \supset B^0 \cap (M \cap N)^{\perp}.$$

Hence

$$B^0 \cap N^{\perp} + M^{\perp} \supset \lambda B^0 \cap (M \cap N)^{\perp}.$$

Consequently

$$N^{\perp} + M^{\perp} = \bigcup_{n \in \mathbb{N}} n(B^0 \cap N^{\perp}) + M^{\perp} \supset (M \cap N)^{\perp}.$$

The proof of the lemma is now complete. □

III.3.9 Theorem. *Let M and N be closed subspaces of the Banach space X. Then the following properties are equivalent:*

(i) $M + N$ *closed*
(ii) $M^\perp + N^\perp$ *closed*
(iii) $M^\perp + N^\perp$ *weak*-closed*
(iv) (M, N) *open*
(v) $(M^\perp, N^\perp)$ *open*
(vi) $M + N = (M^\perp \cap N^\perp)^\top$
(vii) $M^\perp + N^\perp = (M \cap N)^\perp$.

PROOF. From Lemmas III.3.3 and III.3.7 we evidently have the equivalences

(1) (i) $\iff$ (iv) $\iff$ (v) $\iff$ (ii).

We shall now verify the chain of implications (i) $\implies$ (ii) $\implies$ (iii) $\implies$ (iv) $\implies$ (v) $\implies$ (vi) $\implies$ (vii) $\implies$ (i).

(i) $\implies$ (ii). This is contained in (1).

(ii) $\implies$ (iii). Assume (ii). Then from (1), (M, N) is open, and so by Lemma III.3.8.

$$M^\perp + N^\perp = (M \cap N)^\perp.$$

Hence

$$\begin{aligned}(M^\perp + N^\perp)^{\top\perp} &= (M \cap N)^{\perp\top\perp} \\ &= (M \cap N)^\perp = M^\perp + N^\perp\end{aligned}$$

and therefore $M^\perp + N^\perp$ is weak*-closed, by the Bipolar Theorem.

(iii) $\implies$ (iv). Assume (iii). Then $M^\perp + N^\perp$ is closed, and so (iv) follows from (1).

(iv) $\implies$ (v). This is contained in (1).

(v) $\implies$ (vi). Assume (v). Then from (1), $M + N$ is closed and so from Lemma III.3.5,

$$M + N = (M + N)^{\perp\top} = (M^\perp \cap N^\perp)^\top.$$

(vi) $\implies$ (vii). Assume (vi). Then $M + N$ is closed. Hence by (1), $\gamma(M, N) > 0$. Consequently from Lemma III.3.8, $M^\perp + N^\perp = (M \cap N)^\perp$.

(vii) $\implies$ (i). Assume that (vii) holds. Then $M^\perp + N^\perp$ is closed. Hence from (1), $M + N$ is closed. □

III.4 CLOSED GRAPH, OPEN MAPPING AND CLOSED RANGE THEOREMS

Let $\tilde{X}$ denote the completion of the normed space X.

III.4.1 Definitions Let $T \in LR(X,Y)$, and let $\tilde{T}$ denote the linear relation in $LR(\tilde{X},\tilde{Y})$ whose graph is the completion of $G(T)$. We call $\tilde{T}$ the **completion** (or **complete closure**) of T.

If $G(T)$ is complete, then T is said to be **completely closed** (or **complete**).

If $\tilde{T}x = Tx$ for all $x \in D(T)$, then T is called **completely closable** (or **completable**).

III.4.2 Theorem. (Closed Graph and Open Mapping theorem)

(a) *$\tilde{T}$ is continuous if and only if $D(\tilde{T})$ is closed.*
(b) *$\tilde{T}$ is open if and only if $R(\tilde{T})$ is closed.*

PROOF. We may clearly suppose that X and Y are complete and that T is closed. Since (b) is the dual statement of (a), it is sufficient to prove (a).

(a) Write $M := G(T)$, $N := \{0\} \times Y$. Then $M + N = D(T) \times Y$. Hence $D(T)$ is closed if and only if $M + N$ is closed, if and only if $\gamma(M,N) > 0$ by III.3.9. It only remains to verify that T is continuous if and only if there exists $\lambda > 0$ for which

$$(1) \qquad \lambda B_{D(T)\times Y} \subset B_{G(T)} + \{0\} \times Y.$$

On $X \times Y$ we shall assign the norm

$$(2) \qquad \|(x,y)\| = \max\{\|x\|, \|y\|\}.$$

We shall first assume that T is single valued.

Suppose that (1) holds. Let $\|x\| = 1$. Then

$$\lambda(x,0) = (\lambda x, \lambda Tx) + (0, -\lambda Tx)$$

where $\lambda \|Tx\| \leq 1$. Hence $\|T\| \leq \lambda^{-1}$, and so T is continuous by Proposition II.3.2 (a).

Conversely, assume that T is continuous. Let $\lambda := \min\{1, \|T\|^{-1}\}$. Then if (x,y) satisfies $\|x\|, \|y\| \leq 1$, we have $\lambda(x,y) = (\lambda x, \lambda Tx) + (0, \lambda y - \lambda Tx)$ where $(\lambda x, \lambda Tx) \in B_{G(T)}$. Therefore (1) is satisfied.

Passing now to the general case, suppose that $D(T)$ is closed. By Proposition II.5.2, $QTJ_{D(T)}$ is a closed linear operator from the Banach space $D(T)$ into the Banach space QY. Hence, from what has been proved, QT is continuous. Therefore T is continuous, by Proposition II.3.2 (a), since $\|T\| = \|QT\|$. The converse is obvious. □

Note A more general version of III.4.2 is derived in Section III.5 (see Corollary III.5.4). Theorem III.4.2 is retained for its simplicity.

III.4.3 Corollary.

(a) *If $D(T)$ is complete then T' is continuous.*
(b) *If $R(T)$ is complete then T' is open.*

PROOF. See III.1.3 and III.1.13. □

III.4.4 Theorem. (Closed Range Theorem).
The following properties are equivalent:

(i) *$R(\tilde{T})$ closed*
(ii) *$R(T')$ closed*
(iii) *$R(\tilde{T}')$ weak*-closed.*

PROOF. (a) Without loss of generality we may assume that X and Y are complete and T is closed. Write $M := G(T)$, $N := X \times \{0\}$. Then $M + N = X \times R(T)$, and $M^\perp + N^\perp = R(T') \times Y'$. Hence the equivalence of (i), (ii) and (iii) follows immediately from Theorem III.3.9. □

III.4.5 Corollary. *The following properties are equivalent:*

(i) *$D(\tilde{T})$ closed*
(ii) *$D(T')$ closed*
(iii) *$D(\tilde{T}')$ weak*-closed.*

Corollary III.4.5 is a generalisation of Brown's Closed Domain theorem (see A. Brown [1], and Kato [2; p.235]).

III.4.6 Proposition.

(a) *T is continuous if and only if $D(T') = T(0)^\perp$.*
(b) *T is open if and only if $R(T') = N(T)^\perp$.*
(c) *If T is continuous, then $\|T\| = \|T'\|$.*
(d) *If T is open, then $\gamma(T) = \gamma(T')$.*

PROOF. Write $J = J^X_{D(T)}$.

(a) Suppose that T is continuous. Then $(QTJ)'$ is continuous by Corollary III.1.13, and by Proposition III.1.2 its domain is the whole space, i.e. $T(0)^\perp$. Hence by Proposition III.1.10 (c), $D(T') = T(0)^\perp$.

Conversely, let $D(T') = T(0)^\perp$. Since T is continuous if and only if QT is continuous, it is enough (by III.1.11) to consider the case when T is single valued. Now we have $D(T') = Y'$. Since for each $y' \in Y'$, $\sup\{|y'Tx| : x \in S_{D(T)}\} \leq \|T'y'\|$, therefore by Corollary II.3.17 (with $W = T(S_{D(T)})$), $\|T\| = \sup\{\|Tx\| : x \in S_{D(T)}\} < \infty$.

The statements (b) and (d) are dual statements of (a) and (c) respectively. It remains only to verify (c).

(c) Let T be continuous. Then

$$\begin{aligned}
\|T\| &= \|QTJ\| = \sup\{\|QTJx\| : x \in B_{D(T)}\} \\
&= \sup\{|y'(QTJx)| : y' \in B_{T(0)^\perp}, x \in B_{D(T)}\} \qquad \text{(see III.1.8)} \\
&= \sup\{|(QTJ)'y'(x)| : y' \in B_{D(T')}, x \in B_{D(T)}\} \\
&\qquad\qquad \text{(since } D(T') = T(0)^\perp \text{ by (a))} \\
&= \sup\{\|(QTJ)'y'\| : y' \in B_{D(T')}\} \\
&= \sup\{\|T'y'\| : y' \in B_{D(T')}\} \qquad \text{(Proposition III.1.10 (c))} \\
&= \|T'\|. \quad \square
\end{aligned}$$

III.4.7 Theorem. *T is closable if and only if $D(T')^\top = T(0)$.*

PROOF. Follows immediately from Propositions II.5.5 and III.1.4(d). □

III.4.8 Corollary. *Let T be a linear operator. Then T is closable if and only if $D(T')$ is weak*-dense in Y'.* (See Goldberg [3; p.54]).

III.4.9 Proposition. *The following properties are equivalent:*

(i) *$\tilde{T}$ is continuous*
(ii) *T' is continuous*
(iii) *$D(T')$ is $\sigma(Y', \tilde{Y})$ closed.*

PROOF. We have

T' is continuous $\iff D(\tilde{T}')$ is closed (Closed Graph Theorem III.4.2)
$\iff D(\tilde{T})$ is closed (Closed Domain Theorem III.4.5)
$\iff \tilde{T}$ is continuous (by III.4.2)
$\iff D(T') = \tilde{T}(0)^{\perp}$ (Proposition III.4.6(a))
$\iff D(T') = D(\tilde{T}')^{\top\perp}$ (Proposition III.1.4(d))
$\iff D(T')$ is $\sigma(Y', \tilde{Y})$ closed (Bipolar Theorem). □

III.4.10 Exercise Let T be completely closable. Show that T is continuous if and only if T' is continuous.

III.4.11 Corollary. *Let E denote the subspace $D(T')^{\top}$. Then $Q_E T$ is single valued and closable. Moreover, T' has the same graph as $(Q_E T)'$.*

PROOF. We have $(Q_E T)' = T' J_{E^{\perp}}$ where $E^{\perp} \supset D(T')$, by III.1.6 and III.1.9. Therefore $D((Q_E T)') = D(T')$. Hence $Q_E T$ is closable by Theorem III.4.7. Furthermore, since $D(T')^{\top} \supset T(0)$ by III.1.4(d), $Q_E T$ is single valued. □

III.4.12 Definition The relation $Q_E T$ of III.4.11 is termed the **regular contraction** of T.

III.4.13 Example Let M, N be a pair of dense linear subspaces of X such that $M \cap N = \{0\}$ and $M + N = X$. Let P be the linear projection with domain X, range M and kernel N. Then P has the following properties:

(1) $G(P') = \{(0,0)\}$
(2) $G(\tilde{P}) = \tilde{X} \times \tilde{X}$
(3) $\tilde{P}(0) = \tilde{X}$
(4) $\|\tilde{P}\| = 0$
(5) $\|P\| = \infty$
(6) $\tilde{P}\tilde{P} = \tilde{P} = \tilde{P}^{-1} = I - \tilde{P}$.

III.4.14 Example Let M, N be a pair of closed subspaces of X such that $M + N$ is dense in X and $M \cap N = \{0\}$. Let P be the linear projection with domain $M + N$, range M and kernel N. Let us determine P' and $\tilde{P}$.

We have

$$N(P') = R(P)^{\perp} = M^{\perp}$$
$$R(P') = N(P)^{\perp} = N^{\perp}$$

If $(y_1', x_1') \in G(P')$, then $y_1'(m) = x_1'(m) + x_1'(n)$ for all $m \in M$, $n \in N$. Hence $y_1' - x_1' \in M^{\perp}$ and $x_1' \in N^{\perp}$. Thus

$$G(P') = \{(y' + x', y') : y' \in N^{\perp}, x' \in M^{\perp}\}.$$

Because $D(P) = M + N$ is dense in X, P' is single valued. Hence $M^\perp \cap N^\perp = 0$. It follows that P' is the linear projection defined on $M^\perp + N^\perp$ with range $N^\perp$ and kernel $M^\perp$.

Now from Proposition III.1.4

$$R(\tilde{P}')^\top = N(\tilde{P}) = \tilde{N}$$
$$D(\tilde{P}')^\top = \tilde{P}(0) = (M^\perp + N^\perp)^\top \subset \tilde{M} \cap \tilde{N}.$$

Also $\tilde{P}$ is open, since $\gamma(\tilde{P}) \geq \gamma(P)$ by II.5.9. Hence $R(\tilde{P}) = \tilde{M}$ (by III.4.2, since $M \subset R(\tilde{P}) \subset R(P)^\sim \subset \tilde{M}$).

By considering a Cauchy sequence in $G(P)$ of the form $(m_k + n_k, m_k)$, we conclude that

$$G(\tilde{P}) = \{(m+n, m) : m \in \tilde{M}, n \in \tilde{N}\}.$$

It follows that

$$\tilde{P}\tilde{P} = \tilde{P}$$
$$\tilde{P}(0) = \tilde{M} \cap \tilde{N}$$

Special case: Let X be complete. Then $\tilde{P} = P$, and so P is continuous if and only if $M + N = X$, if and only if $M^\perp + N^\perp = X'$.

III.5 ALMOST OPEN RELATIONS

III.5.1 Definition T is called **almost open** if $\overline{TB_X} \supset \lambda B_{R(T)}$ for some $\lambda > 0$.

The main result of this section is Theorem III.5.3, for which the following characterization of almost openness is needed.

III.5.2 Proposition. *The following properties are equivalent:*

(i) *T is almost open*
(ii) *$\tilde{T}$ is open*
(iii) *T' is open.*

PROOF. Suppose that for some $\lambda > 0$

$$\text{(1)} \qquad \overline{TB_X} \supset \lambda B_{R(T)} \quad (= \lambda B_Y \cap R(T)).$$

Then from Theorem III.2.2 we have

$$(T')^{-1}(B_{X'}) = (TB_X)^0 \cap D(T') \subset (\lambda^{-1}B_{Y'} + N(T')) \cap D(T') \subset \lambda^{-1}B_{D(T')} + N(T').$$

Hence T' is open, by II.1.10 (a). Therefore (i) $\Longrightarrow$ (iii).

The equivalence of (ii) and (iii) is an obvious consequence of the Open Mapping and Closed Range theorems (III.4.2 (b) and III.4.4).

Finally, the implication (ii) $\Longrightarrow$ (i) follows from the inclusions

$$(TB_X)^\sim \supset \tilde{T}B_{\tilde{X}}, \ \overline{TB_X} \supset (TB_X)^\sim \cap Y. \quad \square$$

III.5.3 Theorem. *(a) Let X be complete. Then $\gamma(\overline{T}) = \gamma(T')$. Moreover, if $\gamma(\overline{T}) > 0$, then $R(\overline{T})$ is closed.*

(b) Let Y be complete. Then $\|\overline{T}\| = \|T'\|$. Moreover if $\|\overline{T}\| < \infty$, then $D(\overline{T})$ is closed.

PROOF. (a) Suppose that T' is open. Then by Proposition III.5.2 there exists $\lambda > 0$ such that

$$\tilde{T}B_X \supset \lambda B_{R(\tilde{T})}. \tag{1}$$

Without loss of generality, we shall assume that T is closed. Plainly, $\tilde{T}B_X \cap Y \subset TB_{D(T)}$. Indeed, if $(x, y) \in G(\tilde{T})$, where $\|x\| \leq 1$ and $y \in Y$, then there is a sequence $(x_n, y_n) \in G(T)$, with $\|x_n\| \leq 1$, such that $(x_n, y_n) \to (x, y)$, and $(x, y) \in X \times Y$ since X is complete. Since T is closed, $x \in B_{D(T)}$ and $y \in TB_{D(T)}$. Thus as $R(\tilde{T})$ is closed by III.4.2, the same is true of $R(T)$, since now

$$R(\tilde{T}) \cap Y \supset R(T) = \bigcup_{n=1}^{\infty} nTB_{D(T)} \supset \left(\bigcup_{n=1}^{\infty} n\tilde{T}B_X \right) \cap Y = R(\tilde{T}) \cap Y.$$

Taking the intersection of both sides of (1) with $R(T)$ now yields

$$TB_{D(T)} \supset \lambda B_{R(T)} \tag{2}$$

showing that T is open. Hence $\gamma(T) = \gamma(T')$ by III.4.6 (d).

If T' is not open, then $\gamma(T') = \gamma(T) = 0$ by III.1.13 and II.2.5.

(b) Substitute T^{-1} for T in (a) and use Theorem II.2.5. □

We have the following generalisation of the Open Mapping and Closed Graph theorems:

III.5.4 Corollary. *Let $T \in LR(X, Y)$ be a closed linear relation.*

(a) *If X is of second category, Y is complete and T is everywhere defined, then T is bounded.*

(b) *If X is complete, Y is of second category and T is surjective, then T is open.*

PROOF. (a) As (a) follows immediately from (b) upon interchanging the roles of T and T^{-1}, it is sufficient to prove (b).

(b) Since $Y = \bigcup_{n\in\mathbb{N}} nTB_X$ is of second category, it follows that T is almost open. Now combine Proposition III.5.2 and Theorem III.5.3(a). □

III.5.5 Exercise Let T be completely closable. Show that $\|T\| = \|T'\|$.

III.6 THE STATE DIAGRAM

This section is concerned with the following question: Given that T has certain properties, to find corresponding properties enjoyed by T', and conversely. The properties considered are various combinations of "states" defined below. The resulting "state diagram" (see III.6.3 and III.6.5) has the same configuration as the State Diagram for Linear Operators in Goldberg [3], widely used in Operator Theory.

In the theory of linear relations, the usefulness of a state diagram is enhanced by the duality present, and the exploitation of such relationships as $(T^{-1})' = (T')^{-1}$, $\overline{T}^{-1} = (T^{-1})^{-}$, and $\overline{T}' = (T')^{-} = T'$.

A feature of the Goldberg state diagram is its stability under finite dimensional perturbations when the underlying states are suitably modified (see Theorems V.7.4 and V.8.5).

III.6.1 Definition: States of a Linear Relation

Consider the following classification:

$I:\ R(T) = Y$
$II:\ R(T) \neq Y$ but $\overline{R(T)} = Y$
$III:\ \overline{R(T)} \neq Y$

1: T^{-1} is single valued and continuous
2: T^{-1} is single valued but not continuous
3: T^{-1} is not single valued.

If, for instance $R(T) = Y$ we say that T is in the state I and we write $T \in I$. The statement "$T \in 3$" means that T is in the state 3, i.e. T is not injective. A similar interpretation applies to all six symbols I, II, III, 1, 2, 3. Each triple, (I, II, III) and (1, 2, 3), describes a partition of $LR(X, Y)$. The intersections $I \cap 1$, $I \cap 2$, $II \cap 1$, are denoted by I_1, I_2, II_1, respectively. The same notation is used for T'. If $T \in II_3$ and $T' \in III_1$, we write $(T, T') \in (II_3, III_1)$.

Note that $T \in 1$ if and only if there exists $c > 0$ such that for all $x \in D(T)$,

$$\|Tx\| \geq c\,\|x\|\,.$$

Such a relation T is said to be **bounded below.**

The classification described above is the same as the one described in Goldberg [3].

The next lemma tabulates the properties needed to construct the state diagram. All these properties can be read off from the state diagram that ensues (III.6.3).

III.6.2 Lemma.

(a) $T \in 1 \iff T' \in I$
(b) $T' \in 1 \implies T' \notin II$
(c) $T \in I_1 \cup II_1 \iff T' \in I_1$
(d) $T \in III \iff T' \in 3$
(e) $T \in 3 \implies T' \in III$
(f) *If Y is complete and $T \in I$, then $T' \in 1$.*

PROOF. (a) Immediate from III.4.6 (b) and III.1.4(c).

(b) $T' \in 1 \implies R(T')$ closed (by III.4.2) $\implies T' \notin II$.

(c) $T \in I_1 \cup II_1 \iff T' \in I$ (by (a)) and $R(T)^{\perp} = \{0\} \iff T' \in I$ and $N(T') = \{0\}$ (III.1.4 (a)) $\iff T' \in I_1$ (III.4.2).

(d) $T \in III \iff R(T)^{\perp} = N(T') \neq \{0\}$ (III.1.4 (a)) $\iff T' \in 3$.

(e) $T \in 3 \implies \{0\} \neq N(T) \subset R(T')^{\top} \cap D(T)$ (III.1.4 (c)) $\implies R(T')^{\top} \neq \{0\} \implies T' \in III$.

(f) Let Y be complete. Then $T \in I \implies \tilde{T} \in I \implies N(T') = \{0\}$ and $R(T')$ closed (by III.4.4, since $R(T') = R(\tilde{T}'))$ $\implies T' \in 1$ (by III.4.2). □

III.6.3 State Diagram for linear relations

III_3	■	■	■	■	■	■	■		
III_2	■	Y	Y	■			■	■	■
III_1	■			■			■	■	■
II_3	■	■	■	■	■	■	■		■
II_2	■	Y	■	■		■	■	■	■
II_1	■	■	■	■	■	■	■	■	■
I_3	■	■	■	■	■	■		■	■
$\uparrow$ I_2	■	■	■	■	■	■	■	■	■
T' I_1		■	■		■	■	■	■	■
	I_1	I_2	I_3	II_1	II_2	II_3	III_1	III_2	III_3
	T	$\rightarrow$							

Y: Cannot occur if Y is complete.

The state diagram for closed linear relations follows from the next lemma:

III.6.4 Lemma. *Let X be complete and let T be a closed linear relation. We have:*

(a) $T' \in I_1 \implies T \in I_1$
(b) $T' \in III_1 \implies T \in I_3$
(c) *If X is reflexive and T is injective, then $R(T')$ is dense in X'.*

PROOF. (a) and (b) are immediate consequences of III.5.3 and the state diagram III.6.3. (See Exercise III.5.5).

(c) By reflexivity, $R(T')^\top = R(T')^\perp$. Therefore $N(T'') = R(T')^\perp = N(T) = \{0\}$. □

III.6.5 State Diagram for closed linear relations

III_3	■	■	■	■	■	■	■	$X-R-c$	
III_2	■	Y $(X-R-c)$	Y	■	$X-R-c$		■	■	■
III_1	■	$X-c$		■	$X-c$	$X-c$	■	■	■
II_3	■	■	■	■	■	■	■		■
II_2	■	Y	■	■		■	■	■	■
II_1	■	■	■	■	■	■	■	■	■
I_3	■	■	■	■	■	■		■	■
$\uparrow$ I_2	■	■	■	■	■	■	■	■	■
T' I_1		■	■	$X-c$	■	■	■	■	■
	I_1	I_2	I_3	II_1	II_2	II_3	III_1	III_2	III_3
	T	$\rightarrow$							

Y: Cannot occur if Y is complete.
$X-c$: Cannot occur if X is complete and T is closed.
$X-R-c$: Cannot occur if X is reflexive and T is closed.

The reader will observe that the state diagrams III.6.3 and III.6.5 are identical with the corresponding ones for linear operators in Goldberg [3; II.3.14 and II.4.11].

Examples are exhibited in Goldberg [3] of continuous densely defined linear operators whose states (T, T') can occur. They show that all the "possible" states in III.6.3 and III.6.5 above do indeed eventuate.

Since III.6.5 contains the state diagram III.6.3, in future "the state diagram" will refer to III.6.5.

III.6.6 Inverse States

The inverse states are obtained from the states defined in III.6.1 by substituting T^{-1} for T. Hence the following dual classification:

$$I: D(T) = X$$
$$II: D(T) \neq X \text{ but } \overline{D(T)} = X$$
$$III: \overline{D(T)} \neq X$$

1 : T is single valued and continuous
2 : T is single valued but not continuous
3 : T is not single valued.

Notice that I_1 is just the class of bounded linear operators from X to Y.

III.6.7 Inverse State Diagram

	III_3	■	■	■	■	■	■	■	$Y-R-c$	
	III_2	■	X $Y-R-c$	X	■	$Y-R-c$		■	■	■
	III_1	■	$Y-c$		■	$Y-c$	$Y-c$	■	■	■
	II_3	■	■	■	■	■	■	■		■
	II_2	■	X	■	■		■	■	■	■
	II_1	■	■	■	■	■	■	■	■	■
	I_3	■	■	■	■	■	■		■	■
$\uparrow$	I_2	■	■	■	■	■	■	■	■	■
T'	I_1		■	■	$Y-c$	■	■	■	■	■
		I_1 T	I_2 $\rightarrow$	I_3	II_1	II_2	II_3	III_1	III_2	III_3

X: Cannot occur if X is complete.
$Y-c$: Cannot occur if Y is complete and T is closed.
$Y-R-c$: Cannot occur if Y is reflexive and T is closed.

Superimposing the state diagram on its inverse, one obtains various propositions about linear operators or relations. For example, let X be reflexive, Y complete and T closed. Then T is densely defined, is not single valued, has dense range, is injective but not open, if and only if T' is not densely defined, is single valued, has dense range, is injective but not open.

Another example: *The relation $T : D(T) \subset X \to Y$ is an injective isomorphism operator if and only if T' is everywhere defined and surjective.*

One can of course reduce the number of possible states by considering the induced relation T_0 defined by

$$T_0 \in LR(D(T), R(T))$$
$$G(T_0) = G(T)$$

so that both T_0 and T_0^{-1} are in the state I.

III.7 SMALL PERTURBATION THEOREM

III.7.1 Definition For $T \in LR(X, Y)$ we define $\overline{\beta}(T) := \dim Y/\overline{R(T)}$ (for the definitions of $\alpha(T)$ and $\beta(T)$ see I.6.6). Write $J_Y := J_Y^{\tilde{Y}}$.

III.7.2 Proposition $\alpha(T') = \overline{\beta}(T) = \overline{\beta}(J_Y T)$.

PROOF. We have $\alpha(T') = \dim N(T') = \dim R(T)^{\perp}$ (by III.1.4(a)) $= \dim(Y/\overline{R(T)})' = \dim Y/\overline{R(T)} = \overline{\beta}(T)$. The proof of the second equality is similar, observing that $T' = (J_Y T)'$. □

We shall assume the following geometric property of normed spaces (see e.g. Goldberg [3])

III.7.3 Lemma. *Let M and N be subspaces of X with $\dim M > \dim N$. Then there exists $m \in M$, $m \neq 0$, such that*

$$\|m\| = d(m, N).$$

III.7.4 Theorem. *Let $\gamma(T) > 0$ and let S satisfy $S(0) \subset \overline{T(0)}$, $D(S) \supset D(T)$ and $\|S\| < \gamma(T)$. Then*

(a) $\alpha(T + S) \leq \alpha(T)$
(a) $\overline{\beta}(T + S) \leq \overline{\beta}(T)$.

PROOF. (a) For $x \neq 0$, $x \in N(T+S)$, we have

$$\begin{aligned}\gamma(T)d(x, N(T)) &\leq \|Tx\| = \|QTx\| \\ &= \|QSx\| \quad (\text{since } QTx = -QSx, \text{ where } Q \text{ denotes } Q_T) \\ &\leq \|S\|\,\|x\| < \gamma(T)\,\|x\|.\end{aligned}$$

Thus $d(x, N(T)) < \|x\|$ for all $x \in N(T+S)$. Hence by Lemma III.7.3, $\alpha(T+S) \leq \alpha(T)$.

(b) Since $\gamma(T) > 0$, we have

$$\gamma(T') = \gamma(T) > \|S\| \geq \|S'\|$$

by III.4.6(d). Also $(T+S)' = T' + S'$ by III.1.5. Replacing X by $D(S)$ if necessary, we may assume S to be everywhere defined, thus making S' single valued. Then by part (a) above and Proposition III.7.2, we have

$$\overline{\beta}(T+S) = \alpha(T'+S') \leq \alpha(T') = \overline{\beta}(T). \quad \square$$

III.7.5 Corollary. *Let T be open with dense range. Then for any relation S satisfying $S(0) \subset \overline{T(0)}$, $D(S) \supset D(T)$ and $\|S\| < \gamma(T)$, $T+S$ has dense range.*

III.7.6 Corollary. *Let T be open and injective. If S is a linear relation such that $S(0) \subset \overline{T(0)}$, $D(S) \supset D(T)$ and $\|S\| < \gamma(T)$, then $T+S$ is open and $\overline{\beta}(T+S) = \overline{\beta}(T)$.*

PROOF. Write $\gamma := \gamma(T)$. Choose $n \in \mathbb{N}$ such that

$$(1) \qquad \|S\|/n \;<\; \gamma - \|S\|.$$

Let $x \in D(T+S)$. Since S is single valued, we have for $0 \leq k \leq n$,

$$\left\|(T+S-\frac{k}{n}S)x\right\| \geq \|Tx\| - \left\|(1-\frac{k}{n})Sx\right\| \geq \gamma\,\|x\| - (1-\frac{k}{n})\,\|S\|\,\|x\| \geq (\gamma - \|S\|)\,\|x\|.$$

Therefore

$$(2) \qquad \gamma(T+S-\frac{k}{n}S) \;\geq\; \gamma - \|S\| \qquad (0 \leq k \leq n).$$

In particular,

$$\gamma(T+S) > 0.$$

Now applying III.7.4 to the relations $T+S-\frac{k}{n}S$ and $\frac{1}{n}S$, using (1) and (2), we obtain

$$\overline{\beta}(T+S-\frac{k+1}{n}S) \;\leq\; \overline{\beta}(T+S-\frac{k}{n}S).$$

Putting $k = n-1, n-2, \ldots, 1, 0$ successively gives

$$(3) \qquad \overline{\beta}(T) \leq \overline{\beta}(T+S).$$

Equality now follows from Theorem III.7.4. $\square$

III.7.7 Corollary. *Let T be open and injective with dense range. Then for any relation S such that $S(0) \subset \overline{T(0)}$, $D(S) \supset D(T)$ and $\|S\| < \gamma(T)$, we have $T + S$ is open and injective with dense range.*

III.7.8 Remarks (1) In Corollary III.7.5, the corresponding statement concerning surjectivity fails (see Labuschagne [7; p.389]).

(2) The requirement that T be injective in III.7.6 and III.7.7 can be replaced with $N(S) \supset N(T)$. Upon perturbation one gets $N(T+S) = N(T)$. See Labuschagne [5; Theorem 2.16] for details. Labuschagne [7; Cor. 13] is a related result.

III.8 THE $D(T')$ SEMINORM AND THE CONTINUITY OF T' AND T''

Given $T \in LR(X,Y)$, useful topologies on Y and Y'' are those induced by the subspace $D(T') \subset Y'$. We shall first consider an arbitrary subspace $V \subset Y'$. Our main aim is to obtain conditions equivalent to the continuity of T' and T''.

III.8.1 Definition Given a linear subspace V of Y', the V-**seminorm** on Y is defined by

$$\|y\|_V = \sup\{|y'y| : y' \in B_V\} \qquad (y \in Y). \tag{1}$$

We shall denote

$$Y_V := (Y, \| \ \|_V)$$

i.e., Y_V is the space Y endowed with the V-seminorm.

We shall first establish some elementary propositions before passing to the case $V = D(T')$.

III.8.2 Proposition. *The following properties are equivalent:*

(i) *The V-seminorm is a norm*
(ii) $V^{\top\perp} = Y'$.

PROOF. We have the following chain of equivalences:

For some $y \neq 0$, $\|y\|_V = 0$
$\iff$ for some $y \neq 0$, $y'y = 0$ for all $y' \in V$
$\iff$ for some $y \neq 0$, $y \in V^{\top}$
$\iff$ $V^{\top} \neq 0$,
$\iff$ $V^{\top\perp} \neq Y'$. □

III.8.3 Proposition. $(Y_V)' \subset Y'$.

(Here both these Banach spaces are considered as linear subspaces of the algebraic dual of Y.)

PROOF. Follows immediately from the definitions. □

III.8.4 Proposition. *Let σ denote $\sigma(Y',Y)$. Then*

$$(Y_V)' = \bigcup_{n=1}^{\infty} n\overline{B}_V^{\sigma} \subset \overline{V}^{\sigma}.$$

(Here $\overline{E}^{\sigma}$ denotes the σ-closure of E.)

PROOF. A local base of neighbourhoods for Y_V is $\{\lambda(B_V)_0 : \lambda \in \mathbb{K}\}$. Hence

$$(Y_V)' = \bigcup_{\lambda\in\mathbb{K}} \lambda((B_V)_0)^0 = \bigcup_{n=1}^{\infty} n((B_V)_0)^0$$

where the second polar is taken in the algebraic dual of Y. Now since $(B_V)_0 \supset B_Y$, it follows that $((B_V)_0)^0 \subset (B_Y)^0 = B_{Y'} \subset Y'$. By the Bipolar Theorem applied to the pair (Y',Y), we now have $((B_V)_0)^0 = \overline{B}_V^{\sigma}$. Thus $(Y_V)' = \bigcup_{n=1}^{\infty} n\overline{B}_V^{\sigma}$. It follows immediately that $(Y_V)' \subset \overline{V}^{\sigma}$. □

III.8.5 Corollary. *If V is $\sigma(Y',Y)$ closed, then $V = (Y_V)'$.*

III.8.6 Proposition. *The following statements are equivalent:*

(i) *V is $\sigma(Y',\tilde{Y})$ closed*
(ii) *B_V is $\sigma(Y',Y)$ closed*
(iii) *B_V is $\sigma(Y',Y)$ compact.*

PROOF. The topology $\sigma(Y',Y)$ is Hausdorff. Hence (iii) $\Longrightarrow$ (ii). Assume (ii). Observing that the $\sigma(Y',Y)$ and $\sigma(Y',\tilde{Y})$ topologies coincide on the norm bounded subsets of Y', it follows that B_V is $\sigma(Y',\tilde{Y})$ closed. Now let y' be a point in the $\sigma(Y',\tilde{Y})$ closure of V, and let $\{y'_\alpha\}$ be a net in V which is $\sigma(Y',\tilde{Y})$ convergent to y'. By the uniform boundedness principle, $\{y'_\alpha\} \subset \lambda B_V$ for some $\lambda > 0$. Hence $y' \in \lambda B_V$. Therefore (ii) $\Longrightarrow$ (i). Finally, assume (i). By the Banach-Alaoglu Theorem (Wilansky [1; p.239]), $B_V = B_{Y'} \cap V$ is $\sigma(Y',\tilde{Y})$ compact, hence $\sigma(Y',Y)$ compact. Thus (i) $\Longrightarrow$ (iii). □

Proposition III.8.6 can be viewed as a generalisation of the Banach-Alaoglu Theorem. Note that $\tilde{Y}$ may replace Y in (ii) and (iii).

III.8.7 Lemma. *Let $D(T) = X$. Then the following properties are equivalent:*

(i) *T is $\sigma(X, X') - D(T')$ seminorm continuous*

(ii) *There exists a finite subset F of X' such that $|\operatorname{co}|(F) \supset T'B_{D(T')}$.*

PROOF. Write $B := B_{D(T')}$. We have the following chain of equivalences:

(i) holds $\Longleftrightarrow$ there exists a $\sigma(X, X')$ neighbourhood of zero in X contained in $T^{-1}(B_0)$ $\Longleftrightarrow$ there exists a finite subset F in X' such that $F_0 \subset T^{-1}(B_0)$.

Now by Corollary III.2.3(a), since $B \subset D(T')$, we have $T^{-1}(B_0) = (T'B)_0 \cap D(T)$. Hence $F_0 \subset T^{-1}(B_0) \Longleftrightarrow F_0 \subset (T'B)_0$ (since $F_0 \subset D(T)$) $\Longleftrightarrow$ $(F_0)^0 \supset ((T'B)_0)^0 \Longleftrightarrow (F_0)^0 \supset T'B$. Now since F is a finite set, $(F_0)^0 = |\operatorname{co}|(F)$. Hence the result follows. □

As a consequence of Lemma III.8.7 we obtain the following characterisation of adjoint relations of finite rank:

III.8.8 Corollary. *Let $D(T) = X$. Then the range of T' is finite dimensional if and only if T is $\sigma(X, X') - D(T')$ seminorm continuous.*

PROOF. The sufficiency is immediate from Lemma III.8.7. Conversely, let $R(T')$ be finite dimensional. Then T' is a closed finite rank linear operator and hence continuous (since its null space is closed). Let B be a ball centred at 0 in X' such that $B \supset T'(B_{D(T')})$.

Let $a_1, \ldots, a_n$ be a normalised basis for $R(T')$. Then there exists $c > 0$ such that

$$\left\| \sum_{k=1}^{n} \lambda_k a_k \right\| \leq \sum_{k=1}^{n} |\lambda_k| \leq c \left\| \sum_{k=1}^{n} \lambda_k a_k \right\|$$

for all scalars $\lambda_1, \ldots, \lambda_n$. Then if d is the radius of $B \subset X'$, it follows that if $x' = \sum_{k=1}^{n} \lambda_k a_k$ belongs to $B \cap R(T')$, then $\sum_{k=1}^{n} |\lambda_k| \leq cd$. Clearly $B \cap R(T')$ is then contained in the absolutely convex hull of $\{cd\, a_1, \ldots, cd\, a_n\}$.

The result now follows from Lemma III.8.7. □

III.8.9 Definition The **bounded weak* topology** of X' is the topology on X' for which a set Λ is closed if and only if $\Lambda \cap B$ is weak* closed whenever B is a norm bounded subset of X'. This topology will be denoted by $b\sigma(X', X)$.

It follows from the definition that a subset Λ of X' is $b\sigma(X', X)$ closed if and only if the limit of every bounded $\sigma(X', X)$ convergent net in Λ is also in Λ (see Day [1; p.41]).

III.8.10 Theorem. *The following properties are equivalent:*

(i) T' *is continuous*
(ii) T *is norm* $-D(T')$ *seminorm continuous*
(iii) $D(T') = (Y_{D(T')})'$
(iv) $D(T')$ *is* $b\sigma(Y',Y)$ *closed*
(v) $D(T')$ *is* $\sigma(Y',\tilde{Y})$ *sequentially closed*
(vi) $D(T'') = D(T)''$
(vii) T'' *is continuous.*

PROOF. Write $B := B_{D(T')}$, $J := J^X_{D(T)}$.

(i) $\iff$ (ii). Clearly T is norm $-D(T')$ seminorm continuous if and only if TJ has the same property. Since $(TJ)' = Q_{T'}T'$, we may suppose that $D(T) = X$. Suppose that T' is continuous. Then for each $y' \in B_{D(T')}$ and $x \in D(T)$, $|y'Tx| \leq \|T'y'\|\,\|x\| \leq \|T'\|\,\|x\|$. Thus for each $x \in D(T)$, $\|Tx\|_{D(T')} \leq \|T'\|\,\|x\|$. Hence (i) $\implies$ (ii). Conversely, assume that (ii) holds. Then there exists a constant $c > 0$ such that for $x \in B_X$ and $y' \in B_{D(T')}$, $|y'Tx| \leq c$, i.e., $|T'y'x| \leq c$, whence $\|T'y'\| \leq c \ \ (y' \in B_{D(T')})$. Hence T' is continuous and the equivalence of (i) and (ii) follows.

(i) $\implies$ (iii). Assume (i). Then $D(T')$ is $\sigma(Y',\tilde{Y})$ closed (Proposition III.3.9). Hence $B_{D(T')}$ is $\sigma(Y',Y)$ closed, by Proposition III.8.6, and hence $(Y_{D(T')})' = D(T')$ by Proposition III.8.4. Therefore (i) $\implies$ (iii).

(iii) $\implies$ (iv). Assume (iii). Let y' belong to the $b\sigma(Y',Y)$ closure of $D(T')$. Then there exists a bounded net $\{y'_\alpha\}$ in $D(T')$ which is $\sigma(Y',Y)$ convergent to y'. Let $c = \sup \|y'_\alpha\|$. Then

$$|y'y| \leq |(y'_\alpha - y')y| + |y'_\alpha y|, \quad \text{so}$$
$$|y'y| \leq c\,\|y\|_{D(T')} \qquad (y \in Y)$$

Hence $y' \in D(T')$. Therefore (iii) $\implies$ (iv).

(iv) $\implies$ (v). Follows immediately.

(v) $\implies$ (vi). Assume (v). Then $D(T')$ is closed in the norm topology. Therefore T' is continuous by the Closed Graph Theorem III.4.2. But then $D(T'') = T'(0)^{\perp}$ (III.4.6). Now $T'(0)^{\perp} = D(T)^{\perp\perp} = D(T)''$. Hence (v) $\implies$ (vi).

(vi) $\implies$ (vii). Follows immediately from Theorem III.4.2.

(vii) $\implies$ (i). Assume (vii). Then $D(T'')$ is closed and hence $D(T')$ is closed by Corollary III.4.5 (Closed Domain Theorem). Therefore T' is continuous (III.4.2). □

III.8.11 Example *There exists a linear operator T such that T' is continuous and $D(T')$ is not $\sigma(Y',Y)$ closed.* Let S be a bounded linear operator such that the pair (S, S') is in the state (II_2, III_1); an example of such an operator is given in Goldberg [3; p.68]. Let $T = S^{-1}$. Then T is a closed linear operator and so $D(T')$ is $\sigma(Y',Y)$ dense in Y'. Also $T' = (S')^{-1}$ is continuous. Since $D(T') = R(S') \neq Y'$, $D(T')$ cannot be $\sigma(Y',Y)$ closed.

III.9 DIFFERENTIAL OPERATORS: AN INTEGRAL BOUNDARY VALUE PROBLEM

An underlying motivation for the introduction of linear relations into operator theory by J. von Neumann [1,2] was to aid the investigation of differential equations governed by non densely defined linear operators. The adjoints of such operators are multivalued linear relations (since $T'(0) = D(T)^{\perp}$, by Proposition III.1.4). With the usual inner product in L_2, the orthogonality formula III.1.1(1) defining the adjoint becomes the familiar "Green's formula" (see III.9.1(17)). Such differential equations arise when certain boundary conditions are imposed, such as the integral boundary conditions (3) of Example III.9.1 below.

III.9.1 Example Consider the partial differential equation

(1) $$u_t = u_{xx} \qquad 0 \leq x \leq 1,\ 0 < t < \infty$$

with the boundary conditions

(2) $$u(0,t) = u(1,t) = 0$$

(3) $$\int_0^1 u(x,t) \sin \pi x \ dx = 0$$

and given initial condition

(4) $$u(x,0) = \phi(x).$$

The special feature of this example is the boundary condition (3), known as an "integral boundary condition".

We shall seek (and find) solutions of the form $u = \Sigma f_n(x) h_n(t)$. To this end suppose that

(5) $$u(x,t) := f(x)h(t).$$

The equation (1) now becomes

$$f(x)h'(t) = f''(x)h(t)$$

whence

(6) $$f''(x)/f(x) = h'(t)/h(t) = c \qquad (f(x) \neq 0,\ h(t) \neq 0)$$

where c is some real constant. From the boundary conditions (2) we conclude that $c \leq 0$. Consequently we may write the equations (6) in the form

(7) $$h' + \lambda^2 h = 0$$

(8) $$f'' + \lambda^2 f = 0$$

with boundary conditions

(9) $$f(0) = f(1) = 0$$

(10) $$\int_0^1 f(x) \sin \pi x \, dx = 0.$$

Solving for f and h we obtain

(11) $$\begin{cases} f_n(x) & = b_n \sin n\pi x \\ h_n(t) & = c_n e^{-n^2\pi^2 t^2} \end{cases} \qquad (n = 2, 3, \dots)$$

(the value $n = 1$ is rejected on account of (10)). The solutions (11) (termed "fundamental solutions") can now be added to form the function

(12) $$u(x,t) = \sum_{u=2}^{\infty} a_n e^{-n^2\pi^2 t^2} \sin n\pi x$$

where the coefficients a_n are determined in such a way that the initial condition (4) is satisfied. Then the resulting function $u(x,t)$ will be a solution to the system described by (1) – (4).

On account of (5) we have

(13) $$u(x,0) = f(x)h(0).$$

Hence the function $f(x)$ must satisfy the system defined by (8), (9) and (10).

Let T be the differential operator $\dfrac{d^2}{dx^2} + \lambda^2 I$ considered as an operator from $L_2[0,1]$ into itself.

Recall the definition of the differentiation operator $\dfrac{d}{dx}$ in $L_2[0,1]$. If $f(x)$ is the indefinite integral of some function in $L_2[0,1]$, denoted by $f'(x)$, then $f'(x)$ is called the *derivative* of $f(x)$. Such a function $f(x)$ is said to be *absolutely continuous*. With this definition, we have

$$D(T) = \{f \in L_2[0,1] : f' \text{ exists and is absolutely continuous}, f'' \in L_2[0,1]\}.$$

Clearly $D(T)$ is dense in $L_2[0,1]$. Let T_0 to be the restriction of T to those functions f satisfying the boundary conditions $f(0) = f(1) = 0$. Then T_0 is again densely defined. If now T_1 denotes the restriction of T_0 to functions f satisfying the integral boundary condition (10), then $D(T_1)$ is a closed subspace of $D(T_0)$ of codimension one (since it is the kernel of a continuous linear functional defined on $D(T_1)$). In particular, T_1 is not densely defined. Consequently, *the adjoint of the differential operator T_1 is a multivalued linear relation* (by Proposition III.1.4(b)).

We shall now verify that T_1' is an extension of the linear relation S determined by the system

$$g'' + \lambda^2 g - \alpha \sin \pi x = 0 \qquad (\alpha \in \mathbb{R} \text{ arbitrary}) \tag{14}$$

$$g(0) = g(1) = 0. \tag{15}$$

In other words,

$$Sg := g'' + \lambda^2 g + \mathrm{sp}\{\sin \pi x\}$$
$$D(S) = \{g \in L_2[0,1] : g \text{ satisfies (15)}, g' \text{ absolutely continuous}, g'' \in L_2[0,1]\ \}$$

We have

$$S(0) = \mathrm{sp}\{\sin \pi x\}.$$

Hence by Proposition III.1.4(b) and (10),

$$T_1'(0) = D(T_1)^{\perp} = S(0). \tag{16}$$

Now for $(g, g'' + \lambda^2 g + \alpha \sin \pi x) \in G(S)$ and $(f, f'' + \lambda^2 f) \in G(T_1)$, we have

$$\begin{aligned} &\int_0^1 [g(f'' + \lambda^2 f) - f(g'' + \lambda^2 g - \alpha \sin \pi x)]dx \\ &= \int_0^1 (gf'' - fg'')dx + \alpha \int_0^1 f(x) \sin \pi x \; dx \\ &= \int_0^1 (gf'' - fg'')dx \quad \text{by (10)} \\ &= 0. \end{aligned} \tag{17}$$

Hence, according to the definition of the adjoint III.1.1(1),

$$G(S) \subset G(T_1'). \tag{18}$$

From (16) and (18) it follows that T_1' is an extension of S. In particular, any solution of $Sg = S(0)$ is a solution of the adjoint equation $T_1'g = T_1'(0)$.

From (4) and (12) we have

$$\phi(x) = \sum_{n=2}^{\infty} a_n \sin n\pi x. \tag{19}$$

Hence

$$a_n = \frac{1}{2} \int_0^1 \phi(x) \sin n\pi x \; dx \qquad (n = 2, 3, \ldots). \quad \square$$

III.10 NOTES AND REMARKS

The definition of the adjoint of a linear relation (III.1.1(1)) is due to J. von Neumann [1] (see R. Arens [1; p.33]). R. Mennicken and S. Sagraloff [1] give an equivalent definition (see also T. Kato [2; p.168]).

Theorem III.1.5 (except for part (c)) is due to R. Arens [1]. Theorem III.1.6(a) and the complementary Example III.1.7 is from M.J. Kascic [1]. The proof of part (c) of Proposition III.1.4 was suggested by M.K. Kinyon.

The notion of gap (or opening) between subspaces of a Hilbert space was introduced by M.G. Krein and M.A. Krasnoselskii [1]. There the gap is defined by the formula

$$\theta(M,N) := \max\{\sup_{m\in S_M} d(m,N), \ \sup_{n\in S_N} d(n,M)\}$$

(if the sphere of a subspace is empty, i.e., if the subspace is $\{0\}$, then the corresponding supremum is taken to be zero). For Banach spaces, this notion first appears in M.G, Krein, M.A. Krasnoselskii and D.P. Milman in [1], where for instance, it is shown that $\theta(M,N) = \theta(N^\perp, M^\perp)$ (c.f. Proposition III.3.7). The non symmetric gap

$$\theta_0(M,N) = \sup_{m\in S_M} d(m,N)$$

was introduced and investigated by T. Kato in [1] (it also satisfies III.3.7). J.D. Newburgh [1] introduced a metric on the space $G(X)$ of all closed subspaces of a Banach space X which generates the same topology on $G(X)$ as θ does. This metric was investigated and compared with θ by E. Berkson [1]. Other early modifications of the gap concept were introduced by I. Gohberg and A.S. Markus [1], R. Douady [1] and V.I. Gurarii [1]. For a survey of the gap concept, and the related notion of "operator opening", the reader is referred to M.I. Ostrovskii [2].

The gap is used in the theory of (closed) semi-Fredholm operators in Banach spaces by I.C. Gohberg and M.G. Krein [1] (see also I.C. Gohberg and A.S. Markus [1]), and T. Kato [1,2]. This investigation is extended by G. Neubauer [1,2,3], who introduced the notion of closed semi-Fredholm relation in a product $X \times Y$ of Banach spaces. R. Mennicken and B. Sagraloff [2,3] carry this study into the more general context of locally convex spaces.

Theorems III.4.2, III.4.4 and their corollaries are applications of Mennicken and Sagraloff's Theorem III.3.9 and illustrate the usefulness of the geometric methods of Section III.2 related to the gap, when applied to completely closed linear relations. The term "regular contraction" (Definition III.4.12) has been borrowed from G. Köthe [1].

Proposition III.4.6(a) is due to P. Pillay (unpublished).

The State Diagram for linear relations in locally convex spaces, of which III.6.5 is a special case, was derived by M. Möller [1]. The original classification (for linear operators) is due to A.E. Taylor and C.J.A. Halberg [1]. For a survey of state diagrams in functional analysis, see V.M. Onieva [1].

Section III.8 generalises part of the author's article Cross [10]. The single valued version of Theorem III.8.10 (listing properties equivalent to the continuity of the adjoint of T), was obtained by A.I. Gouveia (see Cross [10; Theorem 5.9]).

Example VIII.9.1 was suggested by A. Krall. The paper of A. Krall [1] surveys the field of differential systems with general boundary conditions up to 1975. Interest in the application of the methods of linear relations to these problems has gathered momentum after the appearance of R. Aren's paper (Arens [1]). (See the Bibliography for further references; of particular relevance here are the works of E.A. Coddington, A. Dijksma, H.S.V. de Snoo, H. Langer and B. Textorius, and their collaborators).

CHAPTER IV

Operational Quantities of Linear Relations

IV.1 THE QUANTITIES

In this chapter we shall introduce certain functions $f(T)$, referred to as "quantities". These quantities play a useful role in the study of semi-Fredholm, compact and other related classes of linear relations defined in Chapter V. They are all generalisations of quantities well known in the theory of bounded linear operators. However, some of the results are new even in the particular case of bounded linear operators (for example, the density invariance result IV.4.8).

IV.1.1 Definitions

Let $\mathcal{I}(X)$, $\mathcal{C}(X)$ and $\mathcal{P}(X)$ denote the infinite dimensional, finite codimensional and closed finite codimensional subspaces respectively, of the normed space X. We denote the set $\{x \in X : \|x\| = 1\}$ by S_X.

Let $\mathcal{A}$ denote the set $\{\Delta, \Gamma, \Gamma_0, \overline{\Gamma}_0, \tau, \tau_0, \overline{\tau}_0\}$ where each element of $\mathcal{A}$ is a function defined on the class of all linear relations, taking values in the closed semi-infinite interval $[0, \infty]$, determined as follows: If $\dim D(T) < \infty$, then $f(T) = 0$ for all $f \in \mathcal{A}$. If $\dim D(T) = \infty$, then

$$\begin{aligned}
\Gamma(T) &= \inf_{M \in \mathcal{I}(D(T))} \|T|_M\| \\
\Gamma_0(T) &= \inf_{M \in \mathcal{C}(X)} \|T|_M\| \\
\overline{\Gamma}_0(T) &= \inf_{M \in \mathcal{P}(X)} \|T|_M\| \\
\Delta(T) &= \sup_{M \in \mathcal{I}(D(T))} \Gamma(T|_M) \\
\tau(T) &= \sup_{M \in \mathcal{I}(D(T))} \inf_{m \in S_M} \|Tm\| \\
\tau_0(T) &= \sup_{M \in \mathcal{C}(X)} \inf_{m \in S_M} \|Tm\| \\
\overline{\tau}_0(T) &= \sup_{M \in \mathcal{P}(X)} \inf_{m \in S_M} \|Tm\| .
\end{aligned}$$

We shall see later that $\tau_0 = \overline{\tau}_0$ (Theorem V.4.17).

The following inequalities hold:

(1) $\overline{\tau}_0(T) \leq \tau_0(T) \leq \Gamma(T) \leq \Delta(T) \leq \Gamma_0(T) \leq \overline{\Gamma}_0(T)$ (see IV.2.1 and IV.2.4)

(2) $\tau_0(T) \leq \tau(T) \leq \Delta(T)$ (Corollary IV.2.3).

IV.1.2 Proposition.

(a) *Let $M \in \mathcal{C}(X)$ and $N \in \mathcal{I}(X)$. Then $M \cap N$ is a finite codimensional subspace of N. In particular, $M \cap N$ is infinite dimensional.*

(b) *Let $M \in \mathcal{P}(X)$ and $N \in \mathcal{I}(X)$. Then $M \cap N$ is finite codimensional and closed in N.*

(c) *Let $M \in \mathcal{I}(X)$. Then $\mathcal{P}(M) = \{M \cap N : N \in \mathcal{P}(X)\}$ and $\mathcal{C}(M) = \{M \cap N : N \in \mathcal{C}(X)\}$.*

PROOF. (a) Consider the map $A : N/M \cap N \to X/M$ defined by $A(n + M \cap N) = n + M \quad (n \in N)$. Then A is injective. But X/M is finite dimensional. Therefore $N/M \cap N$ is finite dimensional.

(b) Follows immediately from (a).

(c) Let $N \in \mathcal{P}(X)$. Then by (b), $M \cap N \in \mathcal{P}(M)$. Hence $\{M \cap N : N \in \mathcal{P}(X)\} \subset \mathcal{P}(M)$. For the reverse inclusion, let $V \in \mathcal{P}(M)$. There exists a finite dimensional subspace F of M such that $V + F = M$ and $V \cap F = \{0\}$. Let $x_1, \ldots, x_n$ be a basis for F. By the Hahn-Banach Theorem there exists linear functionals $f_1, \ldots, f_n \in X'$ such that $f_i(x_j) = \delta_{ij}$ and $f_i(v) = 0$ for $v \in V (i, j \leq n)$; this is possible as V is closed. Then $N = \cap_{i \leq n} f_i^{-1}(0)$ is a closed finite codimensional subspace of X with $V = M \cap N$, as required. The proof of the second equality is similar. □

IV.1.3 Proposition. *If either*

(i) *$M \in \mathcal{P}(X)$ and $f \in \mathcal{A}$, or*

(ii) *$M \in \mathcal{C}(X)$ and $f \in \{\Delta, \Gamma, \Gamma_0, \tau, \tau_0\}$,*

then $f(T|_M) = f(T)$.

PROOF. Since $f(T) = 0$ when $D(T)$ is finite dimensional, we may assume $\dim D(T) = \infty$ throughout.

Let $f \in \{\Delta, \Gamma, \Gamma_0, \tau, \tau_0\}$. Fix $M \in \mathcal{C}(X)$.

(a) $f = \Gamma$: Clearly $\Gamma(T) \leq \Gamma(T|_M)$. Given $\epsilon > 0$, choose $E \in \mathcal{I}(D(T))$ so that $\Gamma(T) + \epsilon \geq \|T|_E\|$. Then

$$\Gamma(T|_M) = \inf_{N \in \mathcal{I}(M \cap D(T))} \|T|_N\| \leq \inf_{N \in \mathcal{I}(M \cap E)} \|T|_N\| \leq \|T|_E\| \leq \Gamma(T) + \epsilon.$$

Hence $\Gamma(T|_M) = \Gamma(T)$.

(b) $f = \Gamma_0$: Similar to the preceding.

(c) $f = \Delta$: By Lemma IV.1.2 and (a) above, $\Gamma(T|_N) = \Gamma(T|_{N\cap M})$ for $N \in \mathcal{I}(D(T))$. Hence

$$\Delta(T|_M) = \sup_{N\in\mathcal{I}(M)} \Gamma(T|_N) = \sup_{N\in\mathcal{I}(D(T))} \Gamma(T|_{N\cap M})$$
$$= \sup_{N\in\mathcal{I}(D(T))} \Gamma(T|_N) = \Delta(T).$$

(d) $f = \tau$: Clearly $\tau(T|_M) \leq \tau(T)$. Given $\epsilon > 0$, choose $E \in \mathcal{I}(D(T))$ so that

$$\tau(T) - \epsilon \leq \inf_{x\in S_E} \|Tx\| .$$

Then since $E \cap M \in \mathcal{I}(D(T))$ by IV.1.2,

$$\tau(T|_M) = \sup_{N\in\mathcal{I}(M\cap D(T))} \inf_{x\in S_N} \|Tx\| \geq \inf_{x\in S_{E\cap M}} \|Tx\|$$
$$\geq \inf_{x\in S_E} \|Tx\| \geq \tau(T) - \epsilon.$$

Hence $\tau(T|_M) = \tau(T)$.

(e) $f = \tau_0$: Similar to (d).

This completes the proof for the case (ii). Since $\mathcal{P}(X) \subset \mathcal{C}(X)$, it is only necessary to consider the quantitites $\overline{\Gamma}_0$ and $\overline{\tau}_0$ to establish the sufficiency of (i).

(f) $f = \overline{\Gamma}_0$: Similar to part (a), using the fact that the intersection of two subspaces in $\mathcal{P}(X)$ is again in $\mathcal{P}(X)$.

(g) $f = \overline{\tau}_0$: Similar to (d). □

IV.1.4 Corollary. *Let $F, T \in LR(X,Y)$, $D(F) \supset D(T)$, and let F have finite dimensional range.*

(a) *For $f \in \{\Delta, \Gamma, \Gamma_0, \tau, \tau_0\}$ we have*

(1) $$f(T+F) \leq f(T)$$

with equality if $F(0) \subset \overline{T(0)}$.

(b) *If in addition F is continuous, then (1) holds for all f in $\mathcal{A}$.*

PROOF. (a) Since $\dim R(F) < \infty$, it follows that $\dim D(F)/N(F) = \dim D(F)/N(QF) < \infty$.

Now for $x \in N(F)$ we have (see II.1.13)

$$\|(T+F)x\| \leq \|Tx\| \tag{2}$$

Hence from Proposition IV.1.3,

$$f(T+F) = f(T+F|_{N(F)}) \leq f(T|_{N(F)}) = f(T) \tag{3}$$

If $F(0) \subset \overline{T(0)}$ then we clearly have equality in (2) for $x \in N(F)$, and hence equality in (3).

(b) Since $N(F) \in P(X)$, we have by Proposition IV.1.3

$$\overline{\Gamma}_0(T+F) \leq \overline{\Gamma}_0(T|_{N(F)} + F|_{N(F)}) \leq \overline{\Gamma}_0(T|_{N(F)}) = \overline{\Gamma}_0(T)$$

with equality if $F(0) \subset \overline{T(0)}$. The proof for $\overline{\tau}_0$ is similar. □

The example in II.1.13 shows that strict inequality is possible in (1).

IV.1.5 Exercise Let T be continuous. Show that

$$\overline{\Gamma}_0(T) = \Gamma_0(T) \text{ and } \overline{\tau}_0(T) = \tau_0(T).$$

IV.2 INEQUALITIES

IV.2.1 Proposition $\Delta(T) \leq \Gamma_0(T)$.

PROOF. Let $\epsilon > 0$ be given and choose $M \in \mathcal{C}(X)$ so that $\|T|_M\| \leq \Gamma_0(T)+\epsilon$. Since $\Delta(T|_M) = \Delta(T)$ by Proposition IV.1.3, we have

$$\Delta(T) \leq \|T|_M\| \leq \Gamma(T_0) + \epsilon$$

and the result follows. □

IV.2.2 Proposition $\tau(S+T) \leq \tau(S) + \Delta(T)$.

PROOF. We may clearly assume $\dim D(S+T)) = \infty$. Let $\epsilon > 0$ be given and choose $M \in \mathcal{I}(D(S+T))$ such that

$$\tau(S+T) - \epsilon \leq \inf_{x \in S_M} \|(S+T)x\| \,.$$

Then for $x \in S_M$ we have

$$\tau(S+T) - \epsilon \leq \|(S+T)x\| \leq \|Sx\| + \|Tx\| \,.$$

Let $N \in \mathcal{I}(M)$. Then

$$\tau(S+T) - \epsilon \leq \inf_{x \in S_N} \|Sx\| + \|T|_N\| \leq \tau(S|_N) + \|T|_N\| \leq \tau(S) + \|T|_N\| \,.$$

Consequently

$$\tau(S+T) - \epsilon \leq \tau(S) + \Gamma(T|M) \leq \tau(S) + \Delta(T). \quad □$$

IV.2.3 Corollary. $\tau(T) \leq \Delta(T)$.

IV.2.4 Proposition. $\tau_0(T) \leq \Gamma(T)$.

PROOF. Assume without loss of generality that $\dim D(T) = \infty$. Choose $M \in \mathcal{C}(X)$ such that

$$\tau_0(T) - \epsilon \leq \inf_{m \in S_M} \|Tm\| .$$

Then for $m \in S_M$ we have

$$\tau_0(T) - \epsilon \leq \|Tm\| .$$

Hence for $N \in \mathcal{I}(M)$,

$$\tau_0(T) - \epsilon \leq \|T|_N\| . \tag{1}$$

From (1) and Proposition IV.1.3 we now have

$$\tau_0(T) - \epsilon \leq \Gamma(T|_M) = \Gamma(T)$$

from which the desired inequality follows. □

IV.2.5 Proposition. $\Gamma(S+T) \leq \Delta(S) + \Gamma(T|_{D(S)})$.

PROOF. We may clearly assume $D(S+T)$ is infinite dimensional. Furthermore we may suppose that $D(T) = X$. We have for $M \in \mathcal{I}(X)$.

$$\Gamma(S+T) \leq \|S+T|_M\| \leq \|S|_M\| + \|T|_M\| . \tag{1}$$

Fix $\epsilon > 0$ and choose $N \in \mathcal{I}(M)$ such that

$$\|S|_N\| \leq \Gamma(S|_M) + \epsilon.$$

We have from (1) with N in place of M,

$$\begin{aligned} \Gamma(S+T) &\leq \Gamma(S|_M) + \|T|_N\| + \epsilon \\ &\leq \Delta(S) + \|T|_N\| + \epsilon. \end{aligned}$$

Hence

$$\Gamma(S+T) \leq \Delta(S) + \Gamma(T|_M) + \epsilon.$$

Since $\epsilon > 0$ and $M \in \mathcal{I}(X)$ are arbitrary,

$$\Gamma(S+T) \leq \Delta(S) + \Gamma(T)$$

and the desired inequality follows. □

IV.2.6 Corollary. $\Delta(S+T) \leq \Delta(S) + \Delta(T)$.

PROOF. For each subspace M of $D(S+T)$ we have $\Gamma(S+T)_M) \leq \Delta(S|_M) + \Gamma(T|_M) \leq \Delta(S) + \Delta(T)$. □

IV.2.7 Proposition. *Let M and N be infinite dimensional subspaces of $D(T)$. If $M \subset N$ then*

(a) $\overline{\Gamma}_0(T|_M) \leq \overline{\Gamma}_0(T|_N)$
(b) $\Gamma_0(T|_M) \leq \Gamma_0(T|_N)$
(c) $\overline{\tau}_0(T|_M) \geq \overline{\tau}_0(T|_N)$
(d) $\tau_0(T|_M) \geq \tau_0(T|_N)$.

PROOF. (a) We have

$$\overline{\Gamma}_0(T|_M) = \inf_{V \in \mathcal{P}(M)} \|T|_V\| = \inf_{V \in \mathcal{P}(N)} \|T|_{M \cap V}\| \text{ (by IV.1.2(c))}$$
$$\leq \inf_{V \in \mathcal{P}(N)} \|T|_V\| = \overline{\Gamma}_0(T|_N).$$

The proof of (b) is similar.

(c) We have similarly,

$$\overline{\tau}_0(T|_M) = \sup_{V \in \mathcal{P}(M)} \inf_{v \in V} \frac{\|Tv\|}{\|v\|} = \sup_{V \in \mathcal{P}(N)} \inf_{v \in M \cap V} \frac{\|Tv\|}{\|v\|}$$
$$\geq \sup_{V \in \mathcal{P}(N)} \inf_{v \in V} \frac{\|Tv\|}{\|v\|} = \overline{\tau}_0(T|_N).$$

The proof of (d) is similar. □

IV.2.8 Lemma. *Let $S \in LR(Y, Z)$, $R(T) \subset D(S)$ and let M be a linear subspace of X. If T is single valued then*

$$\Gamma(ST|_M) \leq \Gamma(T|_M)\Delta(S|_{T(M)}).$$

PROOF. It is enough to consider the case when $\dim M \cap D(ST) = \infty$ and $\Gamma(T|_M) > 0$ (the inequality is not defined if the factors on the right hand side are $0, \infty$ or $\infty, 0$). Since $R(T) \subset D(S)$, we have $D(ST) = D(T)$. Choose $N \in \mathcal{I}(M \cap D(T))$ so that

$$\|T|_N\| \leq \Gamma(T|_M) + \epsilon. \tag{1}$$

We shall suppose without loss of generality that $\Delta(S|_{T(M)}) < \infty$ and $\Gamma(T|_M) < \infty$. Then we have

$$\begin{aligned}\Gamma(ST|_N) &= \inf\{\|ST|_V\| : V \in \mathcal{I}(N)\}\\ &\le \inf\{\|S|_{T(V)}\|\,\|T|_V\| : V \in \mathcal{I}(N)\}\end{aligned}$$

(by II.3.13, since T is single valued)

$$\le \|T|_N\| \inf\{\|S|_{T(V)}\| : V \in \mathcal{I}(N)\}$$

Assume first that $\dim T(N) = \infty$, and let $W \in \mathcal{I}(T(N))$. Write $U = N \cap T^{-1}(W)$. Then $T(U) = W$ (indeed $T(U) = (T|_N)(U) = (T|_N)(T|_N)^{-1}W = W$ since $W \subset T(N)$). Hence $\dim U = \infty$ (since $T(U)$ is infinite dimensional as T is single valued). Therefore

$$\inf\{\|S|_{T(V)}\| : V \in \mathcal{I}(N)\} \le \inf\{\|S|_W\| : W \in \mathcal{I}(T(N))\} = \Gamma(S|_{T(N)}) < \infty$$

since $\Delta(S|_{T(M)}) < \infty$. It follows that

$$\Gamma(ST|_N) \le \Gamma(S|_{T(N)})\,\|T|_N\| \tag{2}$$

If $\dim T(N) < \infty$, then from $K := N(T|_N) \in \mathcal{I}(N)$, we have $\Gamma(ST|_N) \le \|ST|_K\| = 0$. Thus in either case, (2) holds.

Hence from (1)

$$\Gamma(ST|_M) \le \Delta(S|_{T(M)})(\Gamma(T|_M) + \epsilon)$$

and the lemma follows. □

IV.2.9 Theorem. *Let $T \in LR(X, Y)$ and $S \in LR(Y, Z)$. If T is single valued, then*

$$\Delta(ST) \le \Delta(S)\Delta(T)$$

whenever the right hand side is defined.

PROOF. Put $S_1 = S|_{R(T)}$. Then from IV.2.8 we have for $M \subset X$, $\Gamma(S_1T|_M) \le \Gamma(T|_M)\Delta(S_1)$. Taking the supremum over $M \in \mathcal{I}(D(ST))$ gives $\Delta(S_1T) \le \Delta(T)\Delta(S_1)$. But $S_1T = ST$, and $\Delta(S_1) \le \Delta(S)$. Hence the stated inequality follows. □

IV.2.10 Proposition. *Let $T \in LR(X,Y)$ and $S \in LR(Y,Z)$ and suppose that T is single valued. If $\dim D(ST) = \infty$, $\Gamma(S) < \infty$ and $\tau(T) < \infty$, then*

$$\Gamma(S)\tau(T) \leq \|ST\|.$$

If $\|ST\| = 0$ and $\tau(T|_{D(ST)}) = \infty$, then $\Gamma(S) = 0$.

PROOF. The inequality is trivial if $\dim D(S) < \infty$ or $\dim D(T) < \infty$. Accordingly assume that both $D(S)$ and $D(T)$ are infinite dimensional. If $\tau(T) = 0$ there is nothing further to prove. Let $\tau(T) > 0$ and let ϵ be arbitarily chosen in the interval $0 < \epsilon < \tau(T)$. Pick $M \in \mathcal{I}(D(ST))$ so that

$$(1) \qquad \|Tx\| \geq (\tau(T) - \epsilon)\,\|x\| \qquad (x \in M)$$

Then we have

$$(2) \quad \|STx\| \leq \|ST\|\,\|x\| \leq \|ST\|\,\|Tx\|\,/(\tau(T) - \epsilon) \qquad (x \in M, Tx \in D(S)).$$

Let $N = TM$. Then from (2),

$$\|Sy\| \leq \|ST\|\,\|y\|\,/(\tau(T) - \epsilon) \qquad (y \in N \cap D(S))$$

Hence

$$\|S|_N\| \leq \|ST\|\,/(\tau(T) - \epsilon)$$

whence

$$(3) \qquad \Gamma(S|_{TM}) \leq \|ST\|\,/(\tau(T) - \epsilon).$$

Since (1) implies that TM is infinite dimensional since T is injective on M, we have

$$\Gamma(S) \leq \Gamma(S|_{TM}).$$

Hence from (3)

$$\Gamma(S)(\tau(T) - \epsilon) \leq \|ST\|$$

and the desired inequality follows.

For the second assertion, let $\tau(T|_{D(ST)}) = \infty$. We may clearly suppose that $\dim D(S) = \infty$.

Choose $N \in \mathcal{I}(D(ST))$ so that $\inf\limits_{n \in S_N} \|Tn\| > 0$. Then $M := TN \in \mathcal{I}(D(S))$ (as T is injective on N) and hence

$$\Gamma(S) \leq \|S|_M\| \leq \sup_{n \in N} \frac{\|STn\|}{\|Tn\|} = \sup_{n \in N} \left(\frac{\|STn\|}{\|n\|} / \frac{\|Tn\|}{\|n\|} \right) \leq \|ST\| \,/ \inf_{n \in N} \frac{\|Tn\|}{\|n\|}. \qquad \square$$

IV.2.11 Proposition. *Let $S \in LR(Y,Z)$ and $T \in LR(X,Y)$. If T is single valued and $D(ST)$ is infinite dimensional, then*

$$\tau_0(ST) \geq \tau_0(S)\tau_0(T).$$

PROOF. We may clearly suppose that $\tau_0(S)$, $\tau_0(T) > 0$; in particular, $N(S)$ and $N(T)$ are finite dimensional (see Proposition V.2.4). Choose $M_1 \in \mathcal{C}(X)$ and $M_2 \in \mathcal{C}(Y)$. Let $M = M_1 \cap T^{-1}(M_2)$. $T^{-1}(M_2)$ has finite codimension in $D(T)$ (see I.6.1). Hence $M \cap D(ST)$ has finite codimension in $D(ST)$. Therefore

$$\begin{aligned}
\inf_{y \in M_2 \cap D(S)} \frac{\|Sy\|}{\|y\|} \inf_{x \in M_1 \cap D(T)} \frac{\|Tx\|}{\|x\|} \\
&\leq \inf_{m \in M \cap D(ST)} \frac{\|STm\|}{\|Tm\|} \inf_{m \in M \cap D(ST)} \frac{\|Tm\|}{\|m\|} \\
&\leq \inf_{m \in M \cap D(ST)} \frac{\|STm\|}{\|m\|} \\
&\leq \sup_{\operatorname{codim} E < \infty} \inf_{x \in E \cap D(ST)} \frac{\|STx\|}{\|x\|} \\
&= \tau_0(ST). \quad \square
\end{aligned}$$

IV.2.12 Proposition.

(a) $\Gamma_0(S+T) \leq \Gamma_0(S) + \Gamma_0(T)$

(b) $\overline{\Gamma}_0(S+T) \leq \overline{\Gamma}_0(S) + \overline{\Gamma}_0(T)$.

PROOF. (a) Suppose $S, T \in LR(X,Y)$ and let $\epsilon > 0$ be given. Choose $M_1, M_2 \in \mathcal{C}(X)$ such that $\Gamma_0(S) \geq \|S|_{M_1}\| - \epsilon/2$ and $\Gamma_0(T) \geq \|T|_{M_2}\| - \epsilon/2$. Let $M = M_1 \cap M_2$. Then $M \in \mathcal{C}(X)$, $\|S|_{M_1}\| \geq \|S|_M\|$ and $\|T|_{M_2}\| \geq \|T|_M\|$. Hence

$$\Gamma_0(S+T) \leq \|(S+T)|_M\| \leq \Gamma_0(S) + \Gamma_0(T) + \epsilon.$$

The proof of (b) is similar. $\square$

IV.2.13 Exercise Let T be single valued. Then $f(S) - f(T) \leq f(T+S)$ for $f \in \{\Delta, \Gamma_0, \overline{\Gamma}_0\}$.

IV.2.14 Corollary.

(a) *Let* $M \in \mathcal{P}(X)$. *If* $T|_M$ *is continuous, then* T *is continuous.*

(b) T *is continuous if and only if* $\overline{\Gamma}_0(T) < \infty$.

PROOF. (a) Let $\|T|_M\| < \infty$ where $M \in \mathcal{P}(X)$. By Proposition II.4.7, there exists a continuous projection P defined on $D(T)$ with range $M \cap D(T)$ (see II.4.3 and Proposition V.1.2). Then for $x \in D(T)$ we have by II.1.5 and II.3.13,

$$\begin{aligned}\|Tx\| &\leq \|TPx\| + \|T(I-P)x\| \\ &\leq \|T|_M\| \, \|P\| \, \|x\| + \left\|T|_{R(I-P)}\right\| \, \|I-P\| \, \|x\|\end{aligned}$$

where $\left\|T|_{R(I-P)}\right\| < \infty$ by the finite dimensionality of $R(I-P) = N(P)$.

(b) If T is continuous, then obviously $\overline{\Gamma}_0(T) < \infty$.

Conversely, let $\overline{\Gamma}_0(T) < \infty$. Then there exists $M \in \mathcal{P}(X)$ such that $\|T|_M\| < \infty$. Hence T is continuous by part (a). □

IV.2.15 Proposition. *Let* $S \in LR(Y,Z)$ *and* $T \in LR(X,Y)$. *If* T *is single valued, then*

(a) $\overline{\Gamma}_0(ST) \leq \overline{\Gamma}_0(S)\overline{\Gamma}_0(T)$

(b) $\Gamma_0(ST) \leq \Gamma_0(S)\Gamma_0(T)$

whenever the inequalities are defined.

PROOF. (a) The inequality is defined unless $\overline{\Gamma}_0(S) = 0$ and $\overline{\Gamma}_0(T) = \infty$, or $\overline{\Gamma}_0(S) = \infty$ and $\overline{\Gamma}_0(T) = 0$. If either $\overline{\Gamma}_0(S) = \infty$ or $\overline{\Gamma}_0(T) = \infty$ then the inequality holds trivially if it is defined. Accordingly we assume that $\overline{\Gamma}_0(S)$ and $\overline{\Gamma}_0(T)$ are both finite, or equivalently, that S and T are continuous (Corollary IV.2.14). Then ST will be continuous by II.3.13. Given $\epsilon > 0$, choose $M \in \mathcal{P}(X)$ so that $\|Tx\| \leq (\overline{\Gamma}_0(T) + \epsilon)\|x\|$ for $x \in M \cap D(T)$, and $N \in \mathcal{P}(Y)$ so that $\|Sy\| \leq (\overline{\Gamma}_0(S) + \epsilon)\|y\|$ for $y \in N \cap D(S)$. Set $W = T^{-1}(N \cap D(S)) \cap M$. By Proposition I.6.1, $T^{-1}(N \cap D(S))$ is finite codimensional in $D(ST)$ and hence $W \in \mathcal{C}(D(ST))$. We have

$$\begin{aligned}\overline{\Gamma}_0(ST) &= \inf_{E \in P(X)} \sup_{x \in E \cap D(ST)} \frac{\|STx\|}{\|x\|} \\ &\leq \sup_{x \in \overline{W} \cap D(ST)} \frac{\|STx\|}{\|x\|} \\ &= \sup_{x \in W \cap D(ST)} \frac{\|STx\|}{\|x\|} \quad \text{(since } ST \text{ is continuous)} \\ &\leq \sup_{x \in W \cap D(ST)} (\overline{\Gamma}_0(S) + \epsilon)\frac{\|Tx\|}{\|x\|} \quad \text{(by II.3.13)} \\ &\leq (\overline{\Gamma}_0(S) + \epsilon)(\overline{\Gamma}_0(T) + \epsilon)\end{aligned}$$

and the inequality (a) follows.

The proof of inequality (b) is similar except that the continuity of ST is not required. □

Example V.1.13 will show that the single valuedness of T is essential in Proposition IV.2.15.

IV.2.16 Proposition. *Let $S \in LR(Y,Z)$, $T \in LR(X,Y)$ where T is single valued, and let* $\dim D(ST) = \infty$. *Then*

(a) $\Gamma(ST) \geq \tau_0(S)\Gamma(T)$

(b) $\Gamma_0(ST) \geq \tau_0(S)\Gamma_0(T|_{D(ST)})$

(c) $\overline{\Gamma}_0(ST) \geq \tau_0(S)\overline{\Gamma}_0(T|_{D(ST)})$ *(provided $\overline{\Gamma}_0(T|_{D(ST)}) < \infty$)*

whenever the right hand sides are defined.

PROOF. (a) We may clearly suppose that $\tau_0(S)$ and $\Gamma(T)$ are positive. Let ϵ be arbitrary in the interval $0 < \epsilon < \tau_0(S)$. Choose $E \in \mathcal{C}(Y)$ such that

$$\inf_{y \in S_E} \|Sy\| \geq \epsilon. \tag{1}$$

Since $\Gamma(T) > 0$, we have $\dim N(T) < \infty$, and hence $R(T)$ is infinite dimensional. Furthermore, by Proposition I.6.1, $T^{-1}(E)$ is a finite codimensional subspace of $D(T)$. Let $S_1 = S|_{E \cap R(T)}$. Then $D(S_1T) = D(T) \cap T^{-1}(E \cap D(S) \cap R(T)) = D(ST) \cap T^{-1}(E)$ (since T is single valued). Thus $D(S_1T)$ is a finite codimensional subspace of $D(ST)$. In particular, $\dim D(S_1T) = \infty$. Let $M \in \mathcal{I}(D(S_1T))$. We have from (1),

$$\|S_1T|_M\| = \sup_{m \in S_M} \|S_1Tm\| \geq \inf_{y \in S_{D(S_1)}} \|Sy\| \sup_{m \in S_M} \|Tm\| \geq \epsilon\Gamma(T).$$

Hence $\Gamma(ST|_{D(S_1T)}) \geq \epsilon\Gamma(T)$; but $\Gamma(ST|_{D(S_1T)}) = \Gamma(ST)$ by Proposition IV.1.3. Hence $\Gamma(ST) \geq \tau_0(S)\Gamma(T)$.

The proofs of (b) and (c) are similar except that M is chosen arbitrarily in $\mathcal{C}(D(S_1T))$ and in case (c) it can be assumed that $\overline{\Gamma}_0(ST) < \infty$, i.e. that ST is continuous (see IV.2.14) and thus $\|ST|_M\| = \sup_{m \in S_{\overline{M}}} \|S_1Tm\|$. □

IV.3 THE GRAPH OPERATOR

The graph operator is used to reduce T to a bounded everywhere defined relation.

IV.3.1 Definitions Given $T \in LR(X,Y)$, let X_T denote the vector space $D(T)$ normed by

$$\|x\|_T := \|x\| + \|Tx\| \qquad (x \in D(T)). \tag{1}$$

Let $G_T \in LR(X_T, X)$ be the identity injection of X_T into X, ie.

$$D(G_T) = X_T, \quad G_T x = x \qquad (x \in X_T). \tag{2}$$

We write $G_T = G$ when T is understood. G_T is called the **graph operator** of T.

If T is single valued, then X_T is norm isomorphic to the subspace $G(T) \subset X \times Y$.

IV.3.2 Proposition. *We have*

$$\|TG\| = \frac{\|T\|}{1 + \|T\|} \qquad \left(\textit{where } \frac{\infty}{\infty} := 1\ \right).$$

In particular, TG is bounded.

PROOF. $\|TG\| = \sup_{x \in X_T} \|TGx\| \left(\|Gx\| + \|TGx\|\right)^{-1}$
$= \sup_{x \in X_T} \|TGx\| \left(\|Gx\|^{-1} (1 + \|TGx\| \, \|Gx\|^{-1})^{-1} = \|T\| \, (1 + \|T\|)^{-1}.$ □

IV.3.3 Proposition. *The following properties are equivalent:*

(i) $T = \tilde{T}$
(ii) *Both X_T and $T(0)$ are complete.*

PROOF. Let $T = \tilde{T}$. Then $G(T)$ is complete. It follows that $G(QT)$ is complete. Clearly X_T, X_{QT} and $G(QT)$ are isomorphic. Hence X_T is complete. By Proposition II.5.3, $T(0)$ is complete.

Conversely, let X_T and $T(0)$ be complete. Then $G(QT)$ is complete. Hence $T = \tilde{T}$ by Proposition II.5.3. □

IV.3.4 Proposition. *For each $f \in \{\Delta, \Gamma, \Gamma_0, \tau, \tau_0\}$ we have*

$$f(TG) = \frac{f(T)}{1+f(T)} \qquad (\textit{where } \frac{\infty}{\infty} := 1\ ).$$

PROOF. We may clearly suppose that $\dim D(T) = \infty$. We have

$$\begin{aligned}
\Gamma(TG) &= \inf\left\{\sup_{m\in M} \frac{\|TGm\|}{\|Gm\| + \|TGm\|} : M \in \mathcal{I}(X_T)\right\} \\
&= \inf_M \sup_m \left(1 + \frac{\|Gm\|}{\|TGm\|}\right)^{-1} = \inf_M \left(1 + \inf \frac{\|Gm\|}{\|TGm\|}\right)^{-1} \\
&= \inf_M (1 + 1/\left\|T|_{GM}\right\|)^{-1} = (1 + 1/\inf_M \left\|T|_{GM}\right\|)^{-1} = \frac{\Gamma(T)}{1+\Gamma(T)}.
\end{aligned}$$

Similarly, $\Gamma_0(TG) = \dfrac{\Gamma_0(T)}{1+\Gamma_0(T)}$.

Next $\Delta(TG) = \sup_M \dfrac{\Gamma(T|GM)}{1+\Gamma(T|GM)} = \dfrac{\Delta(T)}{1+\Delta(T)}$.

We have

$$\begin{aligned}
\tau(TG) &= \sup\left\{\inf_{m\in M} \frac{\|TGm\|}{\|Gm\| + \|TGm\|} : M \in \mathcal{I}(X_T)\right\} \\
&= \sup_M \left(1 + \sup_{m\in M} \frac{\|Gm\|}{\|TGm\|}\right)^{-1} = \sup_M \left(1 + \left(\inf_{m\in M} \frac{\|TGm\|}{\|Gm\|}\right)^{-1}\right)^{-1} \\
&= \left(1 + \left(\sup_M \inf_{m\in M} \frac{\|TGm\|}{\|Gm\|}\right)^{-1}\right)^{-1} = \frac{\tau(T)}{1+\tau(T)}.
\end{aligned}$$

The equality for τ_0 follows similarly. □

IV.3.5 Corollary.

$$f(TG) \le \frac{\|T\|}{1+\|T\|} \le 1 \qquad f \in \{\Delta, \Gamma, \Gamma_0, \tau, \tau_0\}.$$

IV.3.6 Proposition. *We have*

$$\begin{aligned}
\Gamma(G) &= (1+\tau(T))^{-1}\\
\Gamma_0(G) &= (1+\tau_0(T))^{-1}\\
\tau(G) &= (1+\Gamma(T))^{-1}\\
\tau_0(G) &= (1+\Gamma_0(T))^{-1}\\
\Delta(G) &= (1+\inf_{M\in\mathcal{I}(D(T))}\tau(T|_M))^{-1}
\end{aligned}$$

PROOF.

$$\begin{aligned}
\Gamma(G) &= \inf_{M\in\mathcal{I}(X_T)}\|G|_M\|\\
&= \inf_M \sup_{m\in M}\frac{\|Gm\|}{\|Gm\|+\|TGm\|}\\
&= \inf_M \sup_M (1+\frac{\|TGm\|}{\|Gm\|})^{-1}\\
&= (\sup_M \inf_m (1+\frac{\|TGm\|}{\|Gm\|}))^{-1}\\
&= (1+\tau(T))^{-1}
\end{aligned}$$

The remaining computations are similar except for Δ. For the latter we have

$$\begin{aligned}
\Delta(G) &= \sup_{M\in\mathcal{I}(D(T))}\Gamma(G|G^{-1}M)\\
&= \sup_M \Gamma(G_{T|_M})\\
&= \sup_M (1+\tau(T|_M))^{-1} \quad \text{by the preceding}\\
&= (1+\inf_M \tau(T|_M))^{-1}. \quad \square
\end{aligned}$$

Proposition IV.3.6 is useful for characterising for example, linear relations T for which the embedding map G_T is partially precompact (see Exercise V.9.21).

IV.3.7 Proposition. *Let T be single valued and injective, and let $D(T)$ be infinite dimensional. Then*

$$\Gamma(T^{-1}) = \tau(T)^{-1}$$
$$\Gamma_0(T^{-1}) = \tau_0(T)^{-1}$$
$$\tau(T^{-1}) = \Gamma(T)^{-1}$$
$$\tau_0(T^{-1}) = \Gamma_0(T)^{-1}$$
$$\Delta(T^{-1}) = \left(\inf_{M\in\mathcal{I}(D(T))} \tau(T|_M)\right)^{-1}$$

PROOF.

$$\begin{aligned}\Gamma(T^{-1}) &= \inf_{M\in\mathcal{I}(R(T))} \sup_{m\in M} \frac{\|T^{-1}m\|}{\|m\|} \\ &= \inf_{V\in\mathcal{I}(D(T))} \sup_{v\in V} \frac{\|v\|}{\|Tv\|} \\ &= \left(\sup_V \inf_v \frac{\|Tv\|}{\|v\|}\right)^{-1} \\ &= \tau(T)^{-1}.\end{aligned}$$

The remaining equalities, are proved similarly, except for $\Delta(T)$. For the latter we have:

$$\begin{aligned}\Delta(T^{-1}) &= \sup_{V\in\mathcal{I}(R(T))} \Gamma(T^{-1}|_V) \\ &= \sup_{M\in\mathcal{I}(D(T))} \Gamma((T|_M)^{-1}) \\ &= \sup_M \tau(T|_M)^{-1} \quad \text{by the preceding} \\ &= (\inf_M \tau(T|_M))^{-1}. \quad \square\end{aligned}$$

IV.3.8 Corollary. *Let $\dim D(T) = \infty$. Then*

$$\Gamma(G) = \tau(G^{-1})^{-1}$$
$$\Gamma_0(G) = \tau_0(G^{-1})^{-1}$$
$$\tau(G) = \Gamma(G^{-1})^{-1}$$
$$\tau_0(G) = \Gamma_0(G^{-1})^{-1}$$
$$\Delta(G) = (\inf_{M\in\mathcal{I}(D(T))} \tau(G^{-1}|_M))^{-1}.$$

IV.3.9 Corollary. *Let T be single valued and injective with infinite dimensional domain. Then:*

$$(1) \qquad f(G_{T^{-1}}) = f(TG), \qquad f \in \{\Gamma, \Gamma_0, \tau, \tau_0\}.$$

PROOF. We have

$$\begin{aligned}\Gamma(G_{T^{-1}}) &= (1+\tau(T^{-1}))^{-1} && \text{by IV.3.6.}\\ &= (1+\Gamma(T)^{-1})^{-1} && \text{by IV.3.7}\\ &= \Gamma(TG) && \text{by IV.3.4}\end{aligned}$$

The proofs for the other quantities are similar. □

IV.3.10 Proposition.

(a) *If $T = 0$, then $\gamma(T) = \gamma(TG) = \infty$.*

(b) *If $T \neq 0$, then*

$$(1) \qquad \gamma(TG) = \frac{\gamma(T)}{1+\gamma(T)} \qquad \left(\textit{where } \frac{\infty}{\infty} := 1\right).$$

PROOF. (a) Follows immediately from the definition.

(b) Let $T \neq 0$. In that case $N(TG)$ is a closed proper subspace of X_T. For any $x \in X_T$ we have

$$\begin{aligned}d(x, N(TG)) &= \inf_{Gz \in N(T)} \{\|Gx - Gz\| + \|TGx - TGz\|\}\\ &= \inf_{Gz \in N(T)} \|Gx - Gz\| + \|TGx\|.\end{aligned}$$

Hence

$$(2) \qquad d(x, N(TG)) = d(Gx, N(T)) + \|TGx\|.$$

We shall now show that

$$(3) \qquad \gamma(TG) \leq \frac{\gamma(T)}{1+\gamma(T)}.$$

If $\gamma(TG) = 0$, then (3) holds trivially. Hence suppose that $\gamma(TG) > 0$ and select λ arbitrarily in the interval $(0, \gamma(TG))$. Then from (2) and the definition of $\gamma(TG)$ we have

$$\|TGx\| \geq \lambda(d(Gx, N(T)) + \|TGx\|)$$

whence

$$\|TGx\| \geq \frac{\lambda}{1-\lambda} d(Gx, N(T)) \quad (Gx \notin N(T)).$$

Since $\lambda \in (0, \gamma(TG))$ was chosen arbitrarily, it follows that

$$\gamma(T) \geq \frac{\gamma(TG)}{1-\gamma(TG)}.$$

Hence (3) follows.

It only remains to prove the reverse inequality, namely

$$(4) \qquad \gamma(TG) \geq \frac{\gamma(T)}{1+\gamma(T)}.$$

If $\gamma(T) = 0$, then (4) holds trivially. Hence assume that $\gamma(T) > 0$ and select a monotone increasing sequence $\{\delta_n\} \subset (0,1)$ such that

$$\lim_{n\to\infty} \frac{\delta_n}{1-\delta_n} = \gamma(T).$$

Since $\gamma(T) > \dfrac{\delta_n}{1-\delta_n} > 0$ we have

$$\|TGx\| \geq \frac{\delta_n}{1-\delta_n} d(Gx, N(T))$$

for each $x \in X_T$ and each $n \in \mathbb{N}$. Hence

$$\|TGx\| \geq \delta_n(d(Gx, N(T)) + \|TGx\|)$$

for each $x \in X_T$ and $n \in \mathbb{N}$. It follows that $\gamma(TG) \geq \delta_n$ for each $n \in \mathbb{N}$. Thus

$$(5) \qquad \gamma(TG) \geq \delta_n = \frac{\delta_n/(1-\delta_n)}{1+\delta_n/(1-\delta_n)}.$$

Taking the limit as $n \to \infty$ in (5) we obtain (4). □

IV.3.11 Corollary. *T is open if and only if TG is open.*

IV.3.12 Exercise Show that

(i) $X_T = X_{QT}$
(ii) $G_T = G_{QT}$

An alternative (and deeper) method of reducing an arbitrary linear relation to the continuous case will now be considered. More precisely, we shall construct a bounded surjective linear operator H_T, denoted by H, such that HT is continuous, and moreover, such that the adjoint H' is the graph operator $G_{T'}$, with the dual space $H(X)'$ coinciding with $D(G_{T'}) = Y'_{T'}$ (see Theorem IV.3.17).

We have the following extension of Goldstine's Theorem:

IV.3.13 Proposition. (I. Singer [S1]) *Let M be a subset of X' and let $K : X \to (\mathrm{sp}M)'$ be defined by $(Kx)m = m(x)$ $(x \in X,\ m \in M)$. Then $K(B_X)$ is $\sigma((\mathrm{sp}M)', M)$-dense in $B_{(\mathrm{sp}M)'}$.*

PROOF. Let $[M] := \mathrm{sp}M$, the linear span of $M \subset X'$. Observe that for any $x \in X$,

$$\|Kx\| = \sup_{m \in B_{[M]}} |(Kx)m| = \sup_{m \in B_{[M]}} |m(x)| \leq \|x\| .$$

Hence $K(B_X) \subset B_{[M]'}$. Let $m' \in U_{[M]'}$ (the open unit ball of $[M]'$). We need to show that each $\sigma([M]', M)$ neighbourhood of m' contains an element of $K(B_X)$, i.e., for every $\epsilon > 0$ and finite collection of elements $m_1, \ldots, m_n \in M$ there exists $x_0 \in X$ such that

$$(1) \qquad \|x_0\| \leq 1 \text{ and } |m'(m_i) - m_i(x_0)| \leq \epsilon, \qquad (1 \leq i \leq n)$$

(since $m_i(x_0) = (Kx_0)m_i$). Write $N := \max_{1 \leq i \leq n} \|m_i\|$.

Since $[M]' = X''/M^{\perp}$, we can choose $x'' \in X''$ such that $m' = x'' + M^{\perp}$ and $\|x''\| < 1$ (since $\|m'\| < 1$). According to Helly's Theorem (see e.g. Wilansky [1; p.103]), there exists $x \in X$ such that

$$(2) \qquad m'(m_i) = x''(m_i) = m_i x, \|x\| < (1 + \epsilon/N)\,\|x''\| < 1 + \epsilon/N.$$

Let $x_0 = (\frac{N}{N+\epsilon})x$. We shall verify that x_0 satisfies (1). We have from (2) that

$$\|x_0\| = (\frac{N}{N+\epsilon})\,\|x\| \leq 1$$

and

$$\begin{aligned} |m'(m_i) - m_i(x_0)| &= |m_i(x) - m_i(x_0)| \\ &= \left| \left(\frac{N+\epsilon}{N}\right) m_i(x_0) - m_i(x_0) \right| \\ &= \frac{\epsilon\,|m_i(x_0)|}{N} \\ &\leq \epsilon \end{aligned}$$

as required.

Thus $K(B_X)$ is $\sigma([M]', M)$ dense in $U_{[M]'}$, hence in $B_{[M]'}$. □

IV.3.14 Theorem. *Let X be a Banach space, and let M be a total subspace of X' such that B_X is $\sigma(X,M)$-compact. Then X is a dual space which is isometric to M' under the map $K : X \to M'$ defined by $(Kx)m = m(x)$ $(x \in X,\ m \in M)$.*

PROOF. Since

$$Kx = 0 \iff m(x) = 0 \text{ for each } m \in M \iff x \in M^{\top}$$

K is injective. Also (as in Proposition IV.3.13)

$$\|Kx\| = \sup_{m\in B_M} \|(Kx)m\| = \sup_{m\in B_M} \|m(x)\| \le \|x\| \,.$$

Therefore K is bounded, with $\|K\| \le 1$.

Let $\epsilon > 0$ and $m \in M$ be given, and let $U := \{m' \in M' : |m'm| < \epsilon\}$ be a given $\sigma(M',M)$ subbasic neighbourhood of 0. Then

$$K(X) \cap U = \{Kx : x \in X, |(Kx)m| = |mx| < \epsilon\}.$$

Hence $K^{-1}(U) = \{x \in X : |m(x)| < \epsilon\}$ is a subbasic $\sigma(X,M)$ neighbourhood of 0. Therefore K is $\sigma(X,M) - \sigma(M',M)$ continuous. But B_X is $\sigma(X,M)$ compact. Consequently, $K(B_X)$ is $\sigma(M',M)$ compact, hence $\sigma(M',M)$ closed as the topology is Hausdorff. Therefore by Proposition IV.3.13, $K(B_X) = B_{M'}$. Hence K is surjective.

To verify that K is an isometry, let $\|x\| = 1$ and suppose that $\|Kx\| < 1$. Then $(1+\epsilon)Kx \in B_{M'}$ for some $\epsilon > 0$ and therefore

$$(1+\epsilon)x \in K^{-1}B_{M'} = B_X$$

a contradiction. Therefore $\|Kx\| = 1$. □

IV.3.15 Theorem. *$G_{T'}(B_{Y'_{T'}})$ is $\sigma(D(T'),Y)$ compact.*

PROOF. Since the graph operator of $Q_{T'}T'$ is just $G_{T'}$ (see Exercise IV.3.12) and $(TJ_{D(T)})' = Q_{T'}T'$, we may without loss of generality suppose that $X = D(T)$. Write

$$W := Y'_{T'}$$

where $Y'_{T'} = (D(T'), \|\ \|_{T'})$ (Definition IV.3.1). Since

$$\|y'\|_{T'} = \|y'\| + \|T'y'\| \ge \|y'\|$$

we have

$$G_{T'}(B_W) \subset B_{Y'}.$$

As $B_{Y'}$ is $\sigma(Y',Y)$ compact, it is clearly sufficient to show that $G_{T'}(B_W)$ is $\sigma(Y',Y)$ closed. Let $\{y'_\alpha\}$ be a net in $G_{T'}(B_W)$ converging $\sigma(Y',Y)$ to $y'_0 \in Y'$. Since $B_{Y'}$ is $\sigma(Y',Y)$ closed, we have $y'_0 \in B_{Y'}$. Obviously T' is $\sigma(D(T'),Y)-\sigma(X',X)$ continuous and single valued. Hence $\{T'y'_\alpha\}$ is a Cauchy net in $\sigma(X',X)$. Furthermore, since $T'G_{T'}$ is bounded, we have

$$T'y'_\alpha \in T'(G_{T'}B_W) \subset \lambda B_{X'}$$

for some $\lambda > 0$. However $\lambda B_{X'}$ is $\sigma(X',X)$ compact. Therefore there exists $x'_0 \in \lambda B_{X'}$ such that for all $x \in X$

$$y'_\alpha(Tx) = T'y'_\alpha(x) \to x'_0 x.$$

Since $y'_\alpha(Tx) \to y'_0(Tx)$, we have $y'_0(Tx) = x'_0 x$ for all $x \in X$. It follows that $y'_0 \in D(T')$ and $T'y'_0 = x'_0$.

For all $y \in Y$ we have

$$|y'_0 y| = \lim |y'_\alpha y| \leq \liminf \|y'_\alpha\| \, \|y\|$$

and so

$$\|y'_0\| \leq \liminf \|y'_\alpha\| .$$

Similarly

$$\|T'y'_0\| \leq \liminf \|T'y'_\alpha\| .$$

Hence

$$\begin{aligned} \|y'_0\|_{T'} &= \|y'_0\| + \|T'y'_0\| \leq \liminf(\|y'_\alpha\| + \|T'y'_\alpha\|) \\ &= \liminf \|y'_\alpha\|_T \\ &\leq 1. \end{aligned}$$

Thus $G_{T'}^{-1}y'_0 \in B_W$, i.e. $y'_0 \in G_{T'}(B_W)$, showing that $G_{T'}(B_W)$ is $\sigma(Y',Y)$ closed, as required. $\square$

IV.3.16 Proposition. *$Y'_{T'}$ is a dual space.*

PROOF. Write $W := Y'_{T'}$ and define $A : Y \to W'$ by

$$(1) \qquad (Ay)(G_{T'}^{-1}y') := y'y.$$

We have for each $y' \in D(T')$,

$$\begin{aligned}(Ay)(G_{T'}^{-1}y') = 0 &\text{ for every } Ay \subset AY \\ \iff y'y = 0 &\text{ for every } y \in Y \\ \iff y' &= 0 \\ \iff G_{T'}^{-1}y' &= 0.\end{aligned}$$

Hence AY is a total subset of W'.

Notice that for any $\sigma(W, AY)$ subbasic neighbourhood of 0

$$U = \{G_{T'}^{-1}y' : y' \in D(T'), |(Ay)(G_{T'}^{-1}y')| < \epsilon\}$$

in W, the set

$$G_{T'}(U) = \{y' \in D(T') : |y'y| < \epsilon\}$$

is a $\sigma(D(T'), Y)$ subbasic neighbourhood of 0, and vice versa. Hence W and $D(T')$ are $\sigma(W, AY) - \sigma(D(T'), Y)$ homeomorphic. Therefore B_W is $\sigma(W, AY)$ compact if and only if $G_{T'}(B_W)$ is $\sigma(D(T'), Y)$ compact, which is the case by Theorem IV.3.15. Hence according to Theorem IV.3.14, $W = (AY)'$. □

IV.3.17 Theorem. *Let $T \in LR(X, Y)$. There exists a normed space Z and a bounded linear operator H mapping Y onto Z with the following properties:*

(i) $\|H\| \le 1$
(ii) *HT is continuous and $\|HT\| \le 1$*
(iii) $Z' = Y'_{T'}$
(iv) *$H' = G_{T'}$ (and hence $(HT)' = T'G_{T'}$)*
(v) *$N(H) = \overline{T}(0)$; in particular H is injective if and only if T is closable and single valued.*

PROOF. By Proposition IV.3.16, $Y'_{T'}$ is a dual space. Let the operator A be as in IV.3.16(1), let $Z := AY$ and define the relation H from Y to a subspace of E

$$(1) \qquad (G_{T'}^{-1}y')(Hy) := y'y \qquad (y' \in D(T')).$$

(Here we have identified $G_{T'}^{-1}y'$ with its canonical image in W''.)

Obviously the relation H is linear and surjective. Furthermore, putting $y = 0$ in (1) gives

$$H(0) \subset (Y'_{T'})^{\top} = \{0\}.$$

Therefore H is single valued.

(i) We have

$$\begin{aligned} \|H\| &= \sup_y \|Hy\| \,/\, \|y\| = \sup_y \sup_{z' \in Z'} |e'Hy| \,/\, \|z'\| \, \|y\| \\ &= \sup_y \sup_{y' \in D(T')} \left|(G_{T'}^{-1} y')(Hy)\right| \,/\, \left\|G_{T'}^{-1} y'\right\| \, \|y\| \\ &= \sup |y'y| \,/ (\|y'\| + \|T'y'\|) \, \|y\| \\ &\leq \sup \|y'\| \,/ (\|y'\| + \|T'y'\|) \\ &\leq 1. \end{aligned}$$

(ii) Let $x \in D(T)$. Then for $y \in Tx$ we have

$$\begin{aligned} \|Hy\| &= \sup_{z' \in Z'} |z'Hy| \,/\, \|z'\| = \sup_{y' \in D(T')} \left|(G_{T'}^{-1} y')(Hy)\right| \,/\, \left\|G_{T'}^{-1} y'\right\| \\ &= \sup_{y' \in D(T')} |y'y| \,/ (\|y'\| + \|T'y'\|) \\ &= \sup_{y' \in D(T')} |y'Tx| \,/ (\|y'\| + \|T'y'\|) \quad \text{(see III.1.2)} \\ &\leq \sup_{y' \in D(T')} |T'y'x| \,/\, \|T'y'\| \\ &\leq \|x\| \,. \end{aligned}$$

Hence

$$\|HTx\| = d(HTx, 0) \leq \|Hy\| \leq \|x\| \,.$$

(iii) Follows immediately from IV.3.16 and the definition of Z.

(iv) We have for $y \in Y$ and $y' \in D(T')$,

$$H'(G_{T'}^{-1} y')(y) = (G_{T'}^{-1} y')(Hy) = y'y = (G_{T'} G_{T'}^{-1} y')(y)$$

whence

$$H'(G_{T'}^{-1} y') = G_{T'}(G_{T'}^{-1} y').$$

Hence $H' = G_{T'}$.

(v) We have

$$\begin{aligned} y \in N(H) &\iff (G_{T'}^{-1} y')(Hy) = 0 \text{ for all } y' \in D(T') \\ &\iff y'y = 0 \text{ for all } y' \in D(T') \\ &\iff y \in D(T')^{\top} = \overline{T}(0). \end{aligned}$$

The final remark now follows from Proposition II.5.5. □

IV.3.18 Definition The operator H of Theorem IV.3.17 associated with a given linear relation T will be denoted by H_T. It will be denoted by H when T is understood.

IV.3.19 Exercise Verify that $\|HT\| = \left\|\tilde{T}G_{\tilde{T}}\right\|$.

IV.4 DENSITY INVARIANCE OF CERTAIN QUANTITIES

In this section we investigate conditions under which $f(T|_E) = f(T)$ where f is a given quantity and E is a dense linear subspace of $D(T)$. (The invariance result IV.1.3 is somewhat analogous, where E differs from $D(T)$ by only a finite number of dimensions.)

IV.4.1 Definition A linear subspace E of X such that $G(T|_E)$ is dense in $G(T)$ is called a **core** of T.

From the definition it is clear that

(1) *any core subspace of* T *is dense in* $D(T)$,
(2) $D(T)$ *is a core of* $\overline{T}$,
(3) $\|T|_E\| = \|T\|$ *if* E *is a core of* T.

IV.4.2 Proposition. *Let E be a linear subspace of $D(T)$. Then E is a core of T if and only if E is a core of QT.*

PROOF. If $G(T|_E)$ is dense in $G(T)$, then obviously $G(QT|_E)$ is dense in $G(QT)$. Conversely, suppose that E is a core of QT, and let $(x,y) \in G(T)$. There exists a sequence $(e_n, z_n) \in G(QT|_E)$ such that $(e_n, z_n) \to (x, Qy)$ $(n \to \infty)$. Now $z_n \in QTe_n$, so $z_n \in Qy_n$ where $y_n \in Te_n$ $(n \in \mathbb{N})$. We have $d(y_n, T(0)) \to d(y, T(0))$ $(n \to \infty)$. Hence there exists $k_n \in T(0)$ such that $y_n + k_n - y \to 0$ $(n \to \infty)$. We now have

$$(e_n, y_n + k_n) \to (x, y) \qquad (n \to \infty)$$

where $y_n + k_n \in Te_n$. It follows that $G(T|_E)$ is dense in $G(T)$. □

IV.4.3 Corollary. *Let T be continuous. Then $E \subset D(T)$ is a core of T if and only if E is dense in $D(T)$.*

IV.4.4 Proposition. *The following properties are equivalent:*

(i) *E is a core of T*

(ii) $T' = (T|_E)'$.

PROOF. Suppose that E is a core of T. Then by the Bipolar Theorem, we have $G(-T|_E)^{\perp} = G(-T|_E)^{\perp\top\perp} = \overline{G(-T_E)}^{\perp} = G(-T)^{\perp} = G((T')')$. Therefore $T' = (T|_E)'$.

Conversely, if $T' = (T|_E)'$, then $G(-T|_E)^{\perp} = G(-T)^{\perp}$, and so $\overline{G(-T|_E)} = \overline{G(-T)}$, whence $G(T|_E)$ is dense in $G(T)$, i.e. E is a core of T. □

IV.4.5 Definition The quantity f is said to be **densely invariant** if $f(T|_E) = f(T)$ whenever E is a core of T.

IV.4.6 Examples.

(a) *The norm function $f(T) = \|T\|$ is densely invariant.*

(b) *If f is any densely invariant quantity, then for any closable linear relation T, $f(T) = f(\overline{T})$.* However, if T is not closable, then the equality may fail; for instance, the Example III.3.13 in which $\|P\| = \infty$ and $\left\|\tilde{P}\right\| = 0$.

(c) *The minimum modulus function $\gamma(T)$ is not densely invariant.* Consider the continuous linear operator T of Example II.5.10. By the Open Mapping Theorem III.3.2, $\gamma(\tilde{T}) > 0$, whereas clearly $\gamma(T) = 0$ since ℓ_1 and ℓ_2 are not isomorphic. Since T is completely closable, this shows that γ is not densely invariant.

(d) *The index function $\kappa(T)$ is not densely invariant.* Consider again the linear operator T of Example II.5.10. T is continuous and hence closable, while $0 = \kappa(T) \neq \kappa(\overline{T}) = \infty$.

IV.4.7 Lemma. *Let E be a dense linear subspace of X and let $M \in \mathcal{P}(X)$. Then $E \cap M$ is dense in M.*

PROOF. See e.g. Goldberg [3; Lemma IV.2.8]. □

IV.4.8 Theorem. *The quantities $\Delta, \Gamma, \Gamma_0, \overline{\Gamma}_0, \tau, \tau_0, \overline{\tau}_0$ are densely invariant.*

PROOF. We may clearly suppose that $D(T) = X$. Let E be a core of T. Since E is a core of QT, and $f(T) = f(QT)$, $f(T|_E) = f(QT|_E)$, it is enough to consider the case when T is single valued. In the case of the quantities $\Delta, \Gamma, \Gamma_0, \tau, \tau_0$ Proposition IV.3.4 shows that it is sufficient to consider the bounded linear operator TG_T. We shall therefore assume that for these quantities, T is bounded. The quantitites $\overline{\Gamma}_0$ and $\overline{\tau}_0$ will be considered separately.

The case $\dim X < \infty$ being trivial we assume that X is infinite dimensional. For a given infinite dimensional subspace M of X we construct sequences $\{m_n\}$ in M and (x'_n) in X' as follows. Choose $m_1 \in S_M$ arbitrarily and then $x'_1 \in X'$ so that

$$\|x'_1\| = x'_1 m = \|m_1\|$$

and proceed inductively to select m_n and x'_n so that

$$\|m_n\| = \|x'_n\| = x'_n m_n = 1, \qquad 1 \leq n < \infty \tag{1}$$

$$m_n \in \bigcap_1^{n-1} N(x'_i); \text{ i.e. } x'_i(m_k) = 0, \qquad 1 \leq i < k. \tag{2}$$

Let $W = \operatorname{sp}\{m_1, m_2, \ldots\}$. It can be verified using (1) and (2) that the sequence (m_n) is linearly independent. Thus $W \in \mathcal{I}(X)$. For an arbitrary element $w = \sum_1^n a_i m_i$ we shall verify by induction that

$$|a_i| \leq 2^{i-1} \|w\|, \qquad 1 \leq i \leq n. \tag{3}$$

Suppose that (3) is true. Then from (1) and (2),

$$x'_{j+1} w = \sum_{i=1}^{j} a_i x'_{j+1}(m_i) + a_{j+1}.$$

Hence from the induction hypothesis (3),

$$|a_{j+1}| \leq |x'_{j+1} w| + \sum_{i=1}^{j} |a_i| \, \|x'_{j+1}\| \leq \|w\| + \sum_{i=1}^{j} 2^{i-1} \|w\| \leq 2^j \|w\|$$

as required.

Fix $\epsilon, 0 < \epsilon < 1$. The subspace $E \cap \left(\bigcap_1^n N(x'_i) \right)$ is dense in $\bigcap_1^n N(x'_i)$ (by Lemma IV.4.7) and hence there exists a sequence $\{z_n\}$ in E such that

$$\|z_n - m_n\| \leq 2^{1-2n} \min\{\epsilon, \epsilon c / \|T\|\} \qquad \text{and} \tag{4}$$

$$z_n \in \bigcap_1^{n-1} N(x'_i) \tag{5}$$

where c is any given positive constant.

Let $L = \mathrm{sp}\{z_n\}$. From (1), (4) and (5) it is easily seen that $z_1, z_2, \ldots$ are linearly independent. Thus $L \in \mathcal{I}(E)$.

Now for $z = \sum_1^n a_i z_i \in L$ we have

$$\left\|\sum_1^n a_i(z_i - m_i)\right\| \leq \sum_1^n |a_i| \, \|z_i - m_i\| \leq \sum_{i=1}^n 2^{i-1} \left\|\sum_{j=1}^n a_j m_j\right\| \|z_i - m_i\|$$

$$\leq \sum_{i=1}^n 2^{i-1} 2^{1-2i} \epsilon \left\|\sum_{j=1}^n a_j m_j\right\| \leq \epsilon \sum_1^n 2^{-i} \left\|\sum_1^n a_i m_i\right\| < \epsilon \left\|\sum_1^n a_i m_i\right\|$$

by (3) and (4). Consequently

$$(1-\epsilon)\left\|\sum_1^n a_i m_i\right\| \leq \left\|\sum_1^n a_i m_i\right\| - \left\|\sum_1^n a_i(z_i - m_i)\right\| \leq \left\|\sum_1^n a_i z_i\right\|$$

$$\leq \left\|\sum_1^n a_i m_i\right\| + \left\|\sum_1^n a_i(z_i - m_i)\right\| < (1+\epsilon)\left\|\sum_1^n a_i m_i\right\|.$$

Define the linear bijection $A : L \to W$ by setting $Az_i = m_i$. Then for $z \in L$ we have

$$(6) \qquad (1-\epsilon)\|Az\| \leq \|z\| \leq (1+\epsilon)\|Az\|$$

(a) The quantity Δ: We clearly have $\Delta(T|_E) \leq \Delta(T)$. Suppose that $\Delta(T|_E) < \Delta(T)$. Let $\alpha \in \mathbb{R}$ be such that $\Delta(T|_E) < \alpha < \Delta(T)$. Choose $M \in \mathcal{I}(X)$ such that $\Gamma(T|M) \geq \alpha$. We have

$$(7) \qquad M_0 \in \mathcal{I}(M) \Rightarrow \|T|_{M_0}\| \geq \alpha.$$

Putting $c = 1$ in (4) we have from (3) and (4) that

$$\left\|T\sum_1^n a_i z_i\right\| \geq \left\|T\sum_1^n a_i m_i\right\| - \left\|T\sum_1^n a_i(z_i - m_i)\right\| \geq \left\|T\sum_1^n a_i m_i\right\|$$

$$- \|T\| \sum_1^n |a_i| \, \|z_i - m_i\| \geq \left\|T\sum_1^n a_i m_i\right\|$$

$$-\epsilon \sum_1^n 2^{-i} \left\|\sum_1^n a_i m_i\right\| \geq \left\|T\sum_1^n a_i m_i\right\| - \epsilon\left\|\sum_1^n a_i m_i\right\|.$$

Hence from (6)

$$\frac{\|T|_E z\|}{\|z\|} \geq \left(\frac{\|TAz\|}{\|Az\|} - \epsilon\right)(1-\epsilon). \tag{8}$$

Let $V \in \mathcal{I}(L)$. Then taking the supremum on the left hand side of (8) over elements of V gives $\|T|_{E\cap V}\| \geq \left(\frac{\|TAz\|}{\|Az\|} - \epsilon\right)(1-\epsilon)$ for each $z \in V$. Hence $\|T|_{E\cap V}\| \geq (\|T|_{AV}\| - \epsilon)\,(1-\epsilon)$. Since $AV \in \mathcal{I}(M)$, we have from (7) that $\|T|_{E\cap V}\| > (\alpha-\epsilon)\,(1-\epsilon)$ whence $\Delta(T|_E) = \sup_{V\in\mathcal{I}(E)} \Gamma(T|_{E\cap V}) \geq \sup_{V\in\mathcal{I}(L)} \Gamma(T|_{E\cap V}) \geq (\alpha-\epsilon)\,(1-\epsilon)$. Since ϵ is arbitrary it follows that $\Delta(T|_E) \geq \alpha$, a contradiction.

(b) The quantity Γ: We clearly have $\Gamma(T) \leq \Gamma(T|_E)$.
Suppose that $\Gamma(T) < \Gamma(T|_E) - \delta$ where $\delta > 0$. Write $c = \Gamma(T|_E) - \delta$. Select $M \in \mathcal{I}(X)$ such that

$$\|T|_M\| < c \tag{9}$$

From (9) and (4) we have

$$\begin{aligned}\left\|T\sum_1^n a_i z_i\right\| &\leq \left\|T\sum_1^n a_i m_i\right\| + \left\|T\sum_1^n a_i(z_i - m_i)\right\| \leq \left\|T\sum_1^n a_i m_i\right\| \\ &+ \|T\|\sum_1^n |a_i|\,\|z_i - m_i\| \leq \left\|T\sum_1^n a_i m_i\right\| \\ &+\epsilon\, c\sum_1^n 2^{-i}\left\|\sum_1^n a_i m_i\right\| < (1+\epsilon)c\left\|\sum_1^n a_i m_i\right\|.\end{aligned}$$

Hence from (6)

$$\frac{\|T|_E z\|}{\|z\|} < \frac{1+\epsilon}{1-\epsilon}c, \quad \text{whence}$$

$$\Gamma(T|_E) = \inf_{V\in\mathcal{I}(E)} \|T|_{E\cap V}\| \leq \|T|_{E\cap L}\| \leq \frac{1+\epsilon}{1-\epsilon}c.$$

Since ϵ is arbitrary, $\Gamma(T|_E) \leq c$, a contradiction.

(c) The quantity τ: Evidently $\tau(T) \geq \tau(T|_E)$.
Assume that $\tau(T) > \tau(T|_E)$. Then there exists $\delta > 0$ and $M \in \mathcal{I}(X)$ such that $\tau(T) \geq \inf_{m\in S_M} \|Tm\| > \tau(T|_E) + \delta$. Hence writing $c = \tau(T|_E) + \delta$ we have

$$\frac{\|Tm\|}{\|m\|} > c \text{ for } m \in M. \tag{10}$$

Then from (4) and (10)

$$\left\|T\sum_1^n a_i z_i\right\| \geq \left\|T\sum_1^n a_i m_i\right\| - \left\|T\sum_1^n a_i(z_i - m_i)\right\| \geq \left\|T\sum_1^n a_i m_i\right\|$$
$$- \|T\| \sum_1^n |a_i| (\|z_i - m_i\| \geq \left\|T\sum_1^n a_i m_i\right\|$$
$$-\epsilon\, c \sum_1^n 2^{-i} \left\|\sum_1^n a_i m_i\right\| \geq (1-\epsilon)c \left\|\sum_1^n a_i m_i\right\|.$$

Hence from (6)

$$\frac{\|T|_E z\|}{\|z\|} > \frac{1-\epsilon}{1+\epsilon} c.$$

Consequently $\tau(T|_E) \geq \inf_{z\in L} \dfrac{\|T|_E z\|}{\|z\|} \geq \dfrac{1-\epsilon}{1+\epsilon} c$. Since ϵ is arbitrary, $\tau(T_E) \geq c$, a contradiction.

(d) The quantity $\overline{\Gamma}_0$: We have by Proposition IV.1.2(c)

$$\mathcal{P}(E) = \{M \cap E : M \in \mathcal{P}(X)\}. \tag{11}$$

We have

$$\overline{\Gamma}_0(T|_E) = \inf_{V\in\mathcal{P}(E)} \|T|_V\| = \inf_{M\in\mathcal{P}(X)} \|T|_{M\cap E}\| \qquad \text{(by (11))}.$$

But since evidently $G^{-1}(M) \in \mathcal{P}(X_T)$, we have $G^{-1}(M \cap E)$ dense in $G^{-1}(M)$ by Lemma IV.4.7. Thus given $m \in M$ there exists $x \in M \cap E$ for which $\|m - x\| + \|T(m-x)\| < \epsilon$ for arbitrary $\epsilon > 0$. It follows that $\|T|_M\| = \|T|_{M\cap E}\|$ and hence that $\overline{\Gamma}_0(T|_E) = \inf_{M\in\mathcal{P}(X)} \|T|_M\| = \overline{\Gamma}_0(T)$.

(e) The quantity $\overline{\tau}_0$: We have similarly

$$\begin{aligned}
\overline{\tau}_0(T) &= \sup_{M\in\mathcal{P}(X)} \inf_{x\in S_M} \|Tx\| \\
&= \sup_{M\in\mathcal{P}(X)} \inf_{x\in S_{M\cap E}} \|Tx\| \\
&= \sup_{V\in\mathcal{P}(E)} \inf_{x\in S_V} \|Tx\| \\
&= \overline{\tau}_0(T_E).
\end{aligned}$$

Finally, since $\Gamma_0(T) = \overline{\Gamma}_0(T)$ and $\tau_0(T) = \overline{\tau}_0(T)$ for continuous operators, the theorem is proved. □

IV.4.9 Corollary. *Let T be closable and $f \in \{\Delta, \Gamma, \Gamma_0, \overline{\Gamma}_0, \tau, \tau_0, \overline{\tau}_0\}$. Then $f(T) = f(\overline{T})$.*

IV.4.10 Corollary. *Let T be continuous and single valued, let $S \in LR(Y, Z)$ and let $D(ST)$ be dense in $D(T)$. Then*

$$\Gamma(ST) \leq \Gamma(T)\Delta(S|_{R(T)}) \leq \Gamma(T)\Delta(S).$$

PROOF. We have from Lemma IV.2.8,

$$\Gamma(ST) \leq \Gamma(T|_{D(ST)})\Delta(S|_{R(T)}).$$

But by Theorem IV.4.8, $\Gamma(T|_{D(ST)}) = \Gamma(T)$. □

IV.4.11 Corollary. *Let T be single valued and $S \in LR(Y, Z)$ and let $D(ST)$ be dense in a core of T. Then*

$$\Gamma(STG_T) \leq \Gamma(TG_T)\Delta(S|_{R(T)}).$$

IV.4.12 Corollary. *For each of the quantities $f \in \{\Delta, \Gamma, \Gamma_0, \overline{\Gamma}_0, \tau, \tau_0, \overline{\tau}_0\}$ we have*

(a) $f(\overline{T}) = f(\overline{T}|_{D(T)})$

(b) $f(\overline{T}) \leq f(T)$.

PROOF. (a) Follows immediately from Theorem IV.3.8 since $D(T)$ is a core of $D(\overline{T})$ (see IV.4.1(2)).

(b) By (a), $f(\overline{T}) = f(\overline{T}|_{D(T)})$. For each $M \in \mathcal{I}(D(T))$, we have $\|\overline{T}m\| \leq \|Tm\|$ $(m \in M)$. Hence (b) follows. □

Observe that density invariance was used in an essential way in the proof of the inequality IV.4.12(b).

IV.4.13 Theorem. *Let $D(T)$ be infinite dimensional and let E be a core of T. Then*

$$\inf_V \tau(T|_V) = \inf_M \tau(T|_M)$$

where V, M range over the infinite dimensional subspaces of E and $D(T)$ respectively.

PROOF. As in the proof of Theorem IV.4.8 we assume without loss of generality that T is single valued and bounded. Clearly $\inf \tau(T|_V) \geq \inf \tau(T|_M)$. Let $\epsilon > 0$ be given and choose $M_0 \in \mathcal{I}(X)$ so that $\tau(T|_{M_0}) < \inf\{\tau(T|_M) : M \in \mathcal{I}(X)\} + \epsilon = c$ say. Then

$$(1) \qquad M \in \mathcal{I}(M_0) \implies \inf_{m \in S_M} \|Tm\| < c.$$

With the notation of the proof of IV.4.8 we have for $z = \sum_1^n a_i z_i \in L$ that

$$\begin{aligned}
\left\| T \sum_1^n a_i z_i \right\| &\leq \left\| T \sum_1^n a_i m_i \right\| + \left\| T \sum_1^n a_i (z_i - m_i) \right\| \\
&\leq \left\| T \sum_1^n a_i m_i \right\| + \|T\| \sum_1^n |a_i| \, \|z_i - m_i\| \\
&\leq \left\| T \sum_1^n a_i m_i \right\| + \epsilon c \sum_1^n 2^{-i} \left\| \sum_1^n a_i m_i \right\| \\
&= \left\| T \sum_1^n a_i m_i \right\| + \epsilon c \left\| \sum_1^n a_i m_i \right\| .
\end{aligned}$$

Thus $\|Tz\| \leq \|TAz\| + \epsilon c \, \|Az\|$, whence from (6) of IV.4.8

$$(2) \qquad \frac{\|Tz\|}{\|z\|} \leq \left(\frac{\|TAz\|}{\|Az\|} + \epsilon c \right) \frac{1}{1 - \epsilon}.$$

Let $V \in \mathcal{I}(L)$. Then from (2) we have

$$\inf_{z \in V} \frac{\|Tz\|}{\|z\|} \leq \left(\inf_{Az \in AV} \frac{\|TAz\|}{\|Az\|} + \epsilon c \right) \frac{1}{1 - \epsilon}.$$

But $AV \in \mathcal{I}(M) \subset \mathcal{I}(M_0)$. Hence from (1) $\inf_{z \in V} \frac{\|Tz\|}{\|z\|} \leq \left(\frac{1+\epsilon}{1-\epsilon} \right) c$.

Consequently

$$\tau(T|_L) = \sup_{V \in \mathcal{I}(L)} \inf_{z \in V} \frac{\|Tz\|}{\|z\|} \leq \left(\frac{1+\epsilon}{1-\epsilon} \right) c$$

whence

$$\inf\{\tau(T|_V) : V \in \mathcal{I}(E)\} \le \tau(T|_L) \le \left(\frac{1+\epsilon}{1-\epsilon}\right) c.$$

Since $\epsilon > 0$ is arbitrary we obtain $\inf\{\tau(T|_V) : V \in \mathcal{I}(E)\} \le c$. The desired inequality now follows. □

IV.4.14 Theorem. *Let E be a core of T. Then for each of the quantities $f \in \{\Delta, \Gamma, \Gamma_0, \tau, \tau_0\}$ we have*

$$f(G_{T|_E}) = f(G_T).$$

PROOF. If $f \in \{\Gamma, \Gamma_0, \tau, \tau_0\}$ then the equality follows immediately from Theorem IV.4.8 and Proposition IV.3.6. It only remains to consider $f = \Delta$. From Theorem IV.4.13 and Proposition IV.3.6 we have

$$\Delta(G_{T|_E}) = \left(1 + \inf_{M \in \mathcal{I}(E)} \tau(T|_M)\right)^{-1} = \left(1 + \inf_{M \in \mathcal{I}(D(T))} \tau(T|_M)\right)^{-1} = \Delta(G_T). \quad \square$$

A stronger result holds for a subset of these quantities if T is completely closable – see Theorem V.9.17.

IV.4.15 Lemma. *We have*

(a) $\Gamma(T) = \inf_M f(T|_M)$ *if* $f \in \{\Delta, \Gamma, \Gamma_0, \overline{\Gamma}_0\}$

(b) $\tau(T) = \sup_M f(T|_M)$ *if* $f \in \{\tau, \tau_0, \overline{\tau}_0\}$ *where M ranges over the infinite dimensional subspaces of $D(T)$.*

PROOF. (a) Let $f \in \{\Delta, \Gamma, \Gamma_0, \overline{\Gamma}_0\}$. Clearly

$$\Gamma(T) \le \Gamma(T|_M) \le f(T|_M) \qquad (M \in \mathcal{I}(D(T))).$$

Now select $M \in \mathcal{I}(D(T))$ so that $\|T|_M\| - \epsilon \le \Gamma(T)$. Then $f(T|_M) - \epsilon \le \|T|_M\| - \epsilon \le \Gamma(T)$. Hence $\inf f(T|_M) - \epsilon \le \Gamma(T)$ and the result follows.

(b) Let $f \in \{\tau, \tau_0, \overline{\tau}_0\}$. Clearly $\tau(T) \ge \tau(T|_M) \ge f(T|_M)$ $(M \in \mathcal{I}(D(T)))$. Clearly we may assume $\tau(T) > 0$. Now given any ϵ in the interval $0 < \epsilon < \tau(T)$, select $M \in \mathcal{I}(D(T))$ so that $\inf\{\|Tm\| : m \in S_M\} \ge \epsilon$. Then $f(T|_M) \ge \inf\{\|Tm\| : m \in S_M\} \ge \epsilon$. It follows that $\sup f(T|_M) = \tau(T)$. □

IV.4.16 Theorem. *Let E be a core of T. Then*

(a) $\Gamma(T) = \inf_{V \in \mathcal{I}(E)} f(T|_V) = \inf_{M \in \mathcal{I}(D(T))} f(T|_M)$ *if* $f \in \{\Delta, \Gamma, \Gamma_0, \overline{\Gamma}_0\}$

(b) $\tau(T) = \sup_{V \in \mathcal{I}(E)} f(T|_V) = \sup_{M \in \mathcal{I}(D(T))} f(T|_M)$ *if* $f \in \{\tau, \tau_0, \overline{\tau}_0\}$.

PROOF. (a) By Lemma IV.4.15 (a) we have

$$\begin{aligned} \Gamma(T|_E) &= \inf_{V \in \mathcal{I}(E)} f(T|_V) \\ &\geq \inf_{M \in \mathcal{I}(D(T))} f(T|_M) = \Gamma(T). \end{aligned}$$

Hence the equality follows from Theorem IV.4.8.

(b) Similar to part (a), using Lemma IV.4.15 (b). □

IV.4.17 Exercise Let E be a core of T. Then for any densely invariant quantity f,

$$f(T|_E \, G_{T|_E}) = f(TG_T).$$

IV.5 QUANTITIES DETERMINED BY QUOTIENTS

So far the quantities we have considered have been determined by the norms of restrictions of the relation T to a class of subspaces $E \subset D(T)$. Now from the identity

$$(1) \qquad (TJ_E)' = Q_{E^\perp} T'$$

it would be natural to define "dual" quantities derived from the norms $\|Q_M T\|$, where M runs through some class of subspaces M of Y. Two such classes of subspaces M will be selected, motivated by the pair of equivalences:

$$(2) \qquad \dim E = \infty \iff \operatorname{codim} E^\perp = \infty$$

$$(3) \qquad \operatorname{codim} E < \infty \iff \dim E^\perp < \infty$$

(where E is a linear subspace of X). Hence the following definitions:

IV.5.1 Definitions Let $\mathcal{F}(X)$ denote the collection of finite dimensional subspaces of X, and $\mathcal{E}(Y)$ the collection of closed subspaces of X of infinite codimension in X.

For a given linear relation T we define the quantities $\Gamma_0'(T)$, $\Gamma'(T)$ and $\Delta'(T)$ as follows:

If Y is finite dimensional then all the quantities are zero; thus $\Gamma'_0(T) = \Gamma'(T) = \Delta'(T) = 0$.

If Y is infinite dimensional, then

$$\Gamma'_0(T) := \inf_{M \in \mathcal{F}(Y)} \|Q_M T\|$$

$$\Gamma'(T) := \inf_{M \in \mathcal{E}(\tilde{Y})} \|Q_M J_Y T\|$$

$$\Delta'(T) := \sup_{M \in \mathcal{E}(Y)} \Gamma'(Q_M T)$$

where $J_Y := J_Y^{\tilde{Y}}$ and $Q_M := Q_M^Y$.

The following Lemma is from L.E. Labuschagne [6].

IV.5.2 Lemma. *Let M be a closed subspace of X. We have:*

(a) *If N is a closed subspace of X containing M, then*

(1) $$X/N = (X/M)/(N/M)$$

(where the equality is a canonical isometry) and

(2) $$Q_N^X = Q_{N/M}^{X/M} Q_M^X.$$

(b) *For any closed subspace E of X/M, the closed subspace N of X given by*

$$N := (Q_M^X)^{-1}(E)$$

satisfies

(3) $$N \supset M, \text{ and } (X/M)/E = X/N.$$

PROOF. (a) A representative coset of $(X/M)/(N/M)$ is

$$\begin{aligned}&\{Q_M x + Q_M n : n \in N\}\\ &= Q_M\{x + n : n \in N\}\\ &= Q_N x \quad (x \in X), \text{ since } N \supset M.\end{aligned}$$

Essentially the same argument proves the second equality. It remains to show that the canonical identifications (1) and (2) are isometries. Since $N \supset M$, we have for $x \in X$

$$\begin{aligned}\left\|Q_{N/M}^{X/M} Q_N^X x\right\| &= \inf_{n \in N} \|Q_M^X x - Q_M^X n\| = \inf_{n \in N} \|Q_M^X (x-n)\|\\ &= \inf_{n \in N}\big(\inf_{m \in M} \|(x-n) - m\|\big) = \inf_{n \in N} \|x - n\| = \left\|Q_N^X x\right\|,\end{aligned}$$

as required.

(b) Clearly $(Q_M^X)^{-1}(E)$ is a closed subspace of X. Since $0 \in E$ and $(Q_M^X)^{-1}(0) = M$, we trivially have $M \subset N$. But

$$Q_M^X(N) = E.$$

Hence (3) follows, upon substituting into (1). □

IV.5.3 Lemma. *Let $M \in \mathcal{E}(Y)$. Then*

(a) $\Gamma_0'(Q_M T) = \inf\{\|Q_{M+F}T\| : F \in \mathcal{F}(Y)\}$

(b) $\Gamma'(Q_M T) = \inf\{\|Q_N J_Y T\| : N \in \mathcal{E}(\tilde{Y}), N \supset M\}$

(c) $\Delta'(Q_M T) = \sup\{\Gamma'(Q_N T) : N \in \mathcal{E}(Y), N \supset M\}$.

PROOF. (a) We have

$$\Gamma_0'(Q_M T) = \inf\{\left\|Q_K^{Y/M} Q_M^Y T\right\| : K \in \mathcal{F}(Y/M)\}.$$

Hence from Lemma IV.5.2 it follows that

$$\Gamma_0'(Q_M T) = \inf\{\left\|Q_N^Y T\right\| : N \supset M, \dim N/M < \infty\}.$$

Now for each subspace N containing M with $\dim N/M < \infty$ we may select $F \in \mathcal{F}(Y)$ such that $N \in M + F$. It follows that

$$\Gamma_0'(Q_M T) = \inf\{\|Q_{M+F}T\| : F \in \mathcal{F}(Y)\}.$$

(b) By definition we have

$$\Gamma'(Q_M T) = \inf\{\left\|Q_K^{Y/M} J_{Y/M} Q_M T\right\| : K \in \mathcal{E}((Y/M)^\sim)\}.$$

From the canonical equality $\tilde{Y}/\overline{M} = (Y/M)^\sim$, the operator $J_{Y/M}Q_M$ is naturally identified with $Q_{\overline{M}}^{\tilde{Y}} J_Y$. Consequently

$$\Gamma'(Q_M T) = \inf\{\left\|Q_N^{\tilde{Y}} J_Y T\right\| : N \in \mathcal{E}(\tilde{Y}), N \supset \overline{M}\}.$$

(c) By definition

$$\Delta'(Q_M T) = \sup\{\Gamma'(Q_K^{Y/M} Q_M T) : K \in \mathcal{E}(Y/M)\}.$$

It therefore suffices to apply Lemma IV.5.2. □

IV.5.4 Corollary. *Let Y be infinite dimensional and let $M \in \mathcal{E}(Y)$. Then*

(a) $\Gamma_0'(Q_M T) \le \Gamma_0'(T)$

(b) $\Gamma'(Q_M T) \ge \Gamma'(T)$

(c) $\Delta'(Q_M T) \le \Delta'(T)$.

PROOF. (a) For $F \in \mathcal{F}(Y)$ we have

$$\|Q_{M+F}T\| \le \|Q_F T\|$$

since $M + F \supset F$. Hence by Lemma IV.5.3,

$$\Gamma_0'(Q_M T) \le \Gamma_0'(T).$$

Both inequalities (b) and (c) are immediate consequences of IV.5.3(b) and (c). □

IV.5.5 Proposition. *Let Y be infinite dimensional. Then*

(a) $\Gamma'(T) = \inf\{\Gamma_0'(Q_M J_Y T) : M \in \mathcal{E}(\tilde{Y})\}$

(b) $\Delta'(T) = \sup_{M \in \mathcal{E}(Y)} \inf\{\Gamma_0'(Q_N J_Y T) : N \in \mathcal{E}(\tilde{Y}), N \supset M\}$.

PROOF. From Lemma IV.5.3, we have for any $E \in \mathcal{E}(Y)$,

$$\begin{aligned}\inf_{\substack{M\in\mathcal{E}(\tilde{Y})\\ M\supset E}} \Gamma_0'(Q_M J_Y T) &= \inf_M \inf_{F\in\mathcal{E}(\tilde{Y})} \|Q_{M+F} J_Y T\| \\ &= \inf\{\|Q_K J_Y F\| : K \in \mathcal{E}(\tilde{Y}), K \supset E\} \\ &= \Gamma'(Q_E T)\end{aligned}$$

since $M+F \in \mathcal{E}(\tilde{Y})$ for each $M \in \mathcal{E}(\tilde{Y})$ and each F is in $\mathcal{F}(\tilde{Y})$. Setting $E = \{0\}$ yields (a), whilst (b) is obtained by taking the supremum over all $E \in \mathcal{E}(Y)$. □

IV.5.6 Proposition. *Let $f \in \{\Gamma_0', \Gamma', \Delta'\}$ and let $F \in \mathcal{F}(Y)$. Then*

$$f(Q_F T) = f(T).$$

PROOF. Since $f(Q_F T) = f(T) = 0$ when $\dim Y < \infty$, we may assume that $\dim Y = \infty$.

Case 1: $f = \Gamma_0'$. From Lemma IV.5.3,

$$\begin{aligned}\Gamma_0'(Q_F T) &= \inf\{\|Q_M T\| : M \in \mathcal{F}(Y), M \supset F\} \\ &\geq \inf\{\|Q_M T\| : M \in \mathcal{F}(Y)\} = \Gamma_0'(T).\end{aligned}$$

Hence $\Gamma_0'(T) = \Gamma_0'(Q_F T)$.

Case 2: $f = \Gamma'$. Suppose that $\Gamma'(T) < \infty$ and choose $E \in \mathcal{E}(\tilde{Y})$ so that $\|Q_E J_Y T\| \leq \Gamma'(T) + \epsilon$. Clearly $E + J_Y F \in \mathcal{E}(\tilde{Y})$, and we have by Lemma IV.5.3,

$$\begin{aligned}\Gamma'(Q_F T) &= \inf\{\|Q_M J_Y T\| : M \in \mathcal{E}(\tilde{Y}), M \supset J_Y F\} \\ &\leq \|Q_{E+J_Y F} J_Y T\| \leq \|Q_E J_Y T\| \leq \Gamma'(T) + \epsilon.\end{aligned}$$

Hence $\Gamma'(T) = \Gamma'(Q_F T)$.

Case 3: $f = \Delta'$. First we shall verify that

(1) $$\{M + F : M \in \mathcal{E}(Y)\} = \{E : E \in \mathcal{E}(Y), E \supset F\}.$$

Let $E \in \mathcal{E}(Y)$ and $E \supset F$. Let N be a closed complement in E of the finite dimensional subspace F. Then $N \in \mathcal{E}(Y)$ and $N + F = E$, showing that $E \in \{M + F : M \in \mathcal{E}(Y)\}$. The reverse inclusion is obvious.

From Lemma IV.5.3,

$$\Delta'(Q_F T) = \sup\{\Gamma'(Q_E T) : E \in \mathcal{E}(Y), E \supset F\}. \tag{2}$$

Combining (1) and (2) it follows that

$$\begin{aligned} \Delta'(Q_F T) &= \sup\{\Gamma'(Q^Y_{M+F} T) : M \in \mathcal{E}(Y)\} \\ &= \sup_{M \in \mathcal{E}(Y)} \Gamma'(Q^{Y/M}_{(M+F)/M} Q^Y_M T). \end{aligned} \tag{3}$$

But $(M + F)/M \in \mathcal{F}(Y/M)$. Hence from what has been proved (Case 2),

$$\Gamma'(Q^{Y/M}_{(M+F)/M} Q^Y_M T) = \Gamma'(Q^Y_M T).$$

Consequently from (3),

$$\Delta'(Q_F T) = \sup \Gamma'(Q^Y_M T) = \Delta'(T). \quad \square$$

We have the density invariance result:

IV.5.7 Proposition. *Let T be closable (resp. completely closable). Then for $f \in \{\Gamma'_0, \Gamma', \Delta'\}$ we have*

$$f(T) = f(\overline{T}) \qquad (\textit{resp. } f(T) = f(\tilde{T})).$$

PROOF. Without loss of generality assume that $\dim Y = \infty$. We have for $x \in D(T)$, $\overline{T}|_{D(T)}x = Tx$, and so for $M \subset Y$, $\|Q_M T x\| = \|Q_M \overline{T} x\|$, whence

$$\|Q_M T\| = \left\|Q_M \overline{T}|_{D(T)}\right\| \leq \left\|Q_M \overline{T}\right\|.$$

Hence $\Gamma'_0(T) \leq \Gamma'_0(\overline{T})$.

The reverse inequality will be proved by the method used to prove Proposition II.5.8. Suppose that $\left\|Q_M \overline{T}\right\| > \|Q_M T\|$ and choose $\epsilon > 0$ and (x_0, y_0) in $G(\overline{T})$ so that $\|x_0\| \leq 1$ and

$$d(Q_M y_0, Q_M \overline{T}(0)) > \|Q_M T\| + 2\epsilon. \tag{1}$$

Next choose $(x, y) \in G(T)$ such that $\|x - x_0\| + \|y - y_0\| < \epsilon$ and $\|x\| \leq 1$. Since $\|Q_M\| = 1$, we have

$$\|x - x_0\| + \|Q_M(y - y_0)\| < \epsilon.$$

Then since

$$|d(Q_M y_0, Q_M T(0)) - d(Q_M y, Q_M T(0))| \leq \|Q_M(y - y_0)\| < \epsilon$$

we obtain from (1) that

$$\begin{aligned}\|Q_M T\| + 2\epsilon < d(Q_M y_0, Q_M \overline{T}(0)) &\leq d(Q_M y_0, Q_M T(0)) \\ &\leq d(Q_M y, Q_M T(0)) + \epsilon \leq \|Q_M T\| + \epsilon\end{aligned}$$

which is a contradiction, showing that $\|Q_M T\| \geq \|Q_M \overline{T}\|$. Therefore

$$\|Q_M T\| = \|Q_M \overline{T}\| \qquad (M \subset Y). \tag{2}$$

A similar proof shows that

$$\|Q_M J_Y T\| = \|Q_M J_Y \overline{T}\| \text{ for } M \subset \tilde{Y}. \tag{3}$$

Hence the proposition follows. □

IV.5.8 Proposition. *Let $S \in LR(Z, X)$ and let $S(0) \subset D(T)$. Then for $f \in \{\Gamma_0', \Gamma', \Delta'\}$ we have*

$$f(TS) \leq \|S\| \, f(T) \qquad (\infty.0 \text{ excluded}).$$

PROOF. Since $S(0) \subset D(T)$ we have by Corollary II.3.13,

$$\|Q_M TS\| \leq \|Q_M T\| \, \|S\| \qquad \text{for } M \subset Y$$

and

$$\|Q_M J_Y TS\| \leq \|Q_M J_Y T\| \, \|S\| \qquad \text{for } M \subset \tilde{Y}.$$

The desired inequality now follows immediately in case either $\|S\| < \infty$, or $\|S\| = \infty$ and $f(T) > 0$. □

IV.5.9 Proposition. $\Gamma'(T) \le \Delta'(T) \le \Gamma'_0(T)$.

PROOF. Without loss of generality, let $\dim Y = \infty$. Let $M \in \mathcal{E}(Y)$. It is clear from the definition that $\Gamma'(T) \le \Gamma'_0(T)$. Hence upon substituting $Q_M T$ for T, we have

$$\Gamma'(Q_M T) \le \Gamma'_0(Q_M T).$$

Since clearly $\Gamma'_0(Q_M T) \le \Gamma'_0(T)$ by IV.5.4 it follows that

$$\Gamma'(T) \le \Gamma'(Q_M T) \le \Gamma'_0(T).$$

Taking the supremum over $M \in \mathcal{E}(Y)$ now gives the desired result. □

IV.5.10 Proposition. $\Gamma'_0(T+S) \le \Gamma'_0(T) + \Gamma'_0(S)$.

PROOF. Fix $F \in \mathcal{F}(Y)$. We have $Q_F(T+S) = Q_F T + Q_F S$ (by I.4.2(e)) and hence

$$\text{(1)} \qquad \Gamma'_0(T+S) \le \|Q_F(T+S)\| \le \|Q_F T\| + \|Q_F S\| \le \|Q_F T\| + \|S\|\,.$$

Taking the infimum gives

$$\text{(2)} \qquad \Gamma'_0(T+S) \le \Gamma'_0(T) + \|S\|\,.$$

By Proposition IV.5.6 and (2),

$$\begin{aligned}\Gamma'_0(T+S) = \Gamma'_0(Q_F(T+S)) &\le \Gamma'_0(Q_F T) + \|Q_F S\| \\ &= \Gamma'_0(T) + \|Q_F S\|\,.\end{aligned}$$

Taking the infimum over $F \in \mathcal{F}(Y)$ now yields $\Gamma'_0(T+S) \le \Gamma'_0(T) + \Gamma'_0(S)$. □

IV.5.11 Proposition. *We have*

(a) $\Gamma'(T+S) \le \Delta'(J_Y T) + \Gamma'(S)$
(b) $\Delta'(T+S) \le \Delta'(J_Y T) + \Delta'(S)$.

PROOF. Both inequalities are trivial if $\dim Y < \infty$ or if $\Gamma'(Q_M J_Y T) = \infty$ for some $M \in \mathcal{E}(\tilde{Y})$. Hence we shall assume that $\dim Y = \infty$ and that

$$\text{(1)} \qquad \Gamma'(Q_M J_Y T) < \infty \qquad (M \in \mathcal{E}(\tilde{Y})).$$

(a) Let $\epsilon > 0$ and choose $M \in \mathcal{E}(\tilde{Y})$ so that

$$\text{(2)} \qquad \|Q_M J_Y S\| \le \Gamma'(S) + \epsilon/2.$$

Next select $N \in \mathcal{E}(\tilde{Y}/M)$ so that

$$(3) \qquad \left\| Q_N^{\tilde{Y}/M} Q_M^{\tilde{Y}} J_Y T \right\| \le \Gamma'(Q_M J_Y T) + \epsilon/2.$$

We have

$$\begin{aligned}\|Q_N Q_M J_Y(T+S)\| &\le \|Q_N Q_M J_Y T\| + \|Q_N Q_M J_Y S\| \\ &\le \Gamma'(Q_M J_Y T) + \epsilon/2 + \|Q_N\| \, \|Q_M J_Y S\| .\end{aligned}$$

Hence

$$\Gamma'(T+S) \le \Gamma'(Q_M J_Y T) + \Gamma'(S) + \epsilon.$$

Taking the supremum over $M \in \mathcal{E}(\tilde{Y})$ gives

$$\Gamma'(T+S) \le \Delta'(J_Y T) + \Gamma'(S) + \epsilon$$

from whence the result follows.

(b) Let $M \in \mathcal{E}(Y)$. Writing $J := J_Y^{\tilde{Y}}$, we have from Proposition IV.5.3,

$$(4) \qquad \Delta'(Q_M T) \le \Delta'(Q_{\overline{JM}} JT) \le \Delta'(JT).$$

Hence from part (a) of the proposition,

$$(5) \qquad \Gamma'(Q_M(T+S)) \le \Delta'(JT) + \Gamma'(Q_M S).$$

Taking the supremum in (5) over $M \in \mathcal{E}(Y)$ gives

$$\Delta'(T+S) \le \Delta'(JT) + \Delta'(S). \quad \square$$

IV.5.12 Corollary. *We have*

(a) $\Gamma_0'(J_Y T) \le \Gamma_0'(T)$
(b) $\Gamma'(J_Y T) = \Gamma'(T)$
(c) $\Delta'(J_Y T) \ge \Delta'(T)$.

PROOF. (a) Follows from the fact that the infirmum is taken over a larger set for $\Gamma_0'(J_Y T)$.

(b) Follows immediately from the definition.

(c) Put $S = 0$ in IV.5.11. $\square$

IV.5.13 Theorem. *Let* $S \in LR(Y, Z)$ *and let* $\dim S(0) < \infty$.

(a) *If* $D(S) = Y$, *then* $\Gamma_0'(ST) \le \Gamma_0'(S)\Gamma_0'(T)$.

(b) *If* $\overline{D(S)} = Y$ *and* $T(0) \subset D(S)$, *then* $\Gamma_0'(J_Z ST) \le \Gamma_0'(J_Z S)\Gamma_0'(T)$.

PROOF. (a) Let $D(S) = Y$. Without loss of generality we may suppose that $\Gamma_0'(T) < \infty$ and $\Gamma_0'(S) < \infty$. Let $\epsilon > 0$. Choose $K \in \mathcal{F}(Y)$ such that

$$\|Q_K T\| \le \Gamma_0'(T) + \epsilon. \tag{1}$$

Since $\dim S(0) < \infty$, we have $S(K) \in \mathcal{F}(Z)$ (see I.6.4). Hence by Proposition IV.5.6,

$$\Gamma_0'(S) = \Gamma_0'(Q_{S(K)}S).$$

Hence by Lemma IV.5.3, there exists $W \in \mathcal{F}(Z)$ such that

$$\left\|Q_{W+S(K)}S\right\| \le \Gamma_0'(S) + \epsilon. \tag{2}$$

The relation $Q_{W+S(K)}S$ is single valued (as $S(0) \subset W + S(K)$). Accordingly, it has a canonical factorization $(Q_{W+S(K)}S)^\frown Q_N$, where $N := N(Q_{W+S(K)}S) \supset K$ (see Goldberg [3; p.63]). We conclude from (1) and (2) that

$$\begin{aligned}
\Gamma_0'(ST) &= \inf\{\|Q_F ST\| : F \in \mathcal{F}(Z)\} \\
&\le \left\|Q_{W+S(K)}ST\right\| \\
&= \left\|(Q_{W+S(K)}S)^\frown Q_N T\right\| \\
&\le \left\|(Q_{W+S(K)}S)^\frown\right\| \, \|Q_N T\| \\
&\le \left\|Q_{W+S(K)}S\right\| \, \|Q_K T\| \\
&\le (\Gamma_0'(S) + \epsilon)(\Gamma_0'(T) + \epsilon) \quad \text{by (1) and (2).}
\end{aligned}$$

Since $\epsilon > 0$ is arbitrary, it follows that $\Gamma_0'(ST) \le \Gamma_0'(S)\Gamma_0'(T)$.

(b) Let $\overline{D(S)} = Y$ and $\dim S(0) < \infty$. As before we shall assume without loss of generality that $\Gamma_0'(J_Z S) < \infty$. In order to simplify the notation we shall suppose that Z is complete. The appropriate minor modifications for the incomplete case are straightforward.

Select $W \in \mathcal{F}(Z)$ such that $\|Q_W S\| < \infty$. Then we have

$$\left\|Q_{W+S(0)}S\right\| < \infty. \tag{3}$$

By Proposition IV.5.6 we have

$$\Gamma_0'(ST) = \Gamma_0'(Q_{W+S(0)}ST) \tag{4}$$

and since $Q_{W+S(0)}S$ is continuous by (3) with $\dim Q_{W+S(0)}S(0) \leq \dim S(0) < \infty$, it is closable by Proposition II.5.7 and hence by Proposition IV.5.7,

$$\Gamma_0'(S) = \Gamma_0'(Q_{W+S(0)}S) = \Gamma_0'(\overline{Q_{W+S(0)}S}). \tag{5}$$

Denote $T_1 := T|_{D(ST)}$. Then as $T(0) \subset D(S)$

$$\overline{Q_{W+S(0)}S}T_1 = Q_{W+S(0)}ST. \tag{6}$$

Since $D(\overline{Q_{W+S(0)}S}) = \overline{D(S)} = Y$, it follows from part (a) and (4), (5) and (6) that

$$\begin{aligned}\Gamma_0'(ST) &= \Gamma_0'(Q_{W+S(0)}ST) = \Gamma_0'(\overline{Q_{W+S(0)}S}T_1) \\ &\leq \Gamma_0'(\overline{Q_{W+S(0)}S})\Gamma_0(T_1) = \Gamma_0'(S)\Gamma_0'(T_1).\end{aligned} \tag{7}$$

Clearly, for each $F \in \mathcal{F}(Y)$, $\|Q_F T_1\| \leq \|Q_F T\|$. Therefore $\Gamma_0'(T_1) \leq \Gamma_0'(T)$and the result follows. □

IV.6 NOTES AND REMARKS

Various authors have considered the quantities described in this chapter in the context of bounded linear operators in Banach spaces. Their application to unbounded operators was initiated in R.W. Cross [5]. Our notation (suitably modified for the unbounded case) has been inherited from B. Gramsch [1], M. Schechter [1], A. Lebow and M. Schechter [1], and L. Weis [1,4]. All these quantitites are limits of sequences of s-numbers (see A. Pietsch [2]).

Here is the chronology for bounded linear operators in Banach spaces: B. Gramsch [1] defines Γ (see M. Schechter [1]) (1969), and A.A. Sedaev [1] (1969) treats the analogous quantity Γ_0 which also appears in A. Lebow and M. Schechter [1] (1971). The quantities Δ and τ were introduced by M. Schechter [1] (1972). Next, Γ' amd Δ' appear in the 1974 dissertation of L. Weis [1]. The quantity Γ_0' was introduced by A.S. Fajnshtejn and V.S. Shulman (see A.S. Fajnshtejn [1])(1983), and also J. Zemanek [3] (1984).

A quantity f is called a **measure of noncompactness for (bounded) linear operators** if for every bounded linear operator $T : X \to Y$ (X and Y Banach spaces), T is compact if and only if $f(T) = 0$. As will be seen in Chapter V, the quantities Γ_0 and Γ_0' are measures of noncompactness. In 1965, L.S. Goldenstein and A.S. Markus [1] defined a measure of noncompactness $h(T)$ as the Hausdorff measure of noncompactness of the set $T(B_X)$, namely

$$h(T) := \inf\{\epsilon > 0 : \exists\{y_1, \ldots, y_n\} \subset Y, \forall x \in B_X, \|Tx - y_i\| < \epsilon, 1 \leq i \leq n\}.$$

By interchanging the roles of $h(T)$ and the norm $\|T\|$, new quantities may be derived. Such quantities were introduced in 1983 by V. Rakocevic [1]. The 1985 paper of H.-O. Tylli [1] treats these same quantities and their duals. H.-O. Tylli [2] recently introduced another measure of noncompactness, namely

$$t(T) := \inf\{\epsilon > 0 : \exists Z, \exists K : X \to Z, \forall x \in X, \|Tx\| \leq \|Kx\| + \epsilon \|x\|\}$$

where Z is a Banach space. It is shown by Gonzalez and Martinon [2] that $t = \Gamma_0$. It is known that $h = \Gamma_0'$ (Fajnstejn [1]; see also D.E. Edmunds and H.-O Tylli [1]).

Other authors who have contributed to the study of these and other related quantities include: K. Astala, J. Banas, J. Banas and K. Goebel, J. Banas and J. Rivero, J. Banas and Z. Knap, O.J. Beucher, R.W. Cross and L.E. Labuschagne, R.W. Cross and P. Pillay, W. Damm, J. Danes, F.S. de Blasi, D.E. Edmunds and W.D. Evans, G. Emmanuelle, J. Ewert, V.I. Istratescu, L.E. Labuschagne, L.E. Labuschagne, A. Ströh and J. Swart, A. Martinon, M. Schechter and R.J. Whitley, A. Ströh, A. Ströh and J. Swart, G. West, and others (see Bibliography for further details).

The quantities have been applied to the study of unbounded linear operators in the following papers: T. Alvarez and R.W. Cross [1], T. Alvarez, R.W. Cross and M. Gonzalez [1], R.W. Cross [5, 14, 16], R.W. Cross and L.E. Labuschagne [2], L.E. Labuschagne [1, 4, 6, 8]. Applications to linear relations appear in R.W. Cross [20], and R.W. Cross and P. Pillay [1].

Sections IV.1 and IV.2 are generalisations of corresponding results of Cross [5]. Proposition IV.3.10 is from Labuschagne [5]. Theorem IV.3.17 is a generalisation of S. Goldberg [2] (see Goldberg [3; II.5.1]).

The definition of density invariance first appears in Cross [16]. Our definition of **core** (IV.4.1) coincides for the closed linear operators with Kato's use of the term (Kato [2; p.166]).

Section IV.5 contains results (suitably generalised) from Cross and Labuschagne [2], and Labuschagne [6].

Other works treating the abovementioned and related quantities include the following: J. Alvarez, T. Alvarez and M. Gonzalez [1], Astala [1], Astala and Tylli [1, 2], Banas [1], Banas and Goebel [1], Banas and Rivero [1], Banas and Knap [1], Beucher [1], Danes [1]. De Blasi [1], Edmunds and Evans [1], Edmunds and Tylli [1], Emmanuelle [1], Ewert [1], Fajnshtejn [1], Galaz-Fontes [1], Gonzalez and Martinon [1, 2, 3, 4, 5, 7], Gramsch [1], Istratescu [1], Kinyon [1], Labuschagne, Ströh and Swart [1], Martinon [1], Rakocevic [1], Schechter [1], Schechter and Whitley [1], Sedaev [1], Ströh [1], Ströh and Swart [1], Weis [1, 4], and West [1].

CHAPTER V

Semi–Fredholm Linear Relations

In this chapter we define and investigate the following classes of linear relations: Precompact, compact, strictly singular, strictly cosingular, upper and lower semi–Fredholm, partially continuous, and nowhere continuous linear relations.

V.1 THE MAIN CLASSES: PRELIMINARY RESULTS

V.1.1 Definitions Let $T \in LR(X,Y)$. Then T is called *compact, precompact, strictly singular, upper semi–Fredholm, lower semi–Fredholm, Fredholm, partially continuous (precompact,* or *compact)*, or *nowhere continuous*, respectively, if the linear operator QT has the corresponding property.

In other words:

T is **compact** if and only if $\overline{QTB_X}$ is compact.

T is **precompact** if and only if QTB_X is totally bounded.

T is **strictly singular** if and only if there is no infinite dimensional subspace M of $D(T)$ for which $T|_M$ is injective and open.

T is **upper semi–Fredholm** if and only if there exists a finite codimensional subspace M of X for which $T|_M$ is injective and open. (This is equivalent to the corresponding property for QT, by Corollary II.3.9). T is **lower semi–Fredholm** if and only if T' (or equivalently, $(QT)'$) is upper semi–Fredholm.

T is **partially continuous** (respectively **partially precompact, partially compact**) if there exists a finite codimensional subspace M of X such that $T|_M$ is continuous (respectively precompact, compact).

T is **nowhere continuous** if and only if $T|_M$ is discontinuous whenever $M \in \mathcal{I}(D(T))$.

Note that if $T \in LR(X,Y)$ is precompact and Y is complete, then T is compact.

We shall denote the class of upper semi–Fredholm linear relations by $F_+(X,Y)$, which we abbreviate as F_+. Likewise, $F_-(X,Y)$ (or F_-) will denote the class of lower semi–Fredholm linear relations.

From the definitions we have the equivalences

(1) $T \in F_- \iff T' \in F_+$

(2) $T \in F_+ \iff QT \in F_+$

and it follows that

(3) $T \in F_- \iff QT \in F_-$

From the State Diagram III.6.5 we have the example:
(4) *Y complete, T surjective* $\Longrightarrow T \in F_-$

The relation T is called **Fredholm** if it is both in F_+ and F_-. We write $F(X,Y)$ (or F) for the class of Fredholm relations.

V.1.2 Proposition. *The relation T is upper semi–Fredholm if and only if $T|_M$ is bounded below for some $M \in \mathcal{C}(X)$, i.e.*

$$\|Tm\| \geq c\,\|m\| \qquad (m \in M)$$

where $c > 0$.

PROOF. Suppose the stated inequality holds. Then $T|_M$ is clearly injective and open, i.e. $T \in F_+$.

Conversely, let $T \in F_+$. Then for some $M \in \mathcal{C}(X)$, $T|_M$ is injective and open, i.e.

$$\|Tm\| \geq \gamma(T|_M)d(m, N(T|_M)) = \gamma(T|_M)\,\|m\|$$

where $\gamma(T|_M) > 0$. □

V.1.3 Proposition. *Let T be a continuous linear relation with finite dimensional range. Then T is compact.*

PROOF. Let $T \in LR(X,Y)$. We have

$$QTB_X \subset \|T\|\,B_{R(QT)}$$

where $B_{R(QT)}$ is a compact subset of QY. Therefore T is compact. □

V.1.4 Proposition. *Let T and T_n $(n \in \mathbb{N})$ be everywhere defined and single valued and such that $\lim\limits_{n\to\infty} \|T_n - T\| = 0$, where each T_n $(n \in \mathbb{N})$ is precompact. Then T is precompact.*

PROOF. Let $\epsilon > 0$ be given and choose $N \in \mathbb{N}$ such that

(1) $$\|T - T_N\| < \epsilon/3$$

Since T_N is precompact, $Q_{T_N}T_N B_{D(T)}$ is totally bounded and so there exist points $x_1, x_2, \ldots, x_m$ in $B_{D(T)}$ such that for each $x \in B_{D(T)}$ there exists x_i satisfying

(2) $$\|T_N x - T_N x_i\| < \epsilon/3.$$

Hence from (1) and (2),

$$\|Tx - Tx_i\| \leq \|Tx - T_N x\| + \|T_N x - T_N x_i\| + \|T_N x_i - Tx_i\| < \epsilon$$

showing that $QTB_{D(T)}$ is totally bounded, as required. □

V.1.5 Exercise Show that Proposition V.1.4 holds more generally, for sequences (T_n) such that $\overline{T(0)} \supset \bigcup_{n=1}^{\infty} T_n(0)$. On the other hand the constant sequence, defined by $G(T_n) = X \times X$, satisfies $\|T - T_n\| = 0$ for every $n \in \mathbb{N}$ and every $T \in LR(X,Y)$.

The following theorem is fundamental to the results of this chapter.

V.1.6 Theorem. *Let $T \in LR(X,Y)$. The following properties are equivalent:*

(i) *T is not upper semi–Fredholm.*
(ii) *There is no closed finite codimensional subspace M of X for which $T|_M$ is injective and open.*
(iii) *There exists an infinite dimensional subspace M of $D(T)$ such that $T|_M$ is precompact. (Moreover if T is closed, we may assume M to be a closed subspace of X.)*

PROOF. Since $T(0) = (T|_M)(0)$, it is enough to consider the operator QT. We shall therefore assume, without loss of generality, that T is single valued.

The implication (i) $\Longrightarrow$ (ii) is trivial.

(ii) $\Longrightarrow$ (iii): Suppose that (ii) holds. Then, by Proposition V.1.2 and the Hahn Banach theorem, a pair of sequences $\{x_n\}$ in $D(T)$, and $\{x'_n\}$ in X', can be constructed as follows:

Choose $x_1 \in D(T)$ such that $\|x_1\| = 1$ and $\|Tx_1\| \leq 3^{-1}$. Next choose $x'_1 \in X'$ such that

$$\|x'_1\| = 1 \text{ and } x'_1 x_1 = \|x_1\| = 1.$$

Since $N(x'_1)$ is closed and of codimension 1 in X, there exists $x_2 \in N(x'_1)$ such that

$$\|x_2\| = 1 \text{ and } \|Tx_2\| \leq 3^{-2}.$$

Next choose $x'_2 \in X'$ such that

$$\|x'_2\| = 1 \text{ and } x'_2 x_2 = \|x_2\| = 1.$$

Now $N(x'_1) \cap N(x'_2)$ is closed and of finite codimension in X'. Hence there exists $x_3 \in N(x'_1) \cap N(x'_2)$ such that

$$\|x_3\| = 1 \text{ and } \|Tx_3\| \leq 3^{-3}.$$

Continuing in this way, we obtain

(1) $$\|x_k\| = \|x'_k\| = x'_k x_k = 1, \ \|Tx_k\| \leq 3^{-k} \quad (1 \leq k < \infty)$$

(2) $$x_k \in \bigcap_{i=1}^{k-1} N(x'_i), \text{ i.e. } x'_i x_k = 0 \quad (1 \leq i < k).$$

The set $\{x_k : 1 \leq k < \infty\}$ is clearly linearly independent; let M denote the linear span. Select $x = \sum_{i=1}^{m} \alpha_i x_i$ arbitrarily. Then from (1) and (2)

$$|\alpha_1| = |x'_1 x| \leq \|x'_1\| \, \|x\| = \|x\| .$$

It can be verified as in the proof of (3) of Theorem IV.4.8, that

(3) $$|\alpha_k| \leq 2^{k-1} \|x\| \quad (1 \leq k \leq m).$$

Let P_n denote the linear projection of M onto the linear span of $\{x_1, \ldots, x_n\}$ with kernel the linear span of $\{x_{n+1}, x_{n+2}, \ldots\}$. We shall show that $T|_M$ is the uniform limit of the sequence of operators $\{TP_n\}$. For $x = \sum_{i=1}^{m} \alpha_i(x) x_i \in B_M$ we have

$$\|Tx - TP_n x\| = \left\| \sum_{i=n+1}^{\infty} \alpha_i T x_i \right\| \leq \sum_{i=n+1}^{\infty} |\alpha_i(x)| \, \|Tx_i\|$$

$$\leq \sum_{i=n+1}^{\infty} 2^{i-1} 3^{-i} \|x\| \leq \frac{1}{2} \|x\| \sum_{i=n+1}^{\infty} (2/3)^i \to 0 \text{ uniformly as } n \to \infty$$

as required.

Now TP_n is continuous since

$$\|TP_n x\| = \left\| \sum_{i=1}^{n} \alpha_i T x_i \right\| \leq \frac{1}{2} \|x\| \sum_{i=1}^{\infty} (2/3)^i = \|x\| .$$

Hence, since the range of TP_n is finite dimensional, TP_n is compact by Proposition V.1.3.

Therefore $T|_M$ is precompact, by Proposition V.1.4.

(iii) $\Longrightarrow$ (i): Suppose that (iii) holds. Let E be a finite codimensional subspace of X such that $T|_E \in 1$. The subspace $E \cap M$ is infinite dimensional by Proposition IV.1.2. Since $T|_E$ is injective and open, so is $T|_{E \cap M}$. Hence there exists $\lambda > 0$ such that

$$\lambda B_{T(E \cap M)} \subset T B_{E \cap M}$$

where the right hand side is totally bounded, since $T|_M$ is precompact. Since the unit ball of a normed space is totally bounded if and only if the space is finite dimensional, we conclude that $E \cap M$ is finite dimensional, a contradiction. Therefore (iii) $\Longrightarrow$ (i). $\square$

V.1.7 Corollary. *Let X and Y be Banach spaces and let $T \in LR(X,Y)$ be closed. Then the following properties are equivalent:*

(i) $T \in F_+$

(ii) *T has closed range and finite dimensional kernel.*

PROOF. Since T is closed, we have

$$T(0) \text{ is closed and } QT \text{ is closed} \qquad \text{(by II.5.3)}$$
$$N(T) = N(QT) \qquad \text{(by II.3.4)}$$

and it is easy to see that

$$R(T) \text{ complete} \iff R(QT) \text{ complete}.$$

We may therefore suppose that T is single valued.

Assume (i). By theorem V.1.6 there exists subspaces M and F of $D(T)$, with M closed and F finite dimensional such that

$$D(T) = M + F + N(T)$$

where $M \cap F = F \cap N(T) = M \cap N(T) = \{0\}$ and moreover, such that $(T|_M)^{-1}$ is continuous. Let $Tm_n \to y$ ($m_n \in M$). Then $\{m_n\}$ is Cauchy, so $m_n \to x$ for some $x \in X$ (since X is complete). Thus (m_n, Tm_n) converges to $(x,y) \in X \times Y$. Since T is closed, we have $x \in D(T)$ and $y \in Tx$. Therefore TM is closed, and so $R(T) = T(M) + T(F)$ is closed, since $T(F)$ is finite dimensional (see e.g. Goldberg [3; p.16]).

Conversely, assume (ii). Let M be any closed subspace such that

$$D(T) = M + N(T), \ M \cap N(T) = \{0\}.$$

By hypothesis, $T(M) = R(T)$ is closed. Now $(T|_M)^{-1}$ is a closed linear operator mapping the Banach space $T(M)$ onto the Banach space M. Hence $(T|_M)^{-1}$ is continuous by the Closed Graph theorem. Therefore $T \in F_+$. □

V.1.8 Definition A linear relation T is called a ϕ_+**–relation** if it has finite dimensional kernel and closed range, and a ϕ_-**–relation** if its range is closed and finite codimensional.

Corollary V.1.7 asserts that for closed linear relations between Banach spaces

(1) $$T \in F_+ \iff T \in \phi_+.$$

A sharpened form of the equivalence (1) will be derived later (see Corollary V.7.6).

From the Closed Range theorem III.4.4 and Proposition III.1.4(a) we obtain the following:

V.1.9 Proposition. $T \in F_- \iff \tilde{T} \in \phi_-$.

V.1.10 Proposition. *Let $T \in F_+$. Then any bounded sequence (x_n) in $D(T)$ such that Tx_n is Cauchy has a Cauchy subsequence.*

PROOF. Replacing T by QT if necessary, we may suppose that T is single valued. There exists by Theorem V.1.6 and Proposition II.3.18, a bounded projection P defined on $D(T)$ with $\dim R(I-P) < \infty$ and $T|_{R(P)}$ injective and open. Let (x_n) be a sequence in $D(T)$ satisfying $\|T(x_n - x_m)\| = f(n,m)$ where $f(n,m) \to 0$ $(n,m \to \infty)$. Since $T|_{R(I-P)}$ is continuous (II.3.2), the sequence $T(I-P)x_n$ is bounded in the finite dimensional space $R(T(I-P))$. Select a subsequence $\{x_{k_n}\}$ of $\{x_n\}$ for which $T(I-P)x_{k_n}$ is Cauchy. Then

$$\|TP(x_{k_n} - x_{k_m})\| \leq \|T(x_{k_n} - x_{k_m})\| + \|T(I-P)(x_{k_n} - x_{k_m})\| \to 0 \quad (n,m \to \infty)$$

(see (I.4.2(e)). Hence (TPx_{k_n}) is Cauchy. We have

$$\|TP(x_{k_n} - x_{k_m})\| \geq \gamma(T|_{R(P)})\, \|P(x_{k_n} - x_{k_m})\|$$

since $T|_{R(P)}$ is injective. Therefore (Px_{k_n}) is Cauchy. Choose a subsequence $\{w_n\}$ of $\{x_{k_n}\}$ for which $(I-P)w_n$ is Cauchy. Then $w_n = Pw_n + (I-P)w_n$ is Cauchy. □

V.1.11 Theorem. *The following properties are equivalent:*

(i) $T \notin F_+$.

(ii) *There exists a non precompact bounded subset W of $D(T)$ such that $QT(W)$ is precompact.*

(iii) *T has a singular sequence, i.e. a sequence $\{x_n\}$ of norm one elements of $D(T)$ such that (x_n) has no Cauchy subsequence and* $\lim\limits_{n\to\infty} \|Tx_n\| = 0$.

PROOF. The implication (iii) $\implies$ (ii) is trivial, while (ii) $\implies$ (i) is clear from Proposition V.1.10.

(i) $\implies$ (iii): From (2) of V.1.1, we may suppose T single valued. Assume $T \notin F_+$. Consider the case when T is bounded. In that case, $\tilde{T}$ is bounded and single valued, and by (1) of V.1.8, $\tilde{T} \notin \phi_+$. We first show the existence of a singular sequence for $\tilde{T}$. By Theorem V.1.6, there exists $M \in \mathcal{I}(\tilde{X})$ for which $\tilde{T}|_M$ is compact. Moreover M may be supposed closed (as $\tilde{T}$ is continuous). If $M \cap N(\tilde{T})$ is infinite dimensional, then clearly a singular sequence exists for $\tilde{T}$. Otherwise, there exists a closed infinite dimensional subspace N of M with $N \cap N(\tilde{T}) = \{0\}$. Let $\{x_n\}$ be a norm one sequence in N such that $\left\|\tilde{T}x_n\right\| \to 0$ (see V.1.2). Then $\{x_n\}$ clearly has no Cauchy (hence convergent) subsequence.

Passing to T, select $\{z_n\} \subset D(T)$ so that $\|z_n\| = 1$ and $\|z_n - x_n\| < \frac{1}{n}$ for every n and observe that in this case $\{x_n\}$ is a singular sequence for $\tilde{T}$ if and only if $\{z_n\}$ is a singular sequence for T. Thus (i) $\implies$ (iii) if T is bounded.

Finally observe that by Proposition IV.3.10, $T \notin F_+$ if and only if $TG \notin F_+$ and so TG has a singular sequence $\{z_n\}$ by what we have just verified. Since now $TGz_n \to 0$ it is clear that since $\{z_n\}$ has no Cauchy subsequences (with respect to $\|\cdot\|_T$), the same is true for $\{Gz_n\}$, with in addition $\|Gz_n\| \to 1$. Since then $(\|Gz_n\|^{-1})TGz_n \to 0$, it follows that $\{\|Gz_n\|^{-1} Gz_n\}$ is a singular sequence for T. □

V.1.12 Corollary. *Let P be a linear projection. Then P has an infinite dimensional injective precompact restriction if and only if both the range and kernel of P are infinite dimensional.*

PROOF. Let M be an infinite dimensional subspace of $D(P)$ for which $P|_M$ is both injective and precompact. If $\dim N(P) < \infty$ then $R(P)$ has finite codimension in $D(P)$ and hence $M \cap R(P)$ is an infinite dimensional subspace (IV.1.2) upon which P is both an isomorphism and precompact, which is impossible. Therefore $\dim N(P) = \infty$. On the other hand, if $\dim R(P) < \infty$ then P does not have an infinite dimensional injective restriction, a contradiction. Hence necessity follows.

Conversely, suppose that both $N(P)$ and $R(P)$ are infinite dimensional. Choose a linearly independent sequence (a_n) in $R(P)$ such that $\|a_n\| \to 0$. Next construct a sequence (b_n) in $N(P)$ such that $\|b_n\| = 1$ and $\|b_n - b_k\| \geq 1$ $(n \neq k)$ (for example, by the Gramm–Schmidt construction). Write $x_n = a_n + b_n$ and $E = sp\{x_n\}$. Then $P|_E$ is injective; indeed if $e = \Sigma\alpha_i x_i$, and $Pe = 0$, then $\Sigma\alpha_i Px_i = \Sigma\alpha_i a_i = 0$, which implies $e = 0$ by the linear independence of $\{a_i\}$. Furthermore $Px_n = a_n \to 0$ and (x_n) has no Cauchy subsequence since $\|x_n - x_k\| \geq \|b_n - b_k\| - \|a_n - a_k\| > 1/2$ eventually. Therefore $P|_E$ is not an F_+-operator, by Theorem V.1.11 and hence has an infinite dimensional precompact restriction by Theorem V.1.6. □

Using Corollary V.1.12 we can easily construct a counterexample for Corollary II.3.13, namely a pair of continuous linear relations S, T whose composition ST is not continuous.

V.1.13 Example *There exists a pair of continuous and injective linear relations S and T, with S single valued, such that ST is discontinuous.* Let P be a bounded linear projection defined on a Banach space X, having infinite dimensional kernel and range. According to Corollary V.1.12, there exists an infinite dimensional subspace E of X such that $E \cap N(P) = \{0\}$ and $P|_E$ is a compact operator. Put $S = (J_E^X)^{-1}$ and $T = P^{-1}$. Then S and T are injective and continuous, with S single valued. We have $\|T\| = \|P^{-1}\| = \gamma(P)^{-1} < \infty$ since

P is an open map (see II.3.2, II.3.3). Writing $J := J_E^X$, we clearly have $\|S\| = \|J^{-1}\| = 1$. However

$$\|ST\| = \|J^{-1}P^{-1}\| = \|(PJ)^{-1}\| = \gamma(PJ)^{-1} = \gamma(P|_E)^{-1} = \infty$$

since $P|_E$ is compact and hence not open. □

V.1.14 Exercise Let T be closable and nowhere continuous. Then $\overline{T}$ is nowhere continuous; in particular, $N(\overline{T})$ is finite dimensional.

V.1.15 Exercise Suppose that a scalar λ can be found such that $(\lambda I - T)^{-1}$ is a strictly singular operator. Show that T is nowhere continuous.

V.1.16 Exercise Let $T \in LR(\ell_1, \ell_2)$ be everywhere defined and surjective. Then $\tilde{T}$ is a continuous and strictly singular relation from ℓ_1 onto ℓ_2. Moreover $\tilde{T}$ is compact if and only if $\ell_2/\tilde{T}(0)$ is finite dimensional.

V.1.17 Exercise Let $X = C[0,1] = Y$ and let T be defined by $G(T) := \{(f,g) : g$ is an antiderivative of $f\}$. Show that T is compact, $T(0)$ is one dimensional, and T^{-1} is a closed densely defined upper semi–Fredholm operator.

V.1.18 Exercise Let T be defined by $G(T) := G(I_X) + G(O_X)$. Then $T(0) = X$ and thus $QT = O_X$. Here T is precompact, yet T^{-1} is continuous since by Proposition II.3.4 $\gamma(T) = \gamma(QT) = \gamma(O_X) = \infty$.

V.1.19 Proposition. *Let T be open and let M be a closed finite codimensional subspace of $D(T)$ for which $(T|_M)^{-1}$ is precompact. Then T^{-1} is precompact.*

PROOF. Let $\{y_n\}$ be a sequence in $B_{R(T)}$. Since T is open, $cB_{R(T)} \subset T(B_{D(T)})$ for some $c > 0$, and so there exists a bounded sequence $\{x_n\}$ in $D(T)$ such that

$$y_n \in Tx_n \qquad (n \in \mathbb{N}).$$

Let P be a bounded projection defined on $D(T)$ with range M. Then

$$TPx = (T|_M)Px$$

and

$$(1) \qquad (T_M^{-1})TPx = Px + N(T|_M) \qquad (x \in D(T)).$$

Since $I-P$ is a bounded finite rank projection, and hence $T|_{R(I-P)}$ is continuous by Proposition II.3.2, $\{x_n\}$ has a subsequence, assumed for simplicity to be itself, such that $\{T(I-P)x_n\}$ is norm Cauchy, i.e.

$$(2) \qquad \|T(I-P)(x_n - x_k)\| \to 0 \text{ as } n, k \to \infty.$$

In particular, $\{\|T(I-P)x_n\|\}$ is bounded, as so since

(3) $$d(TPx_n, 0) \le d(y_n, T(0)) + d(T(I-P)x_n, 0),$$

$\{\|TPx_n\|\}$ is a bounded sequence. Hence by the precompactness of $(T|_M)^{-1}$, there exists a subseqence of $\{x_n\}$, for simplicity assumed to be itself, such that $(T|_M)^{-1}(TPx_n)$ is Cauchy modulo $N(T|_M)$, i.e.

(4) $$\|(T_M^{-1})(TP(x_n - x_k)\| = d(P(x_n - x_k), N(T|_M)) \to 0 \qquad (n, k \to \infty).$$

Hence from (1), (2) and (4),

$$\|T^{-1}T(x_n - x_k)\| \le \|T^{-1}TP(x_n - x_k)\| + \|T^{-1}T(I-P)(x_n - x_k)\| \to 0\ (n, k \to \infty).$$

But for $z \in T^{-1}Tx$, we have from II.1.4(a),

$$d(z, T^{-1}(0)) = d(x, T^{-1}(0)).$$

Hence

$$\|T^{-1}(y_n - y_k)\| = \|T^{-1}T(x_n - x_k)\| \to 0 \qquad (n, k \to \infty).$$

Therefore T^{-1} is precompact. □

V.1.20 Proposition. *For any linear subspace $M \subset D(T)$, we have*

$$G_T \textit{ precompact} \implies G_{T|_M} \textit{ precompact.}$$

PROOF. Let G_T be precompact, and let $\{m_k\}$ be a bounded sequence in $X_{T|_M}$, i.e.

(1) $$\|m_k\| + \|(T|_M)m_k\| < c < \infty$$

for some $c > 0$ and all $k \in \mathbb{N}$. We have

$$G_{T|_M}(m_k) = m_k = G_T(m_k)$$

and

$$(T|_M)m_k = Tm_k \qquad (k \in \mathbb{N}).$$

Hence since G_T is compact, there is a subsequence $\{m_{i_k}\}$ of $\{m_k\}$ which is Cauchy in $M \subset X$ (i.e., M considered as a normed subspace of X). Therefore $G_{T|_M}$ is precompact. □

V.1.21 Exercise Show that if X is complete and T closed, then

$$T \in F_+ \implies T \in \phi_+.$$

V.1.22 Exercise Let T be completely closed. Verify the equivalence

$$T \in F_+ \iff \alpha(T) < \infty \text{ and } \gamma(T') > 0.$$

V.2 ZERO SETS OF THE QUANTITIES

Various classes of linear relations will now be characterized as the zero sets of certain quantities. These classes are the precompact, partially precompact, strictly singular, and non upper semi-Fredholm linear relations.

V.2.1 Lemma. *Let B be a bounded subset of X, let $\{x'_1, x'_2, \ldots, x'_n\}$ be a finite subset of X', and let $\epsilon > 0$ be given. Then there exists a finite subset $\{x_1, x_2, \ldots, x_m\}$ of B such that for any $x \in B$ there exists an x_k satisfying*

$$|x'_i x - x'_i x_k| \leq \epsilon \qquad (1 \leq i \leq n).$$

PROOF. The map K from X into $\mathbb{K}^n$ defined by

$$Kx = (x'_1 x, x'_2 x, \ldots, x'_n x)$$

is bounded and linear, hence compact by Proposition V.1.3. Consequently there exist $x_1, x_2, \ldots, x_m$ in B such that given $x \in B$ there exists x_k for which

$$\epsilon \geq \|Kx - Kx_k\| \geq |x'_i x - x'_i x_k| \qquad (1 \leq i \leq n). \quad \square$$

V.2.2 Theorem.

(a) *T is precompact if and only if $\overline{\Gamma}_0(T) = 0$.*

(b) *T is partially precompact if and only if $\Gamma_0(T) = 0$.*

PROOF. We may clearly suppose that T is single valued.

(a) Let T be precompact. Given $\epsilon > 0$, there exist $x_1, x_2, \ldots, x_n$ in $B_{D(T)}$ such that given $x \in B_{D(T)}$ there is an x_k for which

$$\|Tx - Tx_k\| < \epsilon/2. \tag{1}$$

Choose vectors $y'_1, y'_2, \ldots, y'_n$ in $B_{Y'}$ so that $y'_i T x_i = \|Tx_i\|$. The closed subspace $N = \bigcap_{i=1}^{n} N(y'_i T)$ has finite codimension in X (see Proposition IV.1.2(b)). Suppose that in (1), x is a element of $N \cap B_{D(T)}$. Then since $x \in N(y'_k T)$,

$$\|Tx_k\| = y'_k T x_k = y'_k(Tx_k - Tx) \leq \|Tx_k - Tx\| < \epsilon/2. \tag{2}$$

Hence from (1) and (2),

$$\|Tx\| \leq \|Tx - Tx_k\| + \|Tx_k\| < \frac{\epsilon}{2} + \frac{\epsilon}{2}. \tag{3}$$

Taking the supremum in (3) over $x \in B_{N \cap D(T)}$ gives $\|T|_N\| < \epsilon$. Hence

$$\overline{\Gamma}_0(T) = \inf_{N \in \mathcal{P}(X)} \|T|_N\| = 0.$$

Conversely, let $\overline{\Gamma}_0(T) = 0$. Let $\epsilon > 0$. Choose $M \in P(X)$ so that $\|T|_M\| < \epsilon$. There exist $v_1, v_2, \ldots, v_n$ in X such that

$$X = M + sp\{v_1, v_2, \ldots, v_n\}, \quad M \cap sp\{v_1, v_2, \ldots, v_n\} = \{0\}.$$

By the Hahn Banach Theorem there exists a set $\{x'_1, x'_2, \ldots, x'_n\}$ of linear functionals in $M^\perp$ satisfying

$$x'_i v_j = \delta_{ij} \qquad (1 \le i, j \le n).$$

Then each $x \in X$ has a unique representation of the form

$$x = u + \sum_{i=1}^{n} x'_i(x) v_i \tag{4}$$

where $u \in M$.

Since $\|T|_M\| < \epsilon$, we have

$$\|Tx\| \le \epsilon \|u\| + \sum_{i=1}^{n} |x'_i(x)| \, \|Tv_i\| \, . \tag{5}$$

By (4),

$$\|u\| \le \|x\| + \sum_{i=1}^{n} |x'_i(x)| \, \|v_i\| \, . \tag{6}$$

From (5) and (6), putting $c = \max\{\epsilon \|v_i\| + \|Tv_i\| : \ 1 \le i \le n\}$,

$$\|Tx\| \le \epsilon \|x\| + c \sum_{i=1}^{n} |x'_i(x)| \, . \tag{7}$$

According to Lemma V.2.1, there exist $x_1, x_2, \ldots, x_m$ in B_X such that for any $v \in B_X$,

$$\sum_{i=1}^{n} |x'_i v - x'_i x_k| < \epsilon.$$

Substituting $v - x_k$ for x in (7), and noting that $\|v - x_k\| \le 2$, we obtain

$$\|Tv - Tx_k\| \le 2\epsilon + c\epsilon.$$

Since $\epsilon > 0$ is arbitrary, it follows that TB_X is totally bounded.

(b) Let T be partially precompact. Then there exists $M \in \mathcal{C}(X)$ such that $T|_M$ is precompact. Hence by part (a), and Proposition IV.1.3,

$$0 = \overline{\Gamma}_0(T|_M) \ge \Gamma_0(T|_M) = \Gamma_0(T).$$

Conversely, suppose that $\Gamma_0(T) = 0$. Then there exists $M \in \mathcal{C}(X)$ such that $\|T|_M\| < \infty$. Since $T|_M$ is continuous, $\Gamma_0(T|_M) = \overline{\Gamma}_0(T|_M)$. But $\Gamma_0(T|_M) = \Gamma_0(T)$ by Proposition IV.1.3. Hence $T|_M$ is precompact by part (a). $\square$

V.2.3 Corollary. *Precompact linear relations are continuous.*

PROOF. Combine Theorem V.2.2(a) with Corollary IV.2.14(b). □

V.2.4 Theorem. *Let $T \in LR(X,Y)$ and $\dim D(T) = \infty$. Then the following properties are equivalent:*

(i) $T \in F_+$
(ii) $\Gamma(T) > 0$
(iii) $\tau_0(T) > 0$
(iv) $\overline{\tau}_0(T) > 0$.

PROOF. We suppose without loss of generality that T is single valued.

By Theorem V.1.6, if $T \in F_+$ then there exists $M \in \mathcal{P}(X)$ for which $\inf_{m \in S_M} \|Tm\| > 0$, and therefore $\overline{\tau}_0(T) > 0$. Hence (i) $\Longrightarrow$ (iv).

The implications (iv) $\Longrightarrow$ (iii) $\Longrightarrow$ (ii) are trivial. It only remains to verify (ii) $\Longrightarrow$ (i).

Let $\Gamma(T) > 0$. Then for $M \in \mathcal{I}(D(T))$ we have $\overline{\Gamma}_0(T|_M) \geq \Gamma(T|_M) \geq \Gamma(T) > 0$. Therefore $T|_M$ is not precompact, by Theorem V.2.2. Hence $T \in F_+$, by Theorem V.1.6 as required. □

V.2.5 Corollary. *$T \in F_+$ if and only if $TG \in F_+$.*

PROOF. We may clearly assume $\dim D(T) = \infty$. We have

$$\begin{aligned} T \in F_+ &\iff \Gamma(T) > 0 && \text{(Theorem V.2.4)} \\ &\iff \Gamma(TG) > 0 && \text{(Proposition IV.3.4)} \\ &\iff TG \in F_+ && \text{(Theorem V.2.4)} \quad \square \end{aligned}$$

V.2.6 Theorem. *T is strictly singular if and only if $\Delta(T) = 0$, if and only if $\tau(T) = 0$.*

PROOF. Assume without loss of generality that T is single valued, and that $\dim D(T) = \infty$.

Suppose that $\Delta(T) = 0$ and let $M \in \mathcal{I}(D(T))$. Then for each $\epsilon > 0$ there exists $N \in \mathcal{I}(M)$ with $\|Tn\| \leq \epsilon \|n\|$ $(n \in N)$. Consequently $T|_M$ does not have a continuous inverse (see Proposition V.1.2). Hence T is strictly singular.

Conversely, let T be strictly singular. Let G denote the graph operator G_T. We claim that TG is strictly singular. For, if TG is not strictly singular, then for some $M \in \mathcal{I}(D(T))$ we have $\|TGm\| = \|Tm\| \geq k(\|m\| + \|Tm\|)$ for all $m \in M$, where $0 < k < 1$. Hence $\|Tm\| \geq k\|m\|/(1-k)$ $(m \in M)$, which implies that $T|_M$ has a continuous inverse, contrary to the assumption.

Now let $\epsilon > 0$ and $M \in \mathcal{I}(D(T))$ be given. Since $TG|_{G^{-1}M}$ is strictly singular, there exists by Theorems V.2.4 and V.2.5, an $N \in \mathcal{I}(M)$ for which $\|TG|_{G^{-1}N}\| < \epsilon/2$. Hence $\|Tn\| = \left\|TGG^{-1}n\right\| \leq (\epsilon/2)(\|n\| + \|Tn\|)$ $(n \in N)$ and so (if $\epsilon < 1$)

$$\|Tn\| < (\epsilon/2)/(1-\epsilon/2)\|n\| < \epsilon\|n\| \qquad (n \in N).$$

Hence $\Gamma(T|_M) < \epsilon$. Consequently $\Delta(T) \leq \epsilon$. Since ϵ is arbitrary in $(0,1)$, it follows that $\Delta(T) = 0$. The second characterisation $\tau(T) = 0$ is a simple consequence of the definitions. □

V.2.7 Corollary. *The sum of two strictly singular operators is strictly singular.*

PROOF. Combine the triangle inequality $\Delta(T+S) \leq \Delta(T)+\Delta(S)$ (Corollary IV.2.6) with Theorem V.2.6. □

V.2.8 Corollary. *Let T be a linear relation.*

(a) *If T is partially precompact, then T is strictly singular.*
(b) *If QT has finite dimensional range, then T is strictly singular.*

PROOF. (a) Follows immediately from V.2.6 and V.2.2.

(b) Let $\dim R(QT) < \infty$. Then $N(QT)$ has finite codimension in $D(T)$. Hence we have

$$\Delta(T) \leq \Gamma_0(T) = 0$$

and therefore T is strictly singular by V.2.6. □

V.2.9 Lemma. *Let $S \in LR(Y,Z)$ be continuous and let $D(S) \supset \overline{T(0)}$. Then*

(1) $$Q_{ST}SQ^{-1}QT = Q_{ST}ST.$$

PROOF. We have for $(x,y) \in G(T)$,

(2) $$SQ^{-1}Qy = Sy + S\overline{T(0)} \qquad \text{(I.3.1)}$$

Since S is continuous and $D(S) \supset \overline{T(0)}$, we have by II.3.1(5),

$$(3) \qquad S\overline{T(0)} \subset \overline{ST(0)}.$$

Hence from (2) and (3),

$$SQ^{-1}Qy \subset Sy + \overline{ST(0)}$$

whence

$$Q_{ST}SQ^{-1}Qy \subset Q_{ST}Sy \subset Q_{ST}(SQ^{-1}Qy).$$

Therefore

$$Q_{ST}SQ^{-1}Qy = Q_{ST}Sy.$$

Hence

$$Q_{ST}SQ^{-1}QTx = Q_{ST}STx$$

from which (1) follows. □

V.2.10 Proposition. *Let $T \in LR(X,Y)$ be partially precompact (or precompact, or strictly singular) and let $S \in LR(Y,Z)$ be continuous. If $\overline{T(0)} \subset D(S)$, then ST is partially precompact (respectively precompact, or strictly singular).*

PROOF. If T is single valued then the result follows from IV.2.15 (or IV.2.9) combined with V.2.2 (respectively V.2.6).

Turning to the general case, if T is partially precompact then so is the single valued relation QT. Also SQ^{-1} is continuous by II.3.13, since $Q^{-1}(0) = \overline{T(0)} \subset D(S)$. Hence, from what has been proved for the single valued case, $SQ^{-1}QT$ is partially precompact. Hence by Lemma V.2.9, $Q_{ST}ST$ is partially precompact. Therefore ST is partially precompact. The other two properties are proved similarly. □

V.2.11 Corollary. *Let T be partially precompact (or precompact, or strictly singular) and let $S \in LR(Y,Z)$ be bounded. Then ST is partially precompact (respectively precompact, or strictly singular).*

A similar application of the inequalities of IV.2.15 and IV.2.9 yield immediately the next result.

V.2.12 Proposition. *Let $T \in LR(X,Y)$ be single valued and continuous and let $S \in LR(Y,Z)$ be partially precompact (or precompact, or strictly singular). Then ST is partially precompact (respectively precompact, or strictly singular).*

V.2.13 Exercise Let E be a core subspace of T such that $T|_E$ is strictly singular. Then T is strictly singular. In particular, if T is a continuous and strictly singular operator, then its closure is strictly singular.

V.2.14 Exercise Let E be a core subspace of T. Show that $T|_E \in F_-$ if and only if $T \in F_-$.

V.2.15 Exercise Give an example of a compact relation T and a subspace $E \subset Y$ such that $I_E T$ is discontinuous (in particular, not compact).

Concerning products involving F_+–relations we have:

V.2.16 Proposition. *Let $S \in LR(Y,Z)$.*

(a) *If T is single valued, then*

$$S, T \in F_+ \implies ST \in F_+.$$

(b) *If S is bounded, then*

$$ST \in F_+ \implies T \in F_+.$$

PROOF. (a) Let $S, T \in F_+$. We may clearly assume that $D(ST)$ is infinite dimensional. Then $D(T)$ is infinite dimensional, and so also is $D(S)$ (as $T^{-1}(0)$ is finite dimensional). Hence by Theorem V.2.4, $\Gamma(T) > 0$ and $\tau_0(S) > 0$. Now by IV.2.16,

$$\Gamma(ST) \geq \tau_0(S)\Gamma(T) > 0.$$

Hence $ST \in F_+$, by V.2.4.

(b) Let $ST \in F_+$. Then by Proposition V.1.2 there exists $M \in \mathcal{C}(D(ST))$ and $c > 0$, such that

$$\|STm\| \geq c\,\|m\| \qquad (m \in M).$$

Since $D(S) = Y$, we have $D(ST) = D(T)$, and furthermore by (3) of II.3.14,

$$\|S\|\,\|T|_M m\| \geq \|STm\| \qquad (m \in M).$$

Since $\|S\| < \infty$, it follows that $T|_M$ is bounded below on M. Hence $T \in F_+$ by V.1.2. □

V.2.17 Exercise The relation T is said to *respect Cauchy sequences* whenever for each bounded sequence $\{x_n\}$ in $D(T)$, if $\{QTx_n\}$ is Cauchy, then $\{x_n\}$ has a Cauchy subsequence. Show that $T \in F_+$ if and only if T respects Cauchy sequences.

V.3 PERTURBATION THEOREMS FOR F_+ RELATIONS

We are interested in the following question: Let $S, T \in LR(X, Y)$ and suppose that $T \in F_+$. Under what conditions do we have $T + S \in F_+$? It is well-known, for example, that if T is a closed upper semi-Fredholm operator between Banach spaces, then $T + S$ has the same property whenever S is either bounded and strictly singular, or bounded with sufficiently small norm. Theorem V.3.2 contains both these perturbation results in more general form and moreover provides estimates for the stability radius. The next example shows why some restriction on $S(0)$ is necessary.

V.3.1 Example Let X be an infinite dimensional normed space and let $S \in LR(X)$ be defined by $G(S) = X \times X$. Then S is compact, and $\|S\| = 0$. Now the identity operator I_X is Fredholm, while $I_X + S$ is not even upper semi-Fredholm; indeed $I_X + S = S$.

Perturbations with $S(0)$ finite dimensional are considered in Section V.7.

V.3.2 Theorem. *Let $S, T \in LR(X, Y)$ and let $S(0) \subset \overline{T(0)}$ (for example, if S is single valued).*

(a) $\Delta(S) < \Gamma(T) \implies T + S \in F_+$
(b) $\tau(S) < \tau_0(T) \implies T + S \in F_+$.

PROOF. Let $T+S \notin F_+$. Then $\dim D(T+S) = \infty$ and according to Theorem V.2.4, $\Gamma(T+S) = 0$. Hence given $\epsilon > 0$ there exists $M \in \mathcal{I}(D(T+S))$ such that

$$\|T|_M + S|_M\| < \epsilon. \tag{1}$$

(a) Let $\epsilon < \Gamma(T) - \Delta(S)$. Since $S(0) \subset \overline{T(0)}$ it follows that $\overline{(T+S)(0)} = \overline{T(0)}$, and hence that

$$\|Tx\| = \|Tx + Sx - Sx\| \leq \|Tx + Sx\| + \|Sx\|.$$

Hence from (1)

$$\|Sm\| > \|Tm\| - \epsilon \qquad (m \in B_M).$$

Hence

$$\|S|_N\| \geq \|T|_N\| - \epsilon \text{ for } N \in \mathcal{I}(M).$$

Consequently

$$\Delta(S) \geq \Gamma(S|_M) \geq \Gamma(T|_M) - \epsilon \geq \Gamma(T) - \epsilon.$$

Therefore $\Gamma(T) - \Delta(S) \leq \epsilon$. This contradiction establishes the implication (a).

(b) Let V be any finite codimensional subspace of X, and let $W = M \cap V$. Then $\dim W = \infty$ (IV.1.2(a)), and for $x \in S_W$ we have (since $S(0) \subset \overline{T(0)}$)

$$\|Sx\| > \|Tx\| - \epsilon.$$

Hence

$$\tau(S) \geq \tau(S|_M) \geq \inf_{x \in S_W} \|Tx\| - \epsilon \geq \inf_{x \in S_V} \|Tx\| - \epsilon$$

whence $\tau(S) \geq \tau_0(T) - \epsilon$. Therefore $S + T \notin F_+$ implies $\tau(S) \geq \tau_0(T)$ and the result follows. □

V.4 UBIQUITOUS PROPERTIES

The classes where $f(T) = \infty$, for $f \in \{\tau_0, \tau, \Gamma, \Delta, \Gamma_0\}$ will be characterised in this section as follows:

$\left.\begin{matrix}\overline{\tau}_0(T) \\ \tau_0(T)\end{matrix}\right\} = \infty :$ Linear relations with a nontrivial finite codimensional restriction whose inverse is precompact and single valued (Theorems V.4.6 and V.4.17)

$\Gamma(T) = \infty :$ The nowhere continuous linear relations (by definition)

$\left.\begin{matrix}\tau(T) = \infty \\ \Delta(T) = \infty \\ \Gamma_0(T) = \infty\end{matrix}\right\} :$ The non partially continuous linear relations (Theorem V.4.3)

We also have from Corollary IV.2.14:

$\overline{\Gamma}_0(T) = \infty :$ The discontinuous linear relations.

V.4.1 Definition Let $\mathbb{P}$ be a property of linear relations (e.g. continuity). Then $T \in LR(X, Y)$ is said to be **ubiquitously** $\mathbb{P}$ if each infinite dimensional subspace M of $D(T)$ contains another such subspace N for which $T|_N$ has the property $\mathbb{P}$.

V.4.2 Theorem. *The following properties are equivalent:*

(i) T *is ubiquitously continuous*
(ii) $G_T \in F_+$
(iii) T *is partially continuous.*

PROOF. Assume that T is ubiquitously continuous. Write $G := G_T$. Suppose $G \notin F_+$. Then by Theorem V.1.6, there exists $M \in \mathcal{I}(X_T)$ such that $G|_M$ is precompact. By assumption, there exists $N \in \mathcal{I}(GM)$ for which $T|_N$ is continuous. We now have for $n \in G^{-1}(N)$

$$\left\|G^{-1}n\right\| = \|n\| + \|Tn\| \leq (1 + \|T|_N\|)\,\|n\|$$

showing that $G^{-1}|_N$ is continuous. But then $G|_{G^{-1}N}$ is both precompact and an isomorphism, and this implies that $G^{-1}N$ is finite dimensional, contradicting $N \in \mathcal{I}(GM)$. Therefore $G \in F_+$. Choose $E \in \mathcal{C}(X_T)$ so that $(G|_E)^{-1}$ is continuous. Then for $e \in E$ we have

$$\|e\| = \left\|G^{-1}Ge\right\| \leq \left\|G^{-1}|_E\right\| \|Ge\|.$$

Thus

$$\|Ge\| + \|TGe\| \leq \left\|G^{-1}|_E\right\| \|Ge\|.$$

Put $x = Ge \in GE$; then

$$\|Tx\| \leq \left(\left\|G^{-1}|_E\right\| - 1\right) \|x\|$$

showing that $T|_{GE}$ is continuous. Evidently GE is a finite codimensional subspace of $D(T)$. Therefore T is partially continuous.

Finally, suppose that $T|_E$ is continuous where E is some finite codimensional subspace of $D(T)$. Let $M \in \mathcal{I}(D(T))$. Then $M \cap E \in \mathcal{I}(D(T))$ by Proposition IV.1.2, and $T|_{M\cap E}$ is continuous. It follows that T is ubiquitously continuous. □

V.4.3 Theorem. *The following properties are equivalent:*

(i) $\tau(T) < \infty$
(ii) $\Delta(T) < \infty$
(iii) $\Gamma_0(T) < \infty$
(iv) T *is partially continuous.*

PROOF. Since T is partially continuous if and only if QT is partially continuous, we shall suppose without loss of generality, that T is single valued.

(i) $\Longrightarrow$ (iv). Suppose that T is not partially continuous. Then by Proposition V.4.2, there exists $M \in \mathcal{I}(D(T))$ such that $\Gamma(T|_M) = \infty$. Since the null space of $T|_M$ is finite dimensional, there exists $N \in \mathcal{C}(M)$ such that $T|_N$ is injective and $\Gamma(T|_N) = \infty$ (Proposition IV.1.3).

It follows from IV.3.7 that

$$\tau_0((T|_N)^{-1}) \leq \tau((T|_N)^{-1}) = \Gamma(T|_N)^{-1} = 0.$$

In particular, $(T|_N)^{-1} \notin F_+$, by V.2.4. By Theorem V.1.6, there exists $W \in \mathcal{I}(TN)$ for which $(T|_W)^{-1}$ is precompact. Then $\overline{\Gamma}_0((T|_W)^{-1}) = 0$ (Theorem V.2.2). Hence $\Gamma_0((T|_W)^{-1}) = 0$ and by Propositions IV.3.7, $\tau_0(T|_W)^{-1} = 0$, whence $\tau(T) \geq \tau_0(T|_W) = \infty$. Hence (i) $\Longrightarrow$ (iv).

(iv) $\Longrightarrow$ (iii): Let T be partially continuous. There exists $M \in \mathcal{C}(X)$ for which $T|_M$ is continuous. Therefore $\Gamma_0(T|_M) \leq \overline{\Gamma}_0(T|_M) < \infty$ by Corollary IV.2.14. But $\Gamma_0(T) = \Gamma_0(T|_M)$ (IV.1.3). Therefore (iv) $\Longrightarrow$ (iii).

Since $\tau(T) \leq \Delta(T) \leq \Gamma_0(T)$, the implications (iii) $\Longrightarrow$ (ii) $\Longrightarrow$ (i) follow trivially.

The theorem is now proved. □

From Theorems V.4.2 and V.4.3 we deduce a uniform boundedness principle for the quantity Γ.

V.4.4 Corollary. *If* $\Gamma(T|_M) < \infty$ *for each* $M \in \mathcal{I}(D(T))$, *then* $\sup_{M \in \mathcal{I}(D(T))} \Gamma(T|_M) < \infty$.

V.4.5 Theorem. *The following properties are equivalent:*

(i) *T is strictly singular.*
(ii) *T is ubiquitously precompact.*
(iii) *T is ubiquitously strictly singular.*

PROOF. Suppose that T is strictly singular. Then from Theorem V.2.6, $\Gamma(T|_M) \leq \Delta(T) = 0$ for every $M \in \mathcal{I}(D(T))$. Therefore by Theorem V.2.4,

$$T|_M \notin F_+ \qquad (M \in \mathcal{I}(D(T))).$$

Consequently, by Theorem V.1.6, there exists $N \in \mathcal{I}(M)$ for which $T|_N$ is precompact. Hence (i) $\Longrightarrow$ (ii).

Since precompact linear relations are strictly singular, (ii) implies (iii) trivially.

Now suppose that T is not strictly singular. Then $\Gamma(T|_M) > 0$ for some $M \in \mathcal{I}(D(T))$ (Theorem V.2.6). Hence by Theorem V.2.4, $T|_M \in F_+$. Then M has a finite codimensional subspace N for which $T|_N$ is injective and open. Since T is bounded below on the infinite dimensional subspace N, $T|_N$ has no strictly singular restriction to any infinite dimensional subspace of N. Therefore (iii) $\Longrightarrow$ (i). □

V.4.6 Proposition. *Let* $\dim D(T) = \infty$. *Then* $\tau_0(T) = \infty$ *if and only if* T *has a finite codimensional restriction whose inverse is precompact and single valued.*

PROOF. If $\dim N(T) = \infty$ then $\tau_0(T) = 0$ and no finite codimensional restriction of T can be injective (see IV.1.2). We may therefore assume $\dim N(T) < \infty$. Then there is a finite codimensional restriction of T which is injective. Hence by Proposition IV.1.3 it is sufficient to consider the case when T is injective. Hence assume that T is injective.

Suppose first that T is single valued. We have

$$\begin{aligned}
\tau_0(T) = \infty &\iff \Gamma_0(T^{-1}) = 0 \qquad \text{(Proposition IV.3.7)}\\
&\iff T^{-1} \text{ is partially precompact} \qquad \text{(Theorem V.2.2)}\\
&\iff T^{-1}|_M \text{ is precompact for some } M \in \mathcal{C}(R(T))\\
&\iff (T|_{T^{-1}M})^{-1} \text{ is precompact,}
\end{aligned}$$

where $T^{-1}M \in \mathcal{C}(D(T))$, since T and T^{-1} are injective.

For the general case, note that T^{-1} is single valued, and that $(QTJ_M)^{-1} = (TJ_M)^{-1}Q^{-1}$ is precompact for some $M \in \mathcal{C}(D(T))$ if and only if $(TJ_M)^{-1}U_Y$ is totally bounded, since $Q^{-1}U_{QY} = U_Y$. □

V.4.7 Example. *A single valued linear relation* T *with* $\overline{\Gamma}_0(T) = \infty$ *and* $\Gamma_0(T) = 0$. Let T be a discontinuous linear functional. Then $\Gamma_0(T) = 0$ (since $\|T|_{N(T)}\| = 0$) while $T|_M$ is discontinuous for every $M \in \mathcal{P}(X)$.

V.4.8 Example. *A single valued linear relation with* $\tau(T) = \infty$ *and* $\Gamma(T) = 0$. Let $X = Y = \ell_2$ and let (e_n) be the unit vector basis in ℓ_2. Let T be the operator with domain the linear span of (e_n) defined by setting $T(e_{2n}) = \dfrac{1}{n}e_{2n}$ and $T(e_{2n-1}) = ne_{2n-1}$. Since T is evidently not an F_+-operator we have $\Gamma(T) = 0$ by Theorem V.2.3. Also T^{-1} is compact on an infinite dimensional subspace of $R(T)$. Therefore T is not partially continuous, whence $\tau(T) = \infty$ by Theorem V.4.3.

V.4.9 Example. *An unbounded and everywhere defined bijection $T : \ell_2 \to \ell_2$ such that for every $M \in \mathcal{P}(\ell_2)$ there exists $N \in \mathcal{I}(M)$, with N closed, such that $T|_N$ is an isomorphism.*

Let M and N_1 be a pair of closed mutually orthogonal infinite dimensional subspaces of ℓ_2 such that $M \dotplus N_1 = \ell_2$ (where $\dotplus$ denotes the orthogonal sum), and select a dense proper subspace N of N_1 so that $N_1 = K + N$ where K is one–dimensional (N will be the null space of a discontinuous linear functional on N_1). On $\ell_2 = M \dotplus (K + N)$ define P to be the projection of ℓ_2 onto $M + K$ with null space N. P is unbounded since its null space $N(P)$ is not closed. Let $T = I + P$. Then T is unbounded. Also $T|_{M+N}$ is an isomorphism; indeed for $m \in M$, $n \in N$ we have $\|m+n\|^2 = \|m\|^2 + \|n\|^2 \leq \|2m\|^2 + \|n\|^2 = \|2m+n\|^2 = \|T(m+n)\|^2 \leq 4\,\|m+n\|^2$ so $\|m+n\| \leq \|T(m+n)\| \leq 2\,\|m+n\|$. Now let $E \in \mathcal{P}(\ell_2)$. Then $M = (M \cap E) + W$, where $E \cap W = \{0\}$, and $\dim W < \infty$ by Proposition IV.1.2(a). Hence $F = M \cap E$ is infinite dimensional and has the required property.

V.4.10 Exercise Show that any two norms defined on an infinite dimensional space are comparable on some infinite dimensional subspace.

V.4.11 Exercise Let $G(T)$ be complete, and let T be ubiquitously continuous. Then T is continuous.

V.4.12 Exercise Let X be complete. If T is an injective bounded linear operator with a partially continuous inverse, then T is an isomorphism.

V.4.13 Exercise Show that T is strictly singular if and only if T is ubiquitously small, i.e. for each $\epsilon > 0$ and $M \in \mathcal{I}(D(T))$, there exists $N \in \mathcal{I}(M)$ such that $\|T|_M\| < \epsilon$. Consequently, strictly singular relations are partially continuous.

V.4.14 Proposition. *Let T be partially continuous and let $S \in LR(X,Y)$ with $D(S)$ dense in a finite codimensional subspace of $D(T)$. Then*

$$\Gamma(S+T) \leq \Delta(S) + \Gamma(T).$$

PROOF. First assume that T is continuous and that $\overline{D(S)} \supset D(T)$. Then by Theorem IV.4.8, $\Gamma(T|_{D(S)}) = \Gamma(T)$ and the desired inequality follows from Proposition IV.2.5.

Now let T be partially continuous. Then there exist finite codimensional subspaces M_1 and M_2 of $D(T)$ such that $T|_{M_1}$ is continuous and $\overline{D(S)} \supset D(T|_{M_2})$. Let $M = M_1 \cap M_2$. Then M is finite codimensional (see Proposition IV.1.2), and it follows from the first part combined with Proposition IV.1.3 that

$$\Gamma(S+T) = \Gamma(S+T|_M) \leq \Delta(S) + \Gamma(T|_M) = \Delta(S) + \Gamma(T). \quad \square$$

V.4.15 Remark *Proposition* V.4.14 *fails without the density requirement for $D(S)$.* Indeed let $X = Y = \ell_2$, $D(T) = X$. Denoting an arbitrary element of ℓ_2 by (a_n), define $T((a_n)) = (a_1, 0, a_3, 0, \dots)$, $D(S) = \{(a_n) : a_{2n-1} = 0,$

$n \in \mathbb{N}\}$ and $S((a_n)) = 0$. Clearly $\Gamma(T) = 0$ (as $N(T)$ is infinite dimensional), $\Gamma(S+T) = \Gamma(T|_{D(S)}) = \Gamma(I|_{D(S)}) = 1$, and $\Delta(S) = 0$.

V.4.16 Exercise Show that T is continuous if and only if $\overline{\Gamma}_0(T) < \infty$, if and only if there exists $M \in \mathcal{P}(X)$ such that $T|_M$ is continuous.

V.4.17 Theorem. *For any $T \in LR(X,Y)$ we have $\tau_0(T) = \overline{\tau}_0(T)$.*

PROOF. We may clearly suppose that T is single valued (from the definition of the quantities τ_0 and $\overline{\tau}_0$), and furthermore that $X = D(T)$, $Y = R(T)$, and $\dim D(T) = \infty$.

Clearly $\overline{\tau}_0(T) \leq \tau_0(T)$. For the converse, suppose that $\tau_0(T) > 0$. Then $T \in F_+$ by V.2.4. Hence there exists $E \in \mathcal{P}(D(T))$ for which $(TJ_E)^{-1}$ is continuous and single valued (Theorem V.1.6). But $\tau_0(TJ_E) = \tau_0(T)$ and $\overline{\tau}_0(TJ_E) = \overline{\tau}_0(T)$ (Proposition IV.1.3). We may therefore assume that T is injective and that its inverse T^{-1} is continuous. We have

$$\tau_0(T) = \sup_{M \in \mathcal{C}(X)} \gamma(T|_M) \tag{1}$$

since T is injective. Now for any subspace $E \subset X$,

$$T\overline{E} = T\overline{T^{-1}TE} \supset TT^{-1}\overline{TE} = \overline{TE}$$

by the continuity of T^{-1}, and hence

$$\begin{aligned}\gamma(T|_E) &= \left\|(T|_E)^{-1}\right\|^{-1} = \left\|T^{-1}|_{TE}\right\|^{-1} = \left\|T^{-1}|_{\overline{TE}}\right\|^{-1} \\ (\text{as } T^{-1} \text{ is continuous}) &\leq \left\|T^{-1}|_{T(\overline{E})}\right\|^{-1} = \left\|(T|_{\overline{E}})^{-1}\right\|^{-1} = \gamma(T|_{\overline{E}}).\end{aligned}$$

Thus

$$\gamma(T|_E) \leq \gamma(T|_{\overline{E}}). \tag{2}$$

It follows from (1) and (2) that

$$\tau_0(T) = \sup_{M \in \mathcal{C}(X)} \gamma(T|_M) \leq \sup_{M \in \mathcal{C}(X)} \gamma(T|_{\overline{M}}) = \overline{\tau}_0(T)$$

as required. □

V.5 LOWER SEMI–FREDHOLM RELATIONS

In this section we shall obtain characterisations of the class of lower semi-Fredholm relations and determine conditions for the stability of the class under precompact and strictly cosingular perturbation. Included also is a condition equivalent to the compactness of the adjoint relation T' (Corollary V.5.15).

V.5.1 Proposition. *Let* $\dim D(T) = \infty$. *Then*

(a) $\alpha(T) < \infty \implies \gamma(T) \leq \Gamma(T)$
(b) $\overline{\beta}(T') < \infty \implies \gamma(T') \leq \Gamma(T)$

PROOF. (a) Let $\alpha(T) < \infty$, and suppose that $\gamma(T) > 0$. Let $M \in \mathcal{I}(D(T))$. By Lemma III.7.3, there exists $m \in M$, $m \neq 0$, such that

$$\|m\| = d(m, N(T)).$$

But if $0 \leq \lambda < \gamma(T)$, then

$$\|Tm\| \geq \lambda d(m, N(T)).$$

Consequently

$$\frac{\|Tm\|}{\|m\|} \geq \gamma(T)$$

Therefore

$$\|T|_M\| \geq \gamma(T).$$

Taking the infimum over $M \in \mathcal{I}(D(T))$ gives $\Gamma(T) \geq \gamma(T)$.

(b) Clearly $G(T)\hat{}\subset G(T'')$ (where the "$\hat{}$" symbol signifies canonical injection in the second dual); in particular, $D(T)\hat{}\subset D(T'')$. Let $M \in \mathcal{I}(D(T))$. Then

$$\begin{aligned}\|T''|_{\hat{M}}\| &= \sup\{d(T''\hat{M}, T''(0)) : m \in B_M\} \leq \sup_m d(Tm, T(0)) \\ &= \|T|_M\|\end{aligned}$$

since $(Tm)\hat{}\subset T''\hat{M}$ (in particular, $T(0)\hat{}\subset T''(0)$). Hence

$$\Gamma(T'') \leq \inf_{M \in \mathcal{I}(D(T))} \|T''|_{\hat{M}}\| \leq \Gamma(T).$$

Now $\overline{\beta}(T') = \alpha(T'')$. Hence by part (a),

$$\gamma(T'') \leq \Gamma(T'') \leq \Gamma(T).$$

But $\gamma(T') = \gamma(T'')$ by Theorem III.5.3. Therefore $\gamma(T') \leq \Gamma(T)$. □

V.5.2 Proposition. *The following properties are equivalent:*

(i) $T \in F_-$
(ii) $\overline{\beta}(T) < \infty$ *and* $\gamma(T') > 0$
(iii) $\tilde{T} \in \phi_-$
(iv) $QT \in F_-$.

PROOF. The equivalence (i) $\iff$ (iv) is obvious from the equality $(QT)' = T'J_{T(0)^\perp}$ (since $D(T') \subset T(0)^\perp$).

(i) $\implies$ (ii): Let $T \in F_-$. Then $T' \in \phi_+$ (Corollary V.1.7) and hence $\alpha(T') < \infty$ and $R(T')$ is closed. Therefore $\overline{\beta}(T) < \infty$, by Proposition III.7.2, and $\gamma(T') > 0$ by II.3.2 and III.4.2.

(ii) $\implies$ (iii): Suppose (ii) holds. Then by III.7.2, $\overline{\beta}(\tilde{T}) = \alpha(\tilde{T}') = \alpha(T') = \overline{\beta}(T) < \infty$. Also by II.3.2(b), T' is open, so $R(T')$ is closed (Theorem III.3.2). Hence $\tilde{T}' \in \phi_+$. Therefore $\tilde{T} \in \phi_-$ by III.4.4.

(iii) $\implies$ (i): By V.1.9, $\tilde{T} \in \phi_- \iff T \in F_-$. □

V.5.3 Proposition. *Let T be a continuous linear relation. Then T is precompact if and only if T' is compact.*

PROOF. Since T is continuous if and only if $QTJ_{D(T)}$ is continuous, and precompact if and only if $QTJ_{D(T)}$ is precompact, and since T' and $(QTJ_{D(T)})'$ have the same graph (by III.1.12), we may thus suppose without loss of generality that T is single valued and everywhere defined. The proof now reduces to the standard one for bounded linear operators (see e.g., Goldberg [3; p.82]). □

Any linear relation T for which $D(T') = \{0\}$ is lower semi–Fredholm by definition. If furthermore T is single valued, then T will not be precompact, or even continuous (except in the trivial case) while T' is compact. Hence the continuity hypothesis for T in Proposition V.5.6 is necessary. A linear operator of the above type is exhibited in the example below.

V.5.4 Example Let $X = Y = L_2(0,1)$ and let $T \in LR(X)$ be defined as follows: $D(T)$ consists of the functions which are almost everywhere differentiable in the interval (0,1) and have square integrable derivative. Let $Tf = \dfrac{df}{dt}$ $(f \in D(T))$. Since $N(T)$ includes the step functions, $N(T)$ is dense in $L_2(0,1)$. Now by III.1.6(a), $0 = (TJ_{N(T)})' = J_{N(T)}T' = T'$. Furthermore, T' is injective since $R(T)$ is clearly dense. Therefore $D(T') = \{0\}$.

V.5.5 Theorem. *The following properties are equivalent for a linear relation $T \in LR(X,Y)$:*

(i) $T \notin F_-$.

(ii) *There is no finite dimensional subspace $F \subset Y$ for which $T'|_{F^\perp}$ is bounded below.*

(iii) *For each $\epsilon > 0$, there exists a compact operator $K \in LR(X, \tilde{Y})$ with norm not exceeding ϵ such that $D(K) \supset D(T)$ and $\overline{\beta}(J_Y T + K) = \infty$.*

PROOF. (i) $\implies$ (ii): Assume (i). Then $T' \notin F_+$. Now if $F \in \mathcal{F}(Y)$, then $F^\perp \in \mathcal{P}(Y')$, and so $T'|_{F^\perp}$ is not bounded below by Theorem V.1.6. Hence (i) $\implies$ (ii).

(ii) $\implies$ (iii): Suppose that (ii) holds. Let $\epsilon > 0$ and $\delta > 0$. Since T' is not bounded below, there exists $y_1' \in D(T')$ such that

$$\|y_1'\| = 1 \quad \text{and} \quad \|T'y_1'\| < \epsilon \cdot 2^{-1}(1+\delta)^{-1}.$$

Next select $y_1 \in Y$ such that

$$y_1' y_1 = 1 \quad \text{and} \quad \|y_1\| \le (1+\delta).$$

Now suppose that $\{y_1, \ldots, y_{n-1}\} \subset Y$ and $\{y_1', \ldots, y_{n-1}'\} \subset D(T')$ have been found satisfying

$$\|y_k\| \le (1+\delta)(2+\delta)^{k-1}, \|y_k'\| = 1, \|T'y_k'\| < \epsilon \cdot 2^{-k}(1+\delta)^{-1}(2+\delta)^{1-k}$$

(1) $\qquad y_k'(y_m) = \delta_{km}$ for all $1 \le k, m \le n-1$ and $n \ge 2$.

Consider the restriction of T' to the subspace $\{y_1, \ldots, y_{n-1}\}^\perp$. Then by condition (ii) there exists $y_n' \in \{y_1, \ldots, y_{n-1}\}^\perp \cap D(T')$ such that

$$\|y_n'\| = 1 \quad \text{and} \quad \|T'y_n'\| < \epsilon \cdot 2^{-n}(1+\delta)^{-1}(2+\delta)^{1-n}.$$

Select $y \in Y$ such that

$$y_n'(y) = 1 \quad \text{and} \quad \|y\| \le 1 + \delta.$$

Let

$$y_n = y - \sum_1^{n-1} y_k'(y) y_k.$$

Then

$$y_n'(y_n) = y_n'(y) = 1 \quad \text{and} \quad y_n'(y_k) = 0 \text{ for } 1 \le k < n.$$

Also

$$y_k'(y_n) = y_k'(y) - y_k'(y) = 0 \quad \text{for} \quad 1 \le k < n.$$

Hence

$$y'_k(y_m) = \delta_{km} \quad \text{for} \quad 1 \le k, m \le n.$$

Furthermore

$$\|y_n\| \le \|y\|\,(1+\sum_1^{n-1}\|y_k\|) \le (1+\delta)(1+\sum_1^{n-1}(1+\delta)(2+\delta)^{k-1}) = (1+\delta)(2+\delta)^{n-1}.$$

Consequently, by induction we may construct sequences $\{y_k\} \subset Y$ and $\{y'_k\} \subset D(T')$ such that (1) is satisfied for all $n \in \mathbb{N}$.

Now define the linear operator $K \in LR(X, \tilde{Y})$ by

$$(2) \qquad Kx = \sum_1^{\infty} T'y'_k(x)y_k \qquad (x \in X).$$

(Note that $T'y'$ is single valued for $y' \in D(T')$, by III.1.2.) Observe that

$$\|Kx\| \le \sum_1^{\infty} \|T'y'_k\|\,\|x\|\,\|y_k\| < \epsilon\,\|x\|\,(\sum_1^{\infty} 2^{-k}) = \epsilon\,\|x\|$$

so that Kx exists for each $x \in X$ by the completeness of $\tilde{Y}$. By Propositions V.1.3 and V.1.4, K is compact, and its norm does not exceed ϵ.

Now for each $x \in D(T)$, we have by (1) that

$$y'_k(Kx) = T'y'_k(x) = y'_k(Tx) \quad \text{(by III.1.1(4))}$$

and so each y'_k annihilates $\overline{R(J_Y T - K)}$. Since the set $\{y'_k\}$ is linearly independent by (1), we conclude that $\overline{\beta}(J_Y T - K) = \infty$. Therefore (ii) $\Longrightarrow$ (iii).

(iii) $\Longrightarrow$ (i): Let $T \in F_-$. Then $J_Y T \in F_-$. We may therefore assume that Y is complete. Now let $K \in LR(X, Y)$ be any compact operator such that $D(K) \supset D(T)$. Since $D(K') = Y'$ by III.4.6, we have $(T + K)' = T' + K'$ by III.1.5(b). Now K' is compact by Proposition V.5.3, and $T' \in F_+$. Therefore since $T'(0) = D(T)^\perp \supset D(K)^\perp = K'(0)$ by III.1.4, $(T + K)' \in F_+$ by the perturbation theorem V.3.2 together with V.2.6. Hence $\overline{\beta}(T + K) = \alpha((T + K)') < \infty$, as required. □

V.5.6 Corollary. *The following properties are equivalent:*

(i) $T \in F_-$

(ii) $\overline{\beta}(J_Y T + K) < \infty$ *for every compact single valued* $K \in LR(X, \tilde{Y})$ *such that* $D(K) \supset D(T)$.

PROOF. Let $K \in LR(X, \tilde{Y})$ be a compact operator with $D(K) \supset D(T)$. Then by III.1.5 and III.4.6,

$$(T + K)' = T' + K'$$

where K' is compact by Proposition V.5.3.

Now if (i) is satisfied, then for every such K we have by V.3.2(a) and V.2.6,

$$\overline{\beta}(J_Y T + K) = \alpha((T + K)') < \infty.$$

Hence (i) $\Longrightarrow$ (ii).

The converse is an immediate consequence of (i) $\Longrightarrow$ (iii) of Theorem V.5.5. □

The next proposition shows that the F_--property can be reduced to the case when the adjoint is single valued (see III.1.10(b)).

V.5.7 Proposition $T \in F_- \iff TJ_{D(T)} \in F_-$.

PROOF. Write $J = J_{D(T)}$. By III.1.6 we have

$$(TJ)' = Q_{T'} T'$$

(since $T'(0) = D(T)^\perp$ by III.1.4(b)).

Hence we have the following chain of equivalences

$$T \in F_- \iff T' \in F_+ \iff Q_{T'} T' \in F_+ \iff (TJ)' \in F_+ \iff TJ \in F_-. \quad \square$$

V.5.8 Exercise Show that the converses of Proposition V.5.1(a) and (b) are false.

If T is a densely defined linear operator with $D(T') = \{0\}$, then T is an F_--operator by definition. Extending an operator linearly can result in a trivial adjoint even if the extension is done in a natural way as the following example shows.

V.5.9 Example Consider the formal differentiation operator D. Denoting $L_p[0,1]$ by L_p $(1 \leq p \leq \infty)$, define $T \in LR(L_p, L_q)$ $(1 \leq p, q < \infty)$, by $D(T) = \{f \in L_p : Df \in L_q\}$, $Tf = Df$. Then $D(T)$ can be regarded as a natural domain for D. It will be shown in the next example that $D(T') = \{0\}$ (so $T \in F_-$) and that T has dense null space (so that $\gamma(T) = \infty$). Let E be the subspace of $D(T)$ consisting of the absolutely continuous functions. *The operator $T|_E \in LR(L_p, L_q)$, is closed, densely defined and surjective and has $\alpha(T|_E) = 1$.* From the State Diagram III.6.7, we see that $T|_E$ is a Fredholm operator.

Observe that in the above example, although E is dense in $D(T)$, we have $D(T') = \{0\}$ and $D((T|_E)')$ weak*-dense in L'_p. Furthermore $D((TJ_E)') = D((T|_E)')$, while $(J_E)'T' = T'$. Hence $(TJ_E)' \neq (J_E)'T'$.

In the next example we look at a more general class of differential operators.

V.5.10 Example Let $\tau = \Sigma_0^n a_k D^k$ where $D = \dfrac{d}{dt}$, $D^0 f = f$, $a_k \in C^k(\Omega)$ for $0 \leq k \leq n$ and $a_0 \in L_\infty(\Omega)$ where Ω is an interval containing one of its end points. Write $L_p := L_p(\Omega)$. Define $T : L_p \to L_q$ as follows:

$$D(T) = \{f : f \in L_p, \tau f \in L_q\}$$
$$Tf = \tau f.$$

(a) *The operator T is F_- with $D(T') = \{0\}$ whenever $1 \leq p, q < \infty$ and either $a_0 = 0$ or $p = q$.*

PROOF. Let $1 \leq p, q < \infty$ and let E denote the set of all functions in L_p for which $Df = 0$ almost everywhere. Then E is a linear subspace of $D(T)$. Furthermore for any step function g, $Dg = 0$ almost everywhere. Thus E contains the step functions and is therefore a dense subspace of L_p Cohn [1; Proposition 3.4.3]. Consequently

$$E' = D(T)' = L_{p'} \qquad (\frac{1}{p} + \frac{1}{p'} = 1)$$

and so T' and $(TJ_E)'$ are both in $LR(L_{q'}, L_{p'})$. Suppose $a_0 = 0$. Then as $TJ_E = 0$, therefore $(TJ_E)' = 0_{Y'}$ with T' a restriction of $(TJ_E)'$. Therefore $T'y' = 0$ for $y' \in D(T')$. Clearly $D(T)$ contains the domain of the maximal operator $T_{\tau,p,q}$ defined as in Goldberg [3; VI.1.6]. Hence $R(T)$ contains $R(T_{\tau,p,q})$ and is moreover dense in L_q by Goldberg [3; VI.2.10]. It follows that T' is injective and thus $D(T') = \{0\}$. Now suppose that $a_0 \neq 0$ and $p = q$. Since $a_0 \in L_\infty$, $a_0 I$ is a bounded linear operator defined on L_p. Hence $T' = (T - a_0 I)' + (a_0 I)'$. But from the first part, $D((T - a_0 I)') = \{0\}$. Hence $D(T') = \{0\}$. □

(b) *Suppose the interval Ω has finite Lebesgue measure, and that $1 \leq p, q < \infty$ and $1/a_0 \in L_\infty$. Then $T \in F_-$. Furthermore, $D(T') = \{0\}$ whenever $p \geq q$.*

PROOF. Let E be as in (a). We consider two cases.

Case 1 $(p < q)$: Observe that the subspace $K = E \cap D(T)$ contains the step functions and is therefore dense in both L_p and L_q. Furthermore $TJ_K = a_0 I_K$ where I_K denotes the identity map from $K \subset L_p$ into L_q. Let $f \in L_q$ be arbitrary. Then $(1/a_0)f \in L_q$. Hence there exists a sequence of step functions $(g_n) \subset L_q$ such that $g_n \to (1/a_0)f$. Consequently $\{a_0 g_n\} \subset R(a_0 I_K)$ with $a_0 g_n \to f$ in L_q, and therefore $R(a_0 I_K)$ is dense in L_q. In addition

$(a_0 I_K)^{-1}$ is continuous by Cohn [1; Exercise 3.3.9] and the fact that $1/a_0 \in L_\infty$. Hence $(TJ_K)' = (a_0 I_K)'$ is injective and open. But T' is a restriction of $(TJ_K)'$. Hence $T \in F_-$.

Case 2 ($p \geq q$): In this case the natural injection of L_p into L_q is continuous Cohn [1; loc.cit.]. Denote this injection by U. By (a) we have $T - a_0 U \in F_-$ with $D((T - a_0 U)') = \{0\}$. As $a_0 \in L_\infty$, $a_0 U$ is bounded and hence $T' = (T - aU)' + (aU)'$. Consequently $D(T') = \{0\}$. □

V.5.11 Proposition. *Let F be a finite dimensional subspace of Y. Then*

$$T \in F_- \iff Q_F T \in F_-.$$

PROOF. We have $(Q_F T)' = T' J_{F^\perp}$ where $F^\perp$ is finite codimensional. But $T' \in F_+$ if and only if $T' J_{F^\perp} \in F_+$. Hence the result follows. □

V.5.12 Theorem. (F_- perturbation theorem)

Let $S, T \in LR(X, Y)$ with $D(S) \supset D(T)$, and let $T \in F_-$.

(a) *If $\dim R(S) < \infty$, then $T + S \in F_-$*

(b) *If S is precompact, then $T + S \in F_-$*

(c) *If $\|S\| < \gamma(T')$, then $T + S \in F_-$.*

PROOF. (a) We have $Q_{R(S)}(T + S) = Q_{R(S)}T$, since $D(S) \supset D(T)$. Hence $T + S \in F_-$ by Proposition V.5.11.

(b) We have $(T + S)' = T' + S'$ by III.1.5(b), where $T' \in F_+$, and S' is compact by Proposition V.5.3. Now by Proposition V.5.7 we may assume that $D(S) = X$. In that case $S'(0) = \{0\}$ and it follows from the perturbation theorem for F_+-relations V.3.2 that $T' + S' \in F_+$. Therefore $T + S \in F_-$.

(c) We may clearly suppose that $\dim D(T) = \infty$ (otherwise $\dim QY < \infty$). As in the proof of (b), we assume without loss of generality that $D(S) = X$, so that S' is single valued, and then infer from the perturbation theorem for F_+-relations V.3.2(a) (combined with V.5.1 for the $\gamma(T')$ estimate) that $(T + S)' = T' + S' \in F_+$. □

We shall now consider closed or closable F_--operators.

V.5.13 Proposition. *Consider the properties:*

(i) $T \in F_+$ *and* $\overline{\beta}(T) < \infty$
(ii) $T \in F_-$ *and* $\alpha(T) < \infty$.

Then (i) $\Longrightarrow$ (ii). *If* X *is complete and* T *is closed, then* (ii) $\Longrightarrow$ (i).

PROOF. Assume (i). Then (ii) follows from Proposition V.5.2.

Now let T be closed and X complete. Assume (ii). Then $T' \in \phi_+$, and so $\gamma(T) = \gamma(T') > 0$ (see III.5.3 and III.4.2). Now apply Proposition V.5.1(a) and Theorem V.2.4. □

V.5.14 Theorem. *The following properties are equivalent:*

(i) *There exists* $F \in \mathcal{F}(Y)$ *such that* $Q_F T$ *is precompact.*
(ii) $\Gamma_0'(T) = 0$.

PROOF. (i) $\Longrightarrow$ (ii). Assume that (i) holds. Since for any $F \in \mathcal{F}(Y)$, $\Gamma_0'(Q_F T) = \Gamma_0'(T)$ by Proposition IV.5.6, it is enough to consider the case when T is precompact. Let $\epsilon > 0$ be given. Then there exists a finite subset $\{x_1, \ldots, x_n\} \in B_{D(T)}$ such that for each $x \in B_{D(T)}$ there is an x_k, $1 \leq k \leq n$, for which $\|Tx - Tx_k\| < \epsilon$. Choose $y_1, \ldots, y_n$ so that $Tx_i = y_i + T(0)$, $1 \leq i \leq n$, and let $F := \text{sp}\{y_1, \ldots, y_n\}$. Then we have

$$\|Q_F Tx\| = \|Q_F(Tx - y_k)\| = \|Q_F(Tx - Tx_k)\| \leq \|Q_F\| \, \|Tx - Tx_k\| < \epsilon.$$

Therefore $\Gamma_0'(T) = 0$.

(ii) $\Longrightarrow$ (i). Assume (ii) and let $\epsilon > 0$ be given. Then there exists $F \in \mathcal{F}(Y)$ such that $\|Q_F(T)\| < \epsilon$. Hence by Corollary III.1.13, $\|(Q_F T)'\| = \|T'|_{F^\perp}\| < \epsilon$. Consequently $\overline{\Gamma}_0(T') = 0$, and so T' is compact by Theorem V.2.2. Hence $(Q_F T)'$ is compact and so $Q_F T$ is precompact by Proposition V.5.3. □

V.5.15 Corollary. $\tilde{T}$ *is compact if and only if* T' *is compact.*

PROOF. Let T' be compact. By Theorem V.5.14, $\inf\{\|Q_F T'\| : F \in \mathcal{F}(X')\} = 0$. But by III.5.5, $\|Q_F T'\| = \left\|(\tilde{T}|_{F^\top})'\right\| = \left\|\tilde{T}|_{F^\top}\right\|$ and $\mathcal{P}(X) = \{F^\top : F \in \mathcal{F}(X')\}$. Therefore $\overline{\Gamma}_0(\tilde{T}) = 0$. Hence $\tilde{T}$ is compact by Theorem V.2.2.

The converse is contained in Proposition V.5.3. □

V.5.16 Theorem. *Let* $\dim Y/\overline{T(0)} = \infty$. *Then*

$$T \in F_- \iff \Gamma'(QT) > 0.$$

PROOF. By the equivalence $T \in F_- \iff QT \in F_-$ (see Proposition V.5.2), it is sufficient to consider the single valued case under the assumption that $\dim Y = \infty$.

Let $T \notin F_+$. Then by Theorem V.5.5 there exists a compact operator $K \in LR(X, \tilde{Y})$ of norm less than $\epsilon > 0$ for which $\overline{\beta}(J_Y T - K) = \infty$. Put $M := \overline{R(J_Y T - K)}$. Then $M \in \mathcal{E}(\tilde{Y})$ and $Q_M J_Y T = Q_M K$ where $Q_M K$ is compact. Hence by Corollary IV.5.4, Proposition IV.5.9 and Theorem V.5.14,

$$\Gamma'(T) \leq \Gamma'(Q_M J_Y T) = \Gamma'(Q_M K) \leq \Gamma'_0(Q_M K) = 0.$$

Conversely, let $\Gamma'(T) = 0$. Let $\epsilon > 0$ and choose $M \in \mathcal{E}(\tilde{Y})$ such that $\|Q_M J_Y T\| \leq \epsilon$. Then $Q_M J_Y T$ is continuous and so its adjoint has domain the whole space $M^\perp \subset D(T')$. Since $M^\perp$ is infinite dimensional, and $\|T' J_{M^\perp}\| \leq \|Q_M J_Y T\| < \epsilon$, this shows that $\Gamma(T') = 0$, whence $T' \notin F_+$, by Theorem V.2.4. Hence $T \notin F_-$. □

V.5.17 Corollary. *Let* $\dim Y = \infty$, *and let* $T(0)$ *be finite dimensional. Then*

$$T \in F_- \iff \Gamma'(T) > 0.$$

V.5.18 Definition A linear relation T is called **strictly cosingular** if $\Delta'(T) = 0$.

V.5.19 Proposition. *If* $F \in \mathcal{F}(Y)$ *and* $Q_F T$ *is precompact, then* T *is strictly cosingular. In particular, linear relations with finite dimensional range, and precompact relations, are strictly cosingular.*

PROOF. Let $Q_F T$ be precompact, where $F \in \mathcal{F}(Y)$. Then $\Gamma'_0(T) = 0$ by Theorem V.5.14. Hence by Proposition IV.5.9.

$$\Delta'(T) \leq \Gamma'_0(T) = 0.$$

Therefore T is strictly cosingular. □

We have the following perturbation result for F_--relations.

V.5.20 Theorem. *Let $T \in F_-$ and let $J_Y S$ be strictly cosingular with $D(S) \supset D(T)$, and $\dim Q_T S(0) < \infty$. Then $T + S \in F_-$.*

PROOF. Since $J_Y T \in F_-$ if and only if $T \in F_-$, we may suppose that Y is complete. First assume that $S(0) = 0$. Then since $T = T + S - S$, and $(T+S)(0) = T(0)$, we have $Q_T T = Q_{T+S}(T + S - S)$ and so from Proposition IV.5.11

$$\begin{aligned}\Gamma'(QT) &= \Gamma'(Q_{S+T}(T+S) - Q_{S+T}S)\\ &\le \Gamma'(Q_{S+T}(T+S)) + \Delta'(QS) \qquad (1)\end{aligned}$$

Let us consider the various cases for $T(0)$:

Case 1: $\overline{T(0)} \in \mathcal{E}(Y)$. Then by Corollary IV.5.4,

$$\Delta'(QS) \le \Delta'(S) = 0.$$

Case 2: $\overline{T(0)}$ finite codimensional. In that case QS is a finite rank operator and by Proposition V.5.19, $\Delta'(QS) = 0$.

Case 3: $\overline{T(0)}$ finite dimensional. This is covered by Case 1 if we assume without loss of generality that $\dim Y = \infty$.

For the general case, let $\dim Q_T S(0) < \infty$. Now for any $F \in \mathcal{F}(Y)$, $T \in F_-$ if and only if $Q_F T \in F_-$ (as $(Q_F T)' = T' J_{F^\perp}$). Hence with $F = Q_T S(0) \in \mathcal{F}(Y/\overline{T(0)})$, we have the chain of implications

$$\begin{aligned}T \in F_- \implies Q_T T \in F_- \implies Q_F Q_T T \in F_- \implies Q_{S+T}T + Q_{S+T}S \in F_-\\ \implies Q_{S+T}(T+S) \in F_- \implies T + S \in F_-.\end{aligned}$$

(Here we have combined Lemma IV.5.2 with the fact that in this setting $\overline{T(0)} + S(0) = \overline{(T+S)(0)}$.) □

V.5.21 Theorem. *Let X be complete and T closed. Then the following properties are equivalent:*

(i) $T \in F_-$

(ii) $\overline{\beta}(T) < \infty$ *and* $\gamma(T) > 0$

(iii) $Q_F T$ *is open and surjective for some finite dimensional subspace* F *of* Y.

(iv) $\overline{\beta}(Q_F T) < \infty$ *and* $\gamma(Q_F T) > 0$ *for some finite dimensional subspace* F *of* Y.

PROOF. We have $\gamma(T) = \gamma(T')$ by Theorem III.5.3(a). Hence (i) $\implies$ (ii), by Proposition V.5.2.

Assume (ii). Then $T \in F_-$ by Proposition V.5.2, since $\gamma(T') \geq \gamma(T) > 0$. Hence by Theorem V.5.5, there exists a finite dimensional subspace F of Y such that $T' J_{F^\perp}$ is bounded below and hence $Q_F T$ is surjective by the State Diagram III.6.5. Now Q_F is a ϕ_+-operator (so $\gamma(Q_F) > 0$) and hence $Q_F T$ is closed, by Proposition II.5.17. Consequently $\gamma(Q_F T) > 0$ by Theorem III.5.3(a). Therefore (ii) $\implies$ (iii).

The implication (iii) $\implies$ (iv) is trivial. Now assume that (iv) holds. Then $(Q_F T)'$ is open and so by Proposition V.5.2, $Q_F T \in F_-$. Hence by Proposition V.5.11, $T \in F_-$. Therefore (iv) $\implies$ (i). □

V.5.22 Example *Let X be an infinite dimensional Banach space. Then there exists a normed space Y and a closed injective and surjective everywhere defined linear operator $T \in LR(X, Y)$ such that $T \notin F_+ \cup F_-$.*

Let $S \in LR(X, Z)$ be an injective everywhere defined precompact linear operator. Let $Y = R(S)$ and let T be the operator S considered as an element of $LR(X, Y)$. Then T is closed, injective and surjective. Clearly $T \notin F_+$. Also, T' is compact, whence $T' \notin F_+$. Therefore $T \notin F_-$. □

V.5.23 Example *The class $F_+ \cap F_-$ may contain non closed single valued linear relations, even if X and Y are complete.* Let X and Y be infinite dimensional Banach spaces and $S \in LR(X, Y)$ an arbitrary Fredholm linear operator. Let M be a dense subspace of $G(S) \subset X \times Y$ of codimension one (the kernel of a discontinuous linear functional) and $(x_0, Sx_0) \in G(S) \backslash M$. Select $0 \neq y_0 \in Y$ arbitrarily. Denoting the bounded canonical surjection of $G(S)$ onto $D(S)$ by P, we define $T \in LR(X, Y)$ as follows:

$TJ_{PM} = SJ_{PM}$ with $Tx_0 = y_0$ if $Sx_0 = 0$, and $Tx_0 = 0$ if $Sx_0 \neq 0$.

Clearly $D(T) = D(S)$. We show that T is not closed and hence that $T \notin \phi_+ \cap \phi_-$. Select $(x_n, Sx_n) \in M$ such that $(x_n, Sx_n) \to (x_0, Sx_0)$ with $Sx_0 \neq Tx_0$ (by definition). Thus T is not closed. As T and S agree on the finite codimensional subspace PM of their domains, it follows that $T \in F_+$. Furthermore denoting $\text{sp}\{Tx_0, Sx_0\}$ by K, it is clear that $Q_K T = Q_K S$. Noting that $S \in F_-$, we conclude from Proposition V.5.11, that $T \in F_-$. □

V.5.24 Proposition. *We have*

$$TG \in F_- \implies T \in F_-.$$

PROOF. We use the characterization of F_--relations of Theorem V.5.2. Let $TG \in F_-$. Then $\overline{\beta}(T) = \overline{\beta}(TG) < \infty$, and $\gamma((TG)') > 0$. But

$$\gamma(T') = \gamma((TGG^{-1})') = \gamma((G^{-1})'(TG)')$$

by III.1.6(a), since $R(G^{-1}) = X_T = D(TG)$, and by Theorem II.3.11, because $G'(0) = \{0\}$, we have

$$\gamma((G^{-1})'(TG)') \geq \gamma((G^{-1})')\gamma((TG)') > 0$$

(as $\gamma((G^{-1})') = \gamma(G^{-1}) = \|G\|^{-1} > 0$). Hence $\gamma(T') > 0$. It follows that $T \in F_-$. □

The converse of Proposition V.5.24 holds for closable T and is contained in Corollary V.5.27 below.

V.5.25 Lemma. *If T is closable, then $(\overline{T}G_{\overline{T}})' = (TG)'$.*

PROOF. Since $\overline{T}$ is an extension of T (as T is closable), X_T is a subspace of $X_{\overline{T}}$; in fact for $x \in D(T)$, $\|x\|_{\overline{T}} = \|x\| + \|\overline{T}x\| = \|x\| + \|Tx\| = \|x\|_T$. Hence if $z \in X_T$, then $\overline{T}G_{\overline{T}}z = \overline{T}(Gz) = TGz$. Therefore

$$\text{(1)} \qquad \overline{T}G_{\overline{T}} \text{ is an extension of } TG.$$

Since $G(\overline{T}G_{\overline{T}}) \subset G(TG)$, we have

$$\text{(2)} \qquad G((\overline{T}G_{\overline{T}})') \subset G((TG)').$$

Let $(y', z') \in G((TG)')$. Then for $(z, y) \in G(TG)$,

$$y'y = z'z, \quad \text{i.e.} \quad y'TGz = z'z.$$

Hence from (1),

$$\text{(3)} \qquad y'\overline{T}G_{\overline{T}}z = z'z \qquad (z \in X_T).$$

But X_T is dense in $X_{\overline{T}}$. Indeed, if $z \in X_{\overline{T}}$, then $G_{\overline{T}}z = x$ where $x \in D(\overline{T})$, and there is a sequence $\{(x_n, y_n)\}$ in $G(T)$ such that $(x_n, y_n) \to (x, y)$, i.e. $\|x_n - x\| + \|y_n - y\| \to 0$. Hence $\|x_n - x\|_{\overline{T}} = \|x_n - x\| + \|\overline{T}(x_n - x)\| \to 0$. Thus $z = \lim G_{\overline{T}}^{-1}x_n = \lim G^{-1}x_n \in \overline{X}_T$. Consequently, by (3) and the continuity of $z' \in X'_{\overline{T}}$, the equality (3) holds for all $z \in X_{\overline{T}}$. Therefore $(y', z') \in G((\overline{T}G_{\overline{T}})')$, showing that both sides of (2) are equal. □

V.5.26 Theorem. *Let X be complete and T closable. Then the following properties are equivalent:*

(i) $T \in F_-$
(ii) $\overline{\beta}(T) < \infty$ *and* $\gamma(T') > 0$
(iii) $\overline{\beta}(\overline{T}) < \infty$ *and* $\gamma(\overline{T}) > 0$
(iv) $TG \in F_-$.

PROOF. (i) $\Longrightarrow$ (ii). Assume (i). Then $\overline{\beta}(T) < \infty$ (Theorem V.5.21) and $T' \in \phi_+$ whence $\gamma(T') > 0$ by the Open Mapping Theorem (III.4.2(b)).

(ii) $\Longrightarrow$ (iii). Assume (ii). Then $\overline{\beta}(\overline{T}) \leq \overline{\beta}(T) < \infty$, and $\gamma(\overline{T}) > 0$ by Theorem III.5.3.

(iii) $\Longrightarrow$ (iv). Assume (iii). Then $\gamma(\overline{T}G_{\overline{T}}) > 0$ by Corollary IV.3.11. Also $\alpha((\overline{T}G_{\overline{T}})') = \overline{\beta}(\overline{T}G_{\overline{T}}) = \overline{\beta}(\overline{T}) < \infty$. Hence $(\overline{T}G_{\overline{T}})' \in F_+$. Therefore $(TG)' \in F_+$ by Lemma V.5.25, i.e., $TG \in F_-$.

Finally, (iv) $\Longrightarrow$ (i) by Proposition V.5.24. □

V.5.27 Corollary. *If T is closable, then*

$$T \in F_- \iff TG \in F_-.$$

PROOF. Clearly T is closable if and only if T is closable when viewed as an element of $LR(\tilde{X}, Y)$. Writing $J := J_X^{\tilde{X}}$, we thus have by Theorem V.5.26,

$$TJ^{-1} \in F_- \iff TJ^{-1}G_{TJ^{-1}} \in F_-.$$

From the definition of adjoint it is obvious that $T' = (TJ^{-1})'$. Furthermore

$$G_{TJ^{-1}} = JG_T$$

so

$$TJ^{-1}G_{TJ^{-1}} = TJ^{-1}JG_T = TG_T.$$

Hence

$$T \in F_- \iff TG \in F_-. \quad \square$$

V.6 OPERATOR RANGES

V.6.1 Definition. A subspace R of a Banach space Y is called an **operator range** if there exists a Banach space X and a bounded linear operator $T : X \to Y$ with domain X and range R.

Not all linear subspaces of a Banach space are operator ranges; for example, the kernel of a discontinuous linear functional is not an operator range. The domain and range of a completely closed linear relation are operator ranges.

V.6.2 Proposition. *Let R_1 and R_2 be operator ranges in the Banach space Y such that $R_1 + R_2$ and $R_1 \cap R_2$ are closed. Then R_1 and R_2 are closed.*

PROOF. There exist Banach spaces X_1 and X_2 and bounded linear operators $T_1 : X_1 \to Y$ and $T_2 : X_2 \to Y$ such that $R(T_1) = R_1$ and $R(T_2) = R_2$. Define $T : X_1 \times X_2 \to Y$ by $T(x_1, x_2) = T_1x_1 + T_2x_2$. Then $R(T) = R_1 + R_2$.

Let P be the projection of $X_1 \times X_2$ onto $X_1 \times \{0\}$ with kernel $\{0\} \times X_2$. Then $TP(x_1, x_2) = T_1x_1$ and $T(I - P)(x_1, x_2) = T_2x_2$ for $(x_1, x_2) \in X_1 \times X_2$. Now $T^{-1}(R_1)$ is closed, since T is continuous, $R_1 \cap R_2$ is closed, and $T^{-1}(R_1) = (T(I - P))^{-1}(R_1 \cap R_2)$. Observing that T has closed range and that $T^{-1}(R_1)$ contains $N(T)$, it follows that R_1 is closed. Similarly, R_2 is closed. □

V.6.3 Corollary. *Let $S \in LR(Z, Y)$ and $T \in LR(X, Y)$ be linear relations for which $R(\tilde{S}) + R(\tilde{T})$ and $R(\tilde{S}) \cap R(\tilde{T})$ are complete. Then $R(\tilde{S})$ and $R(\tilde{T})$ are closed.*

PROOF. We may clearly assume X, Y, Z complete and S and T closed. Furthermore, since $R(T) = D(T^{-1}) = D(Q_{T^{-1}}T^{-1})$, $R(S)$ and $R(T)$ are the domains of closed linear operators. But the domain of a closed linear operator between Banach spaces is an operator range; indeed if $A : D(A) \subset X \to Y$ is such an operator, then with the notation of IV.3.1, $D(A)$ is the range of the bounded linear operator G_A defined on the Banach space X_A. We conclude that $R(S)$ and $R(T)$ are operator ranges, and the result follows from Proposition V.6.2. □

V.6.4 Remark We note the following observation from the proof of Corollary V.6.3 :

$$R(\tilde{T}) \textit{ is an operator range in } \tilde{Y}.$$

Of special interest is the following particular case of Corollary V.6.3:

V.6.5 Corollary. *If $R(\tilde{T})$ has a complementary closed subspace in $\tilde{Y}$, then $R(\tilde{T})$ is complete.*

It is shown by J.W. Calkin [1] that if T is a closed densely defined linear operator between two separable Hilbert spaces, then T is compact if and only if its range contains no closed infinite dimensional subspace. Motivated by this property we introduce the following definition:

V.6.6 Definition A linear relation $T \in LR(X, Y)$ is called a **Calkin** relation if its range contains no infinite dimensional Banach space.

V.6.7 Theorem. *Let X and Y be Banach spaces. If T is a closed Calkin relation, then T is continuous and strictly singular.*

PROOF. Since $D(T^{-1})$ contains no closed infinite dimensional subspace, T^{-1} is a ϕ_+-relation by Theorem V.1.6 and Corollary V.1.7. Consequently $D(T) = R(T^{-1})$ is closed and so T is continuous by the Closed Graph Theorem. Now suppose M is an infinite dimensional closed subspace of $D(T)$ such that $T|_M$ is injective and open. Then $QT|_M$ is an isomorphism and therefore $QT(M)$ is a closed infinite dimensional subspace of QY. But then $T(M)$ is a closed infinite dimensional subspace of Y, contradicting the Calkin property of T. Therefore T is strictly singular. □

Since strictly singular operators in Hilbert space are compact, Theorem V.6.7 covers Calkin's theorem referred to above.

V.6.8 Corollary. *Let X and Y be Banach spaces and T closed.*

(a) *If $D(T)$ contains no infinite dimensional Banach space, then $R(T)$ is closed.*

(b) *If $R(T)$ contains no infinite dimensional Banach space, then $D(T)$ is closed.*

V.6.9 Corollary. *Let X and Y be Banach spaces and T closed. We have:*

(a) *T is Calkin $\Longrightarrow$ T is continuous $\Longrightarrow$ T^{-1} is not Calkin*

(b) *T^{-1} is Calkin $\Longrightarrow$ T is open $\Longrightarrow$ T is not Calkin.*

V.6.10 Remarks Let X and Y be Banach spaces:

(1) The existence of a bounded (and necessarily strictly singular) linear mapping of ℓ_1 onto ℓ_2 shows that the converse of Theorem V.6.7 fails without additional assumptions. However, if T is continuous and strictly singular and the null space $N(\overline{T})$ of $\overline{T}$ is complemented in $D(\overline{T})$, say $D(\overline{T}) = N(\overline{T}) + E$ when E is closed and $N(\overline{T}) \cap E = \{0\}$, then $R(\overline{T})$ (and hence $R(T)$) contains no infinite dimensional closed subspace. Indeed, $\overline{T}$ is strictly singular by Corollary IV.4.12, and it follows from the Closed Graph Theorem that the injective continuous strictly singular linear relation $\overline{T}|_E$ defined everywhere on the Banach space E is a Calkin relation.

(2) Let T be closed and let $D(T)$ contain no infinite dimensional Banach space; any relation with a strictly singular resolvent has this property (for instance, if T has a compact resolvent – see Section VII.2). Then according to Corollary V.6.8, for every infinite dimensional and relatively closed subspace E of $D(T)$, $T(E)$ is a Banach space.

V.7 THE ESSENTIAL STATE DIAGRAM

Our aim in this section is to set up a state diagram in which the states are finite dimensional variants of the standard states of Section III.6, the so–called "essential states". The diagram will show that the classes of upper and lower semi–Fredholm relations bear the same relationship to each other as the corresponding classes of bounded linear operators.

V.7.1 Definition: Essential States

The **essential states** are defined as follows:

I_e: $R(T)$ is closed and $\beta(T) < \infty$
II_e: $R(T)$ is not closed and $\overline{\beta}(T) < \infty$
III_e: $\overline{\beta}(T) = \infty$

1e: $T \in F_+$
2e: $\alpha(T) < \infty$ and $T \notin F_+$
3e: $\alpha(T) = \infty$.

W shall employ a notation analogous to that in Section III.6. Thus, for example, if $T \in II_e$ and $T \in 3e$, then we shall write $T \in II_3e$. The same notation is used for the adjoint.

V.7.2 Lemma. *Let T be a closed linear relation. Then T and QT belong to the same essential states.*

PROOF. By II.3.4, $\alpha(T) = \alpha(QT)$, and by definition, $T \in 1e$ if and only if $QT \in 1e$. Therefore $T \in 1e, 2e$ or $3e$ accordingly as $QT \in 1e, 2e$ or $3e$ respectively.

Clearly $R(T)$ is closed if and only if $QR(T)$ is closed, since $T(0) \subset Q^{-1}(0) \subset \overline{R(T)}$. Moreover, by IV.5.2, $\overline{\beta}(T) = \overline{\beta}(QT)$ since $T(0) \subset \overline{R(T)}$. Thus T belongs to I_e, II_e or III_e according as QT belongs to the same class. □

V.7.3 Proposition.

(a) $T \in 1e \implies T' \in I_e$
(b) $T \in I_e \cup II_e \iff T' \in 1e \cup 2e$
(c) $T \in 3e \implies T' \in III_e$
(d) $T \in I_1e \cup II_1e \implies T' \in I_1e$
(e) $T' \notin II_1e$
(f) $T \in 2e \implies T' \in II_e \cup III_e$
(g) *Y complete and $T \in I_2e \cup I_3e \implies T' \in III_1e$.*

PROOF. (a) We have by III.1.10

$$(1) \qquad R(T') = R((QT)')$$

and also

$$(2) \qquad (TJ_{D(T)})' = Q_{T'}T'.$$

Now let $T \in 1_e = F_+$. Then there exists a closed finite codimensional subspace E of $D(T)$ such that $TJ_E^X \in 1$. Write $J := J_E^X$. We have by III.1.6(b) that

$$(3) \qquad (TJ)' = J'T'.$$

Hence $J'T' \in I$, by Lemma III.6.2. Replacing T by $TJ_{D(T)}$ if necessary, we may according to (2), assume that T' is single valued. Let $x' \in X'$ and choose $y' \in Y'$ such that $x' + E^{\perp} = T'y' + E^{\perp}$. Then there exists $w \in E^{\perp}$ such that $x' = T'y' + w$. Therefore $X' = R(T') + E^{\perp}$ where $E^{\perp}$ is finite dimensional. But $R(T')$ is an operator range. Hence by V.6.5, $R(T')$ is closed. Therefore $T' \in I_e$.

(b) Follows immediately from the equality $\alpha(T') = \overline{\beta}(T)$ (Proposition III.7.2).

(c) Let $T \in 3e$, i.e. $\alpha(T) = \infty$. Since $N(T) \subset R(T')^{\top}$ (see III.1.4), therefore $N(T)^{\perp} \supset \overline{R(T')}$. Hence if $T' \in I_e \cup II_e$, then $\operatorname{codim} N(T)^{\perp} < \infty$, which implies $\dim N(T) < \infty$. Hence $T' \in III_e$.

(d) Let $T \in (I_e \cup II_e) \cap 1e$. Then $T' \in 1e \cup 2e$ by (b), and $T' \in I_e$ by (a). Therefore $T' \in \phi_+$, or equivalently, $T' \in F_+$, by Corollary V.1.7. Hence $T' \in I_e \cap 1e$.

(e) Suppose $T' \in II_e \cap 1e$. Then T' is upper semi–Fredholm and in particular has closed range (by V.1.7), contradicting $T' \in II_e$.

(f) Assume $T \in 2e$. Then there exists an infinite dimensional subspace M of $D(T)$ for which TJ_M^X is precompact, by Theorem V.1.6. Write $J = J_M^X$. Then $(TJ)'$ is precompact (Proposition V.5.3). Now in view of Lemma V.6.2, and the equalities (1) and (2), we may assume that T and T' are single valued (for example, to determine the state of T' we look at $Q_{T'}T'$ obtained by replacing T by $TJ_{D(T)}$). In that case $(TJ)'$ is an extension of $J'T'$, and so $J'T'$ will be compact. Now suppose if possible that $T' \in I_e$. Let $X' = R(T') + W$ where $\dim W < \infty$ and $R(T') \cap W = \{0\}$. Then

$$J'(X') = J'(R(T') + W) \subset R(J'T') + J'W = J'(X')$$

since $J' \in I$. Thus

$$R(J'T') + J'W = J'X' = X'/M^{\perp}.$$

But $R(T')$ is an operator range and so therefore is its continuous linear image $R(J'T')$. Consequently $R(J'T')$ is closed by V.6.5. Since $J'T'$ is a compact operator, it follows that $R(J'T')$ is finite dimensional. Therefore $\dim X'/M^{\perp} < \infty$. Hence $\dim M = \dim M' = \dim X'/M' < \infty$, a contradiction. We conclude that $T' \notin I_e$.

(g) Let Y be complete. First suppose that $T \in I_2e$. Then there exists a closed finite codimensional subspace E of $D(T)$, and a finite dimensional subspace M of Y, such that $E + N(T) = D(T)$, $E \cap N(T) = \{0\}$, and $R(T) + M = Y$, $R(T) \cap M = \{0\}$. Write $Q = Q_M^Y$ and $J = J_E^X$. Clearly $\alpha(TJ) = 0$; furthermore by Corollary II.6.4, $TJ \in 2$. Moreover $R(TJ) = R(T)$. Hence $TJ \in I_e$. It is now easily verified that $QTJ \in I_2$. Now by III.1.6(a) and (b), $(QTJ)' = (TJ)'Q' = J'T'Q'$, and by the State Diagram III.6.3, $(QTJ)' \in III_1$. By Proposition III.1.15, $R((QTJ)') = R(J'T')$. In particular, $R(J'T')$ is closed. But since $TJ \in I_2e$, (b), (e) and (f) give

$$J'T' \in III_1e \cup II_2e \cup III_2e.$$

It follows that $J'T' \in III_1e$. By Proposition V.2.16(b), we now have $T' \in 1e$, while the finite dimensionality of $E^{\perp}$ implies that $T' \in III_e$ (see III.1.14). Thus $T' \in III_1e$.

Now suppose that $T \in I_3e$. Then $Y = R(T) + M$ where $\dim M < \infty$ and $M \cap R(T) = \{0\}$. Write $Q = Q_M^Y$. Then $QT \in I \cap 3e \subset I_3$ (since $\alpha(QT) \geq \alpha(T)$). Since Y is complete, the State Diagram III.6.3 give $(QT)' \in III_1$. But $(QT)' = T'Q' = T'J_{M^{\perp}}^{Y'}$ (Theorem III.1.6), where $\operatorname{codim} M^{\perp} < \infty$. Therefore $T' \in 1e$. Combining with (c) we obtain $T' \in III_1e$. □

The results of Proposition V.7.3 are equivalent to the Essential State Diagram for an arbitrary linear relation. The configuration coincides with that of the standard State Diagram III.6.3, to which we refer the reader. We formalise this observation thus:

V.7.4 Theorem. *The Essential State Diagram coincides with the State Diagram III.6.3 upon substituting "essential states" for the usual "states".*

V.7.5 Corollary. *We have*

$$T \in F_+ \iff T' \in F_-.$$

PROOF. By Theorem V.7.4, $T \in F_+$ if and only if $T' \in I_e$. It now suffices to apply Proposition V.5.2. □

V.7.6 Corollary. *We have*

$$T \in F_+ \iff \tilde{T} \in \phi_+.$$

PROOF. Combine Corollaries V.1.7 and V.7.5. □

V.7.7 Corollary. *Let F be a finite dimensional subspace of Y. Then*

$$T \in F_+ \iff Q_F^Y T \in F_+.$$

In particular, $Q_F^Y \in F_+$.

PROOF. We may clearly suppose that $D(T) = X$. Let $T \in F_+$. We have by III.1.6,

$$(Q_F^Y T)' = T' J_{F^\perp}.$$

Now $\dim Y'/F^\perp = \dim F' = \dim F < \infty$. Therefore $D(T') \cap F^\perp$ has finite codimension in $D(T')$ (by IV.1.2). Hence $T'(F^\perp)$ has finite codimension in $R(T')$ (by I.6.1). But $R(T')$ has finite codimension in X', since $T' \in \phi_-$ (by V.7.5 and V.5.2). Therefore $R(T' J_{F^\perp})$ has finite codimension in X', and is moreover closed, by Corollary V.6.5. Hence $T' J_{F^\perp} \in F_-$ (V.5.2), whence $Q_F^Y T \in F_+$ by Corollary V.7.5.

The converse is a special case of Proposition V.2.16. □

V.7.8 Lemma. *Let $S \in LR(X,Y)$ have $\dim S(0) < \infty$. Then $T+S-S \in F_+$ if and only if $T \in F_+$.*

PROOF. We have

$$Q_S(T + S - S) = Q_S T + Q_S(S - S) = Q_S T$$

by I.4.2(e). Hence $Q_S(T + S - S) \in F_+$ if and only if $T + S - S \in F_+$, if and only if $T \in F_+$, by Corollary V.7.7. □

V.7.9 Definition. Let $\mathcal{S}(X,Y)$ (or simply $\mathcal{S}$) be a subset of $LR(X,Y)$. We denote by $\mathbf{P}(\mathcal{S})$ the class of linear relations in $LR(X,Y)$ defined as follows:

$$\mathbf{P}(\mathcal{S}) = \{S \in LR(X,Y) : \dim S(0) < \infty \text{ and } T \in \mathcal{S} \implies T + S \in \mathcal{S}\}.$$

$\mathbf{P}(\mathcal{S})$ is called the **perturbation class** of $\mathcal{S}$.

The Example V.3.1 shows that some restriction on the singular part of the perturbating relation is necessary for the above definition to be effective.

Let $SS(X,Y)$ (abbreviated SS) denote the class of strictly singular relations in $LR(X,Y)$.

We can now prove our main F_+-perturbation result.

V.7.10 Theorem. *$\mathbf{P}(F_+)$ coincides with the class of strictly singular relations S with $\dim S(0) < \infty$.*

PROOF. Let $T \in F_+$ and let $S \in SS$, $\dim S(0) < \infty$. We have from I.4.3(e),

(1) $$Q_S(T+S) = Q_S T + Q_S S$$

where $Q_S T \in F_+$ by Corollary V.7.7. Since $Q_S S$ is strictly singular and single valued, $Q_S T + Q_S S \in F_+$ by the perturbation result V.3.2, combined with V.2.4 and V.2.6. Hence from (1), $Q_S(T+S) \in F_+$. Consequently by Corollary V.7.7, $T+S \in F_+$. Therefore $SS \subset \mathbf{P}(F_+)$.

For the reverse inclusion let $S \in LR(X,Y)$ with $S \notin SS$. Then $\Delta(S) > 0$ (by V.2.5) and hence there exists $M \in \mathcal{I}(D(S))$ such that $\Gamma(S|_M) > 0$. Define $T \in LR(X,Y)$ by $D(T) = M$ and $Tx = -Sx$ $(x \in M)$. Then $\Gamma(T) = \Gamma(S|_M) > 0$. Hence $T \in F_+$ (Theorem V.2.4), while $T + S = S - S|_M \notin F_+$ (as its norm is zero), showing that $S \notin \mathbf{P}(F_+)$. Therefore $\mathbf{P}(F_+) \subset SS$. □

We remark that the statement of Theorem V.7.10 is know to be false for the restricted class of bounded linear operators in Banach spaces. (Since the operators would be everywhere defined, one would have to construct a bounded extension of T from M to the whole space in the second part of the above proof, and this is not in general possible.)

V.7.11 Corollary. *The following properties are equivalent for a linear operator $T \in LR(X,Y)$:*

(i) $T \in F_+$

(ii) *$N(T+K)$ is finite dimensional whenever $K \in LR(X,Y)$ is precompact and $\dim K(0) < \infty$.*

(iii) *$N(T+K)$ is finite dimensional for every precompact linear operator $K \in LR(X,Y)$.*

PROOF. We may clearly suppose that $\dim D(T) = \infty$.

(i) $\implies$ (ii): Immediate from Theorem V.7.10.

(ii) $\implies$ (iii): Trivial.

(iii) $\implies$ (i): Assume $T \notin F_+$. By Theorem V.1.5, there exists $M \in \mathcal{I}(D(T))$ for which $T|_M$ is precompact. Define K by $D(K) = M$ and $Kx = -Tx$ $(x \in M)$. Then K is precompact and $N(T+K) \supset M$. Thus $N(T+K)$ is infinite dimensional. Since K is single valued, it follows that (iii) $\implies$ (i). □

V.7.12 Exercise Let $T \in F_+$ and let M be a finite dimensional subspace of Y. Show that $T+S \in F_+$ for all linear relations S satisfying $\|S\| \le \gamma(Q_M T)$ and $S(0) = M$.

V.8 ESSENTIAL STATE DIAGRAM FOR CLOSED RELATIONS

V.8.1 Lemma. *Let Z be a linear subspace of Y containing $R(T)$, and let $S \in LR(X, Z)$ be defined by*

$$J_Z^Y S = T.$$

Then $R(S') = R(T')$.

PROOF. We have $T' = S' Q_{Z^\perp}^{Y'}$, by III.1.6. It follows immediately that $R(T') = R(S')$ (see Proposition III.1.2). □

V.8.2 Proposition. *Let X be complete and T closed and let $R(T')$ be closed. Then $R(T)$ is closed.*

PROOF. Let T_0 be the relation T considered as an element of $LR(X, \overline{R(T)})$. Then T_0 is closed, and $R(T_0') = R(T')$ by Lemma V.8.1. But T_0' is injective, and is open by the Closed Graph Theorem. Thus $T_0' \in 1$ and so by the State Diagram for closed linear relations III.6.5, $T_0 \in I$. Therefore $R(T) = R(T_0) = \overline{R(T)}$. □

(Proposition V.8.2 should be compared with Proposition III.5.3(a).)

V.8.3 Proposition. *Let X be complete and T closed. Then*

(a) $T \in I_2 e \cup II_e \implies T' \notin 1e$.

(b) *If X is reflexive, then $T \in 2e \implies T' \notin III_e$.*

PROOF. (a) Let $T \in II_e$. Then $R(T)$ is not closed, and hence $R(T')$ is not closed, by Proposition V.8.2. Hence $T' \notin 1e$.

Now let $T \in I_2 e$. Let Q and J be as in the proof of Proposition V.7.3(g). Then $QTJ \in I_2$ and $QTJ \in LR(E, Y/M)$ where E is complete. Moreover QTJ is closed, by Proposition II.5.13. By the State Diagram for closed linear relations III.6.5, $(QTJ)' \in II_2 \cup III_2$. But $(QTJ)' = J'T'Q'$ by III.1.6, and $R(J'T'Q') = R(J'T')$ by Lemma V.8.1. Therefore $R(J'T')$ is not closed (by the Open Mapping Theorem III.3.2). Therefore $R(T')$ is not closed (as J' is a quotient map with finite dimensional kernel $E^\perp$). In particular, $T' \in 1e$.

(b) Let $T \in 2e$. There exists a finite codimensional subspace E of $D(T)$, closed in $D(T)$, such that $E + N(T) = D(T)$ and $E \cap N(T) = \{0\}$. Write $J = J_E^X$. Then $TJ \in 2$ and hence from the State Diagram for closed linear relations III.6.5, and Theorem III.1.6(b), $J'T' \in I \cup II$. It follows immediately that $\overline{\beta}(T') < \infty$, i.e. $T' \notin III_e$. □

Combining Propositions V.8.2 and V.8.3 with the diagram V.7.4 now yields the following:

V.8.4 Theorem. *The Essential State Diagram for closed linear relations coincides with the State Diagram for linear relations III.6.5 upon substituting "essential states" for the usual "states".*

Since V.8.4 already contains V.7.4 we shall in future refer to V.8.4 only as "the essential state diagram".

We conclude this section with a result showing that the Essential State Diagram for closed linear relations in Banach spaces is stable under bounded finite rank perturbations.

V.8.5 Theorem. *Let X and Y be Banach spaces and $T \in LR(X,Y)$ a closed linear relation. Let $F \in LR(X,Y)$ satisfy $D(F) \supset D(T)$, $\|F\| < \infty$ and $\dim R(F) < \infty$. Then T and $T+F$ are in the same essential states.*

PROOF. Let $\alpha(T) < \infty$. Suppose that $x \in (T-F)^{-1}(0)$. Then $Tx - Fx = T(0) - F(0)$. Hence $Tx + F(0) = Fx + T(0)$. Applying T^{-1} to both sides, noting that $T^{-1}(0) \subset T^{-1}F(0)$, gives

$$x + T^{-1}F(0) = T^{-1}Fx + T^{-1}F(0).$$

Thus

$$x \in T^{-1}F(X)$$

where $\dim T^{-1}F(X) \leq \dim F(X) + \dim T^{-1}(0) < \infty$ since $\dim T^{-1}(0) < \infty$. Therefore $\alpha(T+F) < \infty$.

Let $\gamma(T) > 0$. Then $T(X)$ is closed by the Open Mapping Theorem III.3.2. Now

$$(T+F)(X) \subset T(X) + F(X)$$

and

$$T(X) \subset (T+F)(X) + F(X).$$

Therefore

$$T(X) + F(X) = (T+F)(X) + F(X) \tag{1}$$

where $T(X) + F(X)$ is a Banach space. Therefore $(T+F)(X)$ is closed by Corollary V.6.5. Hence $\gamma(T+F) > 0$ (Theorem III.3.2).

Now let $\overline{\beta}(T) < \infty$. We have from (1),

$$\overline{T(X)} + F(X) = \overline{(T+F)(X)} + F(X).$$

Hence $\overline{\beta}(T+F) \leq \dim F(X) + \overline{\beta}(T) < \infty$.

Let us define the states A, C, D as follows:

$A : \alpha(T) < \infty$

$C : \gamma(T) > 0$

$D : \overline{\beta}(T) < \infty.$

Denoting the complementary states respectively by A', C' and D', it is easily verified that

$$\begin{aligned}
I_1 e &= ACD \\
I_2 e &= \emptyset \\
I_3 e &= A'CD \\
II_1 e &= \emptyset \\
II_2 e &= AC'D \\
II_3 e &= A'C'D \\
III_1 e &= ACD' \\
III_2 e &= AC'D' \\
III_3 e &= A'D'
\end{aligned}$$

Since the states A, C, D have been shown to be stable, the stability of the essential states follows. □

V.9 DENSITY INVARIANCE OF COMPLETELY CLOSABLE LINEAR RELATIONS

This section contains further characterisations of partially continuous linear relations and a density invariance result for completely closable relations.

V.9.1 Definition The linear relation $T \in LR(X, Y)$ is called **semicontinuous** if it is of the form $T = A + F$ where A is continuous, and F has finite dimensional range.

Observe that

(1) T **semicontinuous** $\Longrightarrow$ T **partially continuous.**

The converse of (1) holds if T is single valued and Y is complete. We have:

V.9.2 Proposition. *If T is partially continuous and single valued, then $J_Y^{\tilde{Y}}T$ is semicontinuous.*

PROOF. We may clearly assume $D(T) = X$. Write $J = J_Y^{\tilde{Y}}$. Let T be partially continuous. Then $X = M + E$ where $M \cap E = \{0\}$, $\dim E < \infty$, and $T|_M$ is continuous. Since $JT|_M$ is single valued, it has a continuous and single valued extension $(JT|_M)^-$ to $\overline{M}$. Since $\overline{M}$ is topologically complemented in X (II.3.18), $(JT|_M)^-$ extends linearly to a bounded operator $A \in LR(X, \tilde{Y})$ defined on X. Let $F = T - A$. Then $\dim R(F) < \infty$ and $JT = A + F$; i.e. JT is semicontinuous. □

V.9.3 Theorem. *Let T be single valued. Then the following properties are equivalent:*

(i) *T is semicontinuous.*

(ii) *There exists a finite dimensional subspace E of Y such that $Q_E^Y T$ is continuous.*

(iii) *There exist linear operators A and F, with A bounded and $R(F)$ finite dimensional, and $R(A)$ and $R(F)$ both contained in $\overline{R(T)}$, such that $T = A + F$.*

(iv) *$D(T')$ is $\sigma(Y', Y)$ closed and finite codimensional in Y'.*

PROOF. First we prove the equivalence of (i), (ii) and (iii). We may clearly suppose $D(T) = X$. We have (iii) $\Longrightarrow$ (i) trivially.

(i) $\Longrightarrow$ (ii): Assume (i). Then $T = A + F$ where A is bounded and $\dim R(F) < \infty$. Hence $Q_{R(F)}^Y T = Q_{R(F)}^Y A$ which is continuous.

(ii) $\Longrightarrow$ (iii): Assume that (ii) holds. Let $Y = E + M$ where $E \cap M = \{0\}$ and M is closed. Then E and M are topological complements (Proposition II.4.7). Therefore M is norm isomorphic to Y/E under the isomorphism $Q_E^Y J_M^Y$. Also $\dim R(T)/M \cap R(T) < \infty$ (see Proposition IV.1.2(a)). Hence $\dim X/T^{-1}(M \cap R(T)) < \infty$ by I.6.4. Write $N = T^{-1}(M \cap R(T))$. Then since $(Q_E^Y J_M^Y)T|_N$ is continuous, so is $T|_N$ (as T is single valued). Choose W so that $X = \overline{N} + W$, $\overline{N} \cap W = \{0\}$, and define the operator A to be equal to zero on W and equal to the continuous extension (in $LR(X, \tilde{Y})$) of $T|_N$ on $\overline{N}$. Then $A \in LR(X, \tilde{Y})$ and is clearly bounded and single valued. Let $x \in \overline{N}$ and let (x_n) be a sequence in N converging to x. Then $Ax_n = Tx_n \to Ax \in \tilde{Y}$. But since $Q_E^Y T$ is continuous by hypothesis, we have $Q_E^Y Tx_n \to Q_E^Y Tx$. Select a sequence (e_n) in E so that $Tx_n + e_n \to Tx$. Then $e_n \to Tx - Ax \in E$, since E is finite dimensional. We thus have $Tx - Ax \in Y$, whence $Ax \in Y$. This shows that $R(A) \subset \overline{R(T)} \subset Y$. Now $T - A$ has finite dimensional range. Hence (ii) $\Longrightarrow$ (iii).

Finally, we prove the equivalence of (i) and (iv).

Let $\operatorname{codim} D(T') < \infty$, with $D(T')$ $\sigma(Y',Y)$ closed, i.e. $D(T') = D(T')^{\top\perp}$. Then since $T'J^{Y'}_{D(T')} = T'J^{Y'}_{D(T')^{\top\perp}} = (Q^{Y}_{D(T')^{\top}}T)'$ is everywhere defined on the Banach space $D(T')$ (III.1.10), we have $Q^{Y}_{D(T')^{\top}}T$ continuous (III.3.6). But $\dim D(T')^{\top} = \dim(D(T')^{\top})' = \dim Y'/D(T') < \infty$. Therefore T is semicontinuous, by the equivalence of (i) and (ii) proved above.

Conversely, let T be semicontinuous. Then $T = A + F$ when A is continuous, $D(A) = D(F)$ and F has finite dimensional range. We have $D(F') \supset N(F') = R(F)^{\perp}$ (III.1.4) where the latter is a $\sigma(Y',Y)$ closed subspace of finite codimension in Y'. Therefore $D(F')$ is such a subspace. Since A is continuous, $T' = A' + F'$ (III.1.5) where $D(A') = Y'$ (by III.3.6 as A is single valued). Hence $D(T') = D(F')$ and the theorem is proved. □

V.9.4 Corollary. *Let T be a linear operator. Then T is partially continuous if and only if $D(T')$ is a finite codimensional and $\sigma(Y',\tilde{Y})$ closed subspace of Y'.*

V.9.5 Corollary. *Completely closable partially continuous linear relations are continuous.*

PROOF. Let T be completely closable and partially continuous. Then QJ_YT has the same properties (II.5.7). Observing that a closable linear operator with finite dimensional range is continuous (since its null space is topologically complemented in its domain – see II.3.18), it follows from Proposition V.9.2 that QJ_YT is continuous. Hence T is continuous. □

V.9.6 Corollary. *The adjoint of a partially continuous linear relation is continuous.*

PROOF. We have

T partially continuous
$\Longrightarrow$ QT partially continuous
$\Longrightarrow$ $D((QT)') = D(T')$ complete (V.9.4)
$\Longrightarrow$ T' continuous (III.3.2). □

V.9.7 Exercise Let T be single valued. Then T is continuous if and only if T is closable and semicontinuous.

V.9.8 Example *A closed partially continuous linear operator which is not semicontinuous.* Let f be a discontinuous linear functional on X and let Y be the space X renormed by

$\|x\|_Y = \|x\| + |f(x)|$ $(x \in X)$. Define $T \in LR(X,Y)$ by $Tx = x$ $(x \in X)$. Then T^{-1} is bounded and $T|_{N(f)}$ is an isometry. Thus T *is discontinuous but closed and partially continuous.* From Exercise V.9.7, we see that T is not even semicontinuous.

As an application of semicontinuity we have the following density result for partially continuous linear relations:

V.9.9 Corollary. *Let T be partially continuous, let E be a dense subspace of $D(T)$ and let $f \in \{\Delta, \Gamma, \Gamma_0, \tau, \tau_0\}$. Then $f(T) = f(T|_E)$.*

PROOF. We shall assume without loss of generality that Y is complete and that T is single valued. Then there exists an everywhere defined finite rank operator F making $T + F$ continuous (Proposition V.9.2). Thus $(T + F)|_E$ is closable. Moreover the minimal closed extension of $(T + F)|_E$ is an extension of $T + F$ (indeed it coincides with the extension by continuity of $T + F$). Hence by Corollary IV.4.9 $f((T + F)|_E) = f(T + F)$. Therefore by Proposition IV.1.3(ii) $f(T) = f(T + F) = f(T|_E + F|_E) = f(T|_E)$. □

V.9.10 Theorem. *Let T be completely closable and let E be a dense linear subspace of $D(T)$. Then $f(T|_E) = f(T)$ for $f \in \{\Delta, \Gamma_0, \overline{\Gamma}_0, \tau\}$.*

PROOF. We shall assume without loss of generality that Y is complete. By Proposition II.5.7 we may furthermore suppose that T is single valued.

Case (i): Let f be one of the quantities Δ or τ. If $f(T|_E) = \infty$ then obviously $f(T|_E) = f(T)$. Now let $f(T|_E) < \infty$. Then $T|_E$ is partially continuous by Theorem V.4.3 and hence continuous, by Corollary V.9.5. It follows that T is continuous. Therefore E is a core of T and hence $f(T|_E) = f(T)$ by Theorem IV.4.8.

Case (ii): The quantity Γ_0. Suppose that $\Gamma_0(T|_E) = \infty$. Then if $\Gamma_0(T) = \infty$ there is nothing to prove. Accordingly let $\Gamma_0(T) < \infty$. Then T is partially continuous (V.4.3). Hence $T = A+F$ where A is continuous and F is a finite rank operator (Proposition V.9.2). But by Proposition IV.1.3 (ii), $\Gamma_0(T+F) = \Gamma_0(T)$ (as $\dim D(T)/N(F) < \infty$). We may thus suppose that $F = 0$, i.e. that T is continuous. But then E is a core of T and then $\Gamma_0(T|_E) = \Gamma_0(T)$ by Theorem IV.4.8. Finally suppose that $\Gamma_0(T|_E) < \infty$. Then $T|_E$ is partially continuous and hence continuous by V.9.5. Since T is closable it follows that T coincides with the continuous extension of $T|_E$, i.e. T is continuous. It follows again similarly that $\Gamma_0(T|_E) = \Gamma_0(T)$.

Case (iii): The quantity $\overline{\Gamma}_0$. Suppose that $\overline{\Gamma}_0(T|_E) = \infty$. Then $\overline{\Gamma}_0(T) = \infty$ (see IV.2.14). Now assume $\overline{\Gamma}_0(T|_E) < \infty$, i.e. $T|_E$ is continuous. Since T is closable it coincides with the continuous extension of $T|_E$. Therefore T is continuous. Hence $\overline{\Gamma}_0(T|_E) = \Gamma_0(T|_E) = \Gamma_0(T) = \overline{\Gamma}_0(T)$. □

V.9.11 Corollary. *Let X and Y be Banach spaces, let T be closable, and let E be a dense linear subspace of $D(T)$. If $T|_E$ is compact (or strictly singular), then T is compact (resp. strictly singular).*

That the completeness of Y is essential in Corollary V.9.11 is evident from the next example.

V.9.12 Example *An unbounded closed strictly singular operator defined everywhere on a Hilbert space which is compact and injective on a dense subspace of codimension one.*

Let $X = \ell_2 = Z$ and let T_0 be a restriction of the compact operator $S : (a_n) \to (\frac{1}{n} a_n)$ to a dense subspace of codimension one. Set $Y = R(T_0)$ and select a point $x_0 \in X \setminus D(T_0)$. Define $T \in L(X, Y)$ by $Tx = T_0 x$ $(x \in D(T_0))$ and $Tx_0 = 0$. Clearly T is an unbounded everywhere defined strictly singular operator with one dimensional null space, and its restriction T_0 has the stated property. It only remains to verify that T is closable, hence closed. Let P, Q be complementary linear projections defined on X such that $R(P) = \text{sp}\,\{x_0\}$, $R(Q) = D(T_0)$. Then $Tx = SQx$. Let $(x_n, Sx_n - SPx_n) \to (0, y)$ $(y \in Y = SQX)$. Then $Sx_n \to 0$ as S is continuous. Hence $SPx_n \to -y \in SPX$. Since S is injective, $SPX \cap SQX = 0$. Therefore $y = 0$, showing that T is closable.

V.9.13 Remark For the closed operator $T \in L(\ell_2, Y)$ constructed in Example V.9.12, we have $\bar{\Gamma}_0(T) = \infty$ and $\bar{\Gamma}_0(T|E) = 0$ where E is a dense subspace of ℓ_2 of codimension one. This shows that Theorem V.9.10 fails for closed operators T if Y is incomplete, at least for the quantity $\bar{\Gamma}_0$. We do not have examples for the other quantities Δ, Γ_0 and τ.

The following example is due to V.I. Bogachev, B. Kircheim and W. Schachermeyer [1].

V.9.14 Example *Let Y be a Banach space and let $T : \ell_1 \to Y$ be an everywhere defined linear operator (or relation). Then there exists a dense subspace $E \subset \ell_1$ for which $T|_E$ is continuous.*

Indeed, since $X = \bigcup_{n=1}^{\infty} nT^{-1}B_{R(T)}$, it follows from Baire's Category theorem, as in the proof of Theorem II.3.15, that there exists a constant $c > 0$ for which the set $U = \{x \in X : \|Tx\| \le c\}$ is dense in B_X. Now let $\{e_i\}$ denote the canonical basis for ℓ_1 and select a sequence $\{x_i\}$ in ℓ_1 satisfying $\|x_i - e_i\| < 2^{-i-1}$ and $\|Tx_i\| \le c$ $(i \in M)$. Then $\sum_{i=1}^{\infty} \|x_i - e_i\| < \frac{1}{2}$. By the basis perturbation theorem (Lindenstrauss and Tzafriri [1; Proposition 1.a.a]), there exists $d > 0$ such that $\left\|\sum_{i=1}^{n} \alpha_i x_{j_i}\right\| \ge d \left\|\sum_{i=1}^{n} \alpha_i e_{j_i}\right\| = d \sum_{i=1}^{n} |\alpha_i|$. Denoting the linear span of U by W, it follows that $\|T|_W\| \le c/d < \infty$. □

On the other hand, it is known that given an arbitrary infinite dimensional Banach space X, there exists an everywhere defined and nowhere continuous linear operator $T : X \to \ell_1$ (see Bogachev, Kircheim and Schachermeyer [1]).

V.9.15 Exercise Show that there exists Banach spaces X and Y and an everywhere defined linear operator $T : X \to Y$ such that T nowhere continuous and yet $\tilde{T}$ is continuous.

V.9.16 Exercise Let X and Y be Banach spaces and let T be an everywhere defined discontinuous linear operator with one dimensional range. Show that $\Gamma_0(T) = \infty$ and $\Gamma_0(\overline{T}) < \infty$.

V.9.17 Theorem. *Let T be completely closable and let E be a dense subspace of $D(T)$. Then for $f \in \{\Gamma, \tau_0\}$ we have*

$$f(G_{T|_E}) = f(G_T).$$

PROOF. For $f = \Gamma$ or τ_0 let $\overline{f}$ denote the quantities τ or Γ_0 respectively. We have by Proposition IV.3.6,

$$\begin{aligned} f(G_{T|_E}) &= (1 + \overline{f}(T|_E))^{-1} \\ &= (1 + \overline{f}(T))^{-1} \text{ by Theorem V.9.10} \\ &= f(G_T). \quad \square \end{aligned}$$

V.9.18 Theorem. *Let T be completely closable and let E be dense in $D(T)$. Then for $f \in \{\tau, \tau_0, \overline{\tau}_0\}$ we have*

$$\tau(T) = \sup_{V \in \mathcal{I}(E)} f(T|_V) = \sup_{M \in \mathcal{I}(D(T))} f(T|_M).$$

PROOF. By Lemma IV.4.14 (b) we have

$$\tau(T) = \sup_{M \in \mathcal{I}(D(T))} f(T|_M) \geq \sup_{V \in \mathcal{I}(E)} f(T|_E) = \tau(T|_E).$$

But $\tau(T|_E) = \tau(T)$ by Theorem V.9.10. Hence the equality follows. $\square$

V.9.19 Proposition. *Let T be completely closable and let E be a dense subspace of $D(T)$. Then*

$$f(T|_E G_{T|_E}) = f(TG_T) \text{ for } f \in \{\Delta, \Gamma_0, \overline{\Gamma}_0, \tau\}.$$

PROOF. Combine Proposition IV.3.4 with Theorem V.9.10, noting that for $\overline{\Gamma}_0$ we have

$$\overline{\Gamma}_0(T|_E G_{T|_E}) = \Gamma_0(T|_E G_{T|_E}) = \Gamma_0(TG_T) = \overline{\Gamma}_0(TG_T). \quad \square$$

V.9.20 Theorem. *Let T be an F_+–relation and let E be a subspace of X for which $T(E)$ is dense in $R(T)$. Then*

$$f(T|_E) = f(T) \text{ for } f \in \{\Delta, \Gamma, \Gamma_0, \tau, \tau_0\}$$

PROOF. We may assume without loss of generality that $D(T) = X$ and $\dim X = \infty$, and furthermore (replacing T by QT if necessary) that T is single valued. Moreover by Proposition IV.3.4, $f(T|_E) = f(T)$ if and only if $f(TG) = f(T|_E G_{T|_E}) = f(TG|_{G^{-1}(E)})$, where $T(E) = TGG^{-1}(E)$ is given to be dense in a closed finite codimensional subspace of $R(T) = R(TG)$. Hence we may also suppose, upon replacing T by TG if necessary, that T is bounded. Since T is an F_+–relation, there exists $M \in P(X)$ and a finite dimensional subspace F such that $X = M + F$, $M \cap F = \{0\}$ and $T|_M$ is injective and open.

We have $E = E \cap M + N$ where $\dim N < \infty$ (see Proposition IV.1.2(b)). Then $\overline{T(E)} = \overline{T(E \cap M)} + TN$, and thus $T|_{E\cap M}$ has a continuous single valued inverse with range dense in a closed finite codimensional subspace of $R(T|_M)$. Therefore we may assume without loss of generality that $E \subset M = X$. In that case T is now an isomorphism and $W = T^{-1}\overline{T(E)} = \overline{E}$ has finite codimension in X. But $f(T) = f(T|_W)$ (IV.1.3). Hence $f(T) = f(T|_E)$ by Theorem IV.4.8. $\square$

V.9.21 Exercise Let $\dim D(T) = \infty$. Show that:

(a) G_T is strictly singular if and only if T is nowhere continuous (and then $T \in F_+$))
(b) G_T is precompact if and only if T has a precompact inverse on some finite codimensional subspace.

As an application of the density invariance property we have the following characterisation of closed upper semi-Fredholm relations in Banach spaces.

V.9.22 Theorem. *Let X and Y be Banach spaces and let T be a closed linear relation of infinite dimensional range. Then the following properties are equivalent:*

(i) *T is a ϕ_+-relation.*

(ii) *For each linear subspace M of $D(T)$ we have $\overline{T(M)} \subset T(\overline{M})$.*

PROOF. Assume (i). Let M be a linear subspace of $D(T)$. By III.3.2, T^{-1} is continuous. Hence from II.3.1(5),

$$T^{-1}\overline{TM} \subset \overline{T^{-1}TM} = \overline{M + T^{-1}(0)}.$$

Therefore

$$\overline{TM} \subset \overline{TM} + T(0) = TT^{-1}(\overline{TM}) \subset T(\overline{M + T^{-1}(0)}). \tag{1}$$

Since $T^{-1}(0)$ is finite dimensional, $\overline{M} + T^{-1}(0)$ is closed (see e.g. Goldberg [3; I.4.12]). Hence

$$\overline{M} + T^{-1}(0) \subset \overline{M + T^{-1}(0)} \subset \overline{M} + T^{-1}(0).$$

Therefore

$$\overline{M + T^{-1}(0)} = \overline{M} + T^{-1}(0).$$

Hence

$$T(\overline{M + T^{-1}(0)}) = T(\overline{M} + T^{-1}(0)) = T\overline{M}. \tag{2}$$

From (1) and (2) we obtain

$$\overline{TM} \subset T\overline{M}.$$

Hence (i) implies (ii).

Conversely, assume (ii). Suppose $M \in \mathcal{I}(D(T))$ where $M \cap N(T) = \{0\}$. If $T|_M$ is compact, then by the density invariance result V.9.11, $T|_{\overline{M}}$ is compact. Consider the linear operator $QT|_{\overline{M}}$. We have $QT(\overline{M}) \supset Q\overline{T(M)}$. But $\overline{TM} \supset N(Q) = T(0)$. Therefore $Q\overline{T(M)} = \overline{QT(M)}$ which is an infinite dimensional Banach space, as QT is injective on M. But then $QT|_{\overline{M}}$ cannot be a compact operator, a contradiction. Hence $T \in \phi_+$, by Theorem V.1.6 and Corollary V.7.6. □

V.9.23 Exercise Characterise those bounded linear projections P on a Banach space such that $P(M)$ is closed for every closed subspace M.

V.10 INVERSES OF SEMI–FREDHOLM RELATIONS

Consider the case of a bounded linear relation $T : X \to Y$, where X and Y are Banach spaces. It is obvious that $T \in \phi_+$ if and only if T^{-1} is continuous and $\dim T^{-1}(0) < \infty$. In that case QT is ϕ_+-operator, and there exists a continuous linear operator B such that $BQT = I + F$, where F is a bounded finite rank operator. Then $A := BQ$ is a continuous linear operator and

$$AT = I + F. \tag{1}$$

Conversely, if there exist continuous linear operators A and F, with $D(A) \supset R(T)$, and $\dim R(F) < \infty$, satisfying (1), then T is a ϕ_+-relation (consider the restriction to $N(F)$).

Passing to the general case of an arbitrary linear relation $T \in LR(X,Y)$ (X and Y normed spaces), we shall obtain a characterisation of F_+-relations analogous to (1) above (see Theorems V.10.2 and V.10.3).

V.10.1 Lemma. *Let T be single valued,* $\dim N(T) < \infty$, *and* $\operatorname{codim} R(T) = \infty$.

(a) *There exists a bounded finite rank linear operator F_1 such that $T + F_1$ is injective.*
(b) *If in addition $T \in F_+$, then there exists a bounded finite rank operator F_2 for which $T + F_2$ has the following property: there exists a closed finite codimensional subspace M of $D(T)$ such that $D(T) = M + N(T + F_1)$, $M \cap N(T + F_2) = \{0\}$, and $F_2 + T|_M$ is injective and open.*

PROOF. Choose a basis $x_1, \ldots, x_n$ for $N(T)$, and let $y_1, \ldots, y_n$ be linearly independent elements of Y whose linear span intersects $R(T)$ only in $\{0\}$. Let $x'_1, \ldots, x'_n$ be chosen in X' such that $x'_i(x_j) = \delta_{ij}$ $(i, j \le n)$. Define

$$F_1(x) = \sum_{i=1}^{n} x'_i(x) y_i \qquad (x \in X).$$

Then $T + F_1$ is injective since $(T + F_1)x = 0 \implies F_1 x = -Tx \in R(T) \implies F_1(x) = 0 = Tx \implies x \in N(T) \cap N(F_1) \implies x = 0$. This proves (a).

Now suppose $T \in F_+$. Then there exists a closed finite codimensional subspace M of $D(T)$ and a finite dimensional subspace N of $D(T)$, such that $D(T) = M + N$, $M \cap N = \{0\}$, and $F_1 + T|_M$ is injective and open (V.3.2 and V.1.6). By Proposition II.3.18, there is a bounded projection defined on $D(T)$ with range M and kernel N. Let $F_2 = F_1 P - T(I - P)$. Then $T + F_2 = (T + F_1)P$ has a kernel N and coincides with $T + F_1$ on M. □

V.10.2 Theorem. *The following properties are equivalent:*

(i) *T is upper semi–Fredholm.*

(ii) *There exist linear operators A and F, with A partially continuous, $D(A) = R(T)$, F continuous and of finite dimensional range (or precompact, or strictly singular), such that*

$$AT = I_{D(T)} + F.$$

PROOF.

(i) $\Longrightarrow$ (ii). First assume that T is single valued. Assume (i). By embedding Y in a larger space if necessary, we can suppose that $\operatorname{codim} R(T) = \infty$. By Lemma V.10.1 there exists a bounded finite rank operator F_1 for which $T + F_1$ is injective and open on a closed finite codimensional subspace complementary to $N(T + F_1)$. Hence there exists a bounded projection P defined on $D(T)$ with $N(P) = N(T + F_1)$ such that $F_1 + T|_{R(P)}$ is injective and open (II.4.7). Let S_1 denote the inverse of $F_1 + T|_{R(P)}$. Let S_2 be any linear extension of S_1 with $D(S_2) = R(T) + R(F_1)$. Then

(1) $$S_2TPx = x - (I - P)x - S_2F_1Px.$$

Writing $F = -(I - P) - S_2F_1$, we obtain

(2) $$S_2Tx = x + Fx \qquad (x \in D(T))$$

Clearly F is a bounded finite rank operator. Let

$$S = S_2|_{R(T)}.$$

We shall verify that S is partially continuous. Since $T \in F_+$, there exists a continuous projection P_1 defined on $D(T)$ for which $\dim N(P_1) < \infty$ and $T|_{R(P_1)}$ is open and injective. Since $R(T) = R(TP_1) + R(T(I - P_1))$ where $R(T(I - P_1))$ is finite dimensional, it is sufficient to verify that S_2 is continuous on $R(TP_1)$. We have

$$\begin{aligned}
\sup_{x \in D(T)} \frac{\|S_2TP_1x\|}{\|TP_1x\|} &\leq \sup_{x \in D(T)} \frac{\|S_2TP_1x\|}{\|P_1x\|} \sup_{x \in D(T)} \frac{\|P_1x\|}{\|TP_1x\|} \\
&\leq \sup_{x \in D(T)} \frac{\|P_1x + FP_1x\|}{\|P_1x\|} \left(\inf_{x \in D(T)} \frac{\|TP_1x\|}{\|P_1x\|} \right)^{-1} \\
&\leq (1 + \|F\|)\gamma(T|_{R(P_1)})^{-1}.
\end{aligned}$$

Therefore S is partially continuous. Hence (i) $\Longrightarrow$ (ii).

Now suppose that T is an arbitrary linear relation. Then QT is single valued. Hence from the first part we have

$$A_1QT = I_{D(T)} + F$$

and substituting $A = A_1Q$ yields (ii).

(ii) $\Longrightarrow$ (i): Let A and F be as described in (ii), with F strictly singular, so that $AT = I_{D(T)} + F$.

Let M be a finite codimensional subspace of $R(T) = R(T) \cap D(A)$ for which $A|_M$ is continuous. Since $AT \in F_+$ by Theorem V.3.2, we have

$$\dim N(T) \leq \dim N(AT) < \infty.$$

Hence $T^{-1}(M)$ is finite codimensional in $D(AT) = D(T)$ (I.6.5). Furthermore since $ATT^{-1}m = Am + AT(0) = Am \quad (m \in M)$, we have

$$AT|_{T^{-1}(M)} = (A|_M)(T|_{T^{-1}M})$$

where $A|_M$ is continuous. Suppose $T \notin F_+$. Then clearly $T|_{T^{-1}(M)} \notin F_+$ (as $T^{-1}(M)$ is finite codimensional). By Theorem V.1.11, $T|_{T^{-1}(M)}$ has a singular sequence (x_n). By the continuity of $A|_M$, $\|ATx_n\| \to 0$, making (x_n) a singular sequence of AT. But then $AT \notin F_+$ (V.1.11). This contradiction shows that $T \in F_+$. □

In case X is complete, a stronger version of (i) of Theorem V.10.2 holds.

V.10.3 Theorem. *Let X be a Banach space, Y a normed space and $T \in LR(X,Y)$. Then the following properties are equivalent:*

(i) *T is upper semi-Fredholm.*

(ii) *There exists a bounded linear operator A and a bounded finite rank projection operator P such that*

$$AT = I_{D(T)} - P.$$

PROOF. (i) $\Longrightarrow$ (ii): Let $T \in F_+$. Replacing T by QT if necessary, we may assume that T is single valued. According to Theorem V.10.2, there exists a partially continuous operator S with domain $R(T)$ and a continuous finite rank operator F_1 such that

$$ST = I_{D(T)} + F_1 \tag{1}$$

Since X is complete, $S = C + F_2$ where C is bounded and F_2 is of finite rank (V.9.2). We have from (1)

$$CT = I_{D(T)} + F_3 \tag{2}$$

where $F_3 = F_1 - F_2T$. Let P_1 be any bounded projection defined on X, with kernel $R(F_3)$ (see II.3.18). Then from (2)

$$P_1CT = I_{D(T)} + (P_1 - I_X).$$

Hence writing $A = P_1C$, $P = I - P_1$ we have

$$AT = I_{D(T)} - P,$$

showing the (i) $\implies$ (ii).

The converse is contained in Theorem V.10.2. □

We can now characterize the inverse of an F_+–operator. The following special case is obvious:

$$\tilde{T} \in \phi_+ \iff \tilde{T}^{-1} \textit{ is continuous and } \dim N(\tilde{T}) < \infty$$

(see II.3.2). More generally we have:

V.10.4 Theorem. *The linear relation T is upper semi–Fredholm if and only if T^{-1} is partially continuous and $T^{-1}(0)$ is finite dimensional.*

PROOF. Let $T \in F_+$. Then $T^{-1}(0)$ is finite dimensional. Let P be a continuous projection defined on $D(T)$ with finite codimensional range, such that $T|_{R(P)}$ is injective and open (V.1.6 and II.3.18). We have

$$\|TPx\| \geq c\,\|Px\| \qquad (x \in D(T)) \tag{1}$$

where $c > 0$. Now

$$T^{-1}TPx = Px + T^{-1}(0). \tag{2}$$

We shall verify that if $y \in TPx$ then

$$\|T^{-1}y\| = d(Px, T^{-1}(0)). \tag{3}$$

Indeed, we then have $TPx = y + T(0)$ by Proposition I.2.8 and hence $T^{-1}TPx = T^{-1}y + T^{-1}(0)$ by Corollary II.1.10. Thus by (2) we clearly have

$$\|T^{-1}y\| = d(T^{-1}y, T^{-1}(0)) = d(T^{-1}TPx, T^{-1}(0)) = d(Px, T^{-1}(0)).$$

Now since $\dim T^{-1}(0) < \infty$, there exists a finite rank continuous linear projection P_1 defined on $D(T)$ with kernel $T^{-1}(0)$ (II.3.18). We have

$$\|P_1 x\| \geq \gamma(P_1)d(x, N(P_1)) \qquad (x \in D(T)).$$

Hence

$$\|P_1\| \, \|Px\| \geq \|P_1 Px\| \geq \gamma(P_1)d(Px, T^{-1}(0)). \tag{4}$$

Therefore

$$d(Px, T^{-1}(0)) \leq \|P_1\| \, \|Px\| / \gamma(P_1).$$

Consequently by (1) and (3),

$$\left\|T^{-1}y\right\| \leq \|P_1\| \, \|TPx\| / c\gamma(P_1) \qquad (y \in TPx).$$

Therefore

$$\left\|T^{-1}|_{TP(D(T))}\right\| \leq \|P_1\| / c\gamma(P_1).$$

But $TP(D(T))$ is finite codimensional in $R(T)$ by Proposition I.6.1. Therefore T^{-1} is partially continuous.

Conversely, assume that T^{-1} is partially continuous and that $T^{-1}(0)$ is finite dimensional. Let P be a continuous projection defined on $D(T)$ with kernel $T^{-1}(0)$. We have for $x \in D(T)$

$$(PT^{-1})Tx = Px + PT^{-1}T(0) = Px = x - (I - P)x.$$

Clearly PT^{-1} is partially continuous and single valued. Therefore $T \in F_+$ by Theorem V.10.2. □

V.10.5 Exercise

(a) Show that the theory of F_+-relations in normed spaces contains as a special case the theory of continuous linear operators in normed spaces in the following sense: Every continuous linear operator is the inverse of an injective F_+-relation.

(b) Show that the theory of Fredholm relations in Banach spaces contains the theory of bounded linear operators in Banach spaces in the following sense: If X and Y are complete, then the class of bounded linear operators from X to Y coincides with the class of inverses of closed Fredholm relations in $LR(Y, X)$ which are both surjective and injective. The corresponding inverse states classification is $I \cap (1 \cup 2) \cap 1e$.

V.10.6 Exercise Let T be closed and X and Y complete, and let $T(0)$ be finite dimensional. Show that T is continuous if and only if $T\overline{M} \subset \overline{TM}$ for all linear subspaces M.

V.11 INVERSE STATE DIAGRAMS

V.11.1 Definitions The state diagram obtained by substituting for the original states of T the states of T^{-1} in V.8.4, is called the **inverse essential state diagram**.

The inverse classifications are the following (using Theorem V.10.4 to derive the righthand classification):

	States	Essential States
1:	T is single valued and continuous	$\dim T(0) < \infty$ and T is partially continuous
2:	T is single valued but not continuous	$\dim T(0) < \infty$ but T is not partially continuous
3:	T is not single valued	$\dim T(0) = \infty$
I:	$D(T) = X$	$D(T)$ is closed and $\dim X/D(T) < \infty$
II:	$D(T)$ is not closed but $D(T)$ is dense	$D(T)$ is not closed but $\dim X/\overline{D(T)} < \infty$
III:	$D(T)$ is not dense	$\dim X/\overline{D(T)} = \infty$

V.11.2 Theorem. *The inverse state diagram and the inverse essential state diagram coincide with the state diagram III.6.5.*

From the inverse state diagram we know that $D(T') = Y'$ if and only if T is continuous and single valued.

As a consequence of the inverse essential state diagram we have:

V.11.3 Corollary. *The following properties are equivalent:*

(i) *T is partially continuous and* $\dim T(0) < \infty$.
(ii) *$\tilde{T}$ is continuous and* $\dim \tilde{T}(0) < \infty$.
(iii) *$D(T')$ is closed and of finite codimension in Y'.*

V.11.4 Corollary. *$D(\tilde{T})$ is closed and of finite codimension in $\tilde{X}$ if and only if T' is continuous and $\dim T'(0) < \infty$.*

V.11.5 Corollary. *Let $\tilde{X}$ be reflexive. Then $\tilde{T}(0)$ is finite dimensional if and only if the norm closure of $D(T')$ is finite codimensional.*

V.12 UBIQUITOUSLY OPEN RELATIONS

Upper semi–Fredholm relations, unlike their classical counterparts in operator theory, need not be open. However, they are ubiquitously open, a property which we now investigate. We shall see that a linear relation is ubiquitously open if and only if the inclusion

$$(\tilde{T}M)^- \subset \tilde{T}\overline{M}$$

holds for every linear subspace M of $D(\tilde{T})$ (Corollary V.12.5).

We note some properties enjoyed by ubiquitously open linear relations: The product of two such relations is ubiquitously open, whenever the product is defined. In particular, if T is ubiquitously open, so is any restriction of T. Furthermore, if T is ubiquitously open and F is any finite rank linear operator, then $T + F$ is ubiquitously open (Corollary V.12.4).

The results of this section provide further characterisations of F_+–relations not of finite rank.

V.12.1 Definition The linear relation T is called **partially open** if there exists a finite codimensional subspace M of X (or equivalently, of $D(T)$) such that $T|_M$ is open.

V.12.2 Theorem. *The following properties are equivalent for a linear relation T:*

(i) *T is ubiquitously open.*
(ii) *For each $M \in \mathcal{I}(D(T))$ there exists $N \in \mathcal{I}(M)$ such that $\alpha(T_N) + \Delta(T|_N) > 0$.*
(iii) *$\alpha(T|_M) + \Delta(T|_M) > 0$ for each $M \in \mathcal{I}(D(T))$.*
(iv) *For each subspace $E \subset D(T)$ such that $E \cap N(T)$ is finite dimensional, $T|_E$ is upper semi–Fredholm.*
(v) *No infinite dimensional restriction of T is both precompact and injective.*
(vi) *Each infinite dimensional restriction of T is partially open.*
(vii) *Some (alternatively, every) finite codimensional restriction of T is ubiquitously open.*

(For further characterisations see Corollary V.12.5.)

PROOF. We proceed through the following chain of inplications: (i) $\Longrightarrow$ (iii) $\Longrightarrow$ (ii) $\Longrightarrow$ (i), (i) $\Longrightarrow$ (iv) $\Longrightarrow$ (v) $\Longrightarrow$ (i), (i) $\Longrightarrow$ (vi) $\Longrightarrow$ (i), (i) $\Longrightarrow$ (vii) $\Longrightarrow$ (i).

(i) $\Longrightarrow$ (iii): Assume (i) and let $\alpha(T|_M) + \Delta(T|_M) = 0$ for some $M \in \mathcal{I}(D(T))$. Then $T|_M$ is both injective and strictly singular (V.2.6) and consequently for each $N \in \mathcal{I}(M)$ we have $\gamma(T|_N) = 0$, contradicting (i).

(iii) $\Longrightarrow$ (ii): Immediate.

(ii) $\Longrightarrow$ (i): Suppose T is not ubiquitously open. Then there exists $M \in \mathcal{I}(D(T))$ such that $\gamma(T|_N) = 0$ for $N \in \mathcal{I}(M)$. Now $M \cap N(T)$ cannot have finite codimension in M (for then $T|_M$ would have finite rank, giving $\gamma(T|_M) > 0$ – see II.3.2). Hence by passing to an infinite dimensional subspace of M if necessary, we may suppose that $T|_M$ is injective. Then $\alpha(T|_M) = 0$ and $T|_M$ is strictly singular, i.e. $\Delta(T|_M) = 0$ by V.2.6. Thus $\alpha(T|_M) + \Delta(T|_M) = 0$. But then for $N \in \mathcal{I}(M)$ we have $\alpha(T|_N) + \Delta(T|_N) = 0$, contradicting (ii).

(i) $\Longrightarrow$ (iv): Assume (i). If $D(T)$ is finite dimensional there is nothing to prove. Accordingly let $D(T)$ be infinite dimensional and let $E \cap N(T)$ be finite dimensional, where $E \in \mathcal{I}(D(T))$. Suppose that $T|_E \notin F_+$. Since this is equivalent to $QT|_E \notin F_+$, we may assume that T is single valued. By Theorem V.1.6 there exists $M \in \mathcal{I}(E)$ such that $T|_M$ is precompact. By assumption there exists $N \in \mathcal{I}(M)$ for which $T|_N$ is open. But then $T|_N$ has finite dimensional range. However, $T|_N$ has finite dimensional kernel. Consequently N is finite dimensional, a contradiction. This proves (iv).

(iv) $\Longrightarrow$ (v): Assume (iv) and let $T|_M$ be injective where $M \in \mathcal{I}(D(T))$. Then $T|_M \in F_+$. Therefore $T|_M$ is not precompact (V.1.6).

(v) $\Longrightarrow$ (i): Suppose T is not ubiquitously open. Then by the equivalence (i) $\Longleftrightarrow$ (iii) proved above, there exists $M \in \mathcal{I}(D(T))$ such that $T|_M$ is injective and strictly singular. In that case there exists $N \in \mathcal{I}(M)$ for which $T|_N$ is precompact (V.4.5), contradicting (v).

(i) $\Longrightarrow$ (vi): Assume (i). Let M be any linear complement of $N(T)$ in $D(T)$; thus $D(T) = M + N(T)$ where $M \cap N(T) = \{0\}$. Then $T|_M$ is an F_+–operator (by (i) $\Longrightarrow$ (iv)). Hence there exists linear subspaces E and F of M such that $M = E + F$, where F is finite dimensional and $T|_E$ is injective and open. It only remains to verify that $T|_{E+N(T)}$ is open. We have

$$\begin{aligned}\gamma(T|_{E+N(T)}) &= \sup\{\gamma : \|T(e+n)\| \geq \gamma d(e+n, N(T)) \text{ for all } e \in E,\ n \in N(T)\}\\ &= \sup\{\gamma : \|Te\| \geq \gamma d(e, N(T)) \text{ for all } e \in E\}\\ &\geq \gamma(T|_E) > 0\end{aligned}$$

as required.

(vi) $\Longrightarrow$ (i): Suppose that T is not ubiquitously open. Then there exists $N \in \mathcal{I}(D(T))$ for which $T|_N$ is injective and precompact (by (v) $\Longleftrightarrow$ (i) above), clearly contradicting (vi) (see II.3.9).

(i) $\Longrightarrow$ (vii): Both statements are immediate.

(vii) $\Longrightarrow$ (i): It is clearly sufficient to prove the first statement. We may clearly suppose that $D(T)$ is infinite dimensional. Let E be a finite codimensional subspace of $D(T)$ for which $T|_E$ is ubiquitously open. Let $M \in \mathcal{I}(D(T))$. Then $\Delta(T|_M) = \Delta(T|_{E\cap M})$ by IV.1.3, and $\alpha(T|_M) \geq \alpha(T|_{E\cap M})$. Consequently, $\Delta(T|_M) + \alpha(T|_M) \geq \alpha(T|_{E\cap M}) + \Delta(T|_{E\cap M})$. Hence T is ubiquitously open by the equivalence (i) $\Longleftrightarrow$ (iii). □

V.12.3 Corollary. *Let $\overline{\beta}(T) < \infty$. Then the following properties are equivalent:*

(i) *T is lower semi-Fredholm*
(ii) *T' is ubiquitously open*
(iii) *T is almost open.*

PROOF. We have $\overline{\beta}(T) = \alpha(T') < \infty$. Now T is almost open if and only if T' is open (III.5.2), if and only if $T' \in \phi_+$ (V.1.7 and V.8.4), if and only if $T \in F_-$ (V.8.4). Finally, from Theorem V.12.2, $T' \in \phi_+$ if and only if T' is ubiquitously open. □

V.12.4 Theorem. *Let T have infinite dimensional range. Then the following properties are equivalent:*

(i) *T is upper semi-Fredholm*
(ii) *T is ubiquitously open.*

PROOF. For simplicity of notation let $D(T) = X$.

(i) $\Longrightarrow$ (ii): Assume (i). Then all restrictions of T are F_+ and hence by (iv) $\Longleftrightarrow$ (i) of Theorem V.12.1, T is ubiquitously open.

(ii) $\Longrightarrow$ (i): Assume (ii). It is clearly enough to consider the single valued case. Suppose in the first instance that T is continuous. Write $X = N(T) + E$ where $N(T) \cap E = \{0\}$, $S = (T|_E)^{-1}$ and $P = ST$. Then P is a projection with $N(P) = N(T)$ and $R(P) = E$. Let $M \in \mathcal{I}(X)$. By hypothesis there exists $N \in \mathcal{I}(M)$ for which $\gamma(T|_N) > 0$. We have

$$\gamma(P|_N) = \gamma(ST|_N) = \inf_{n\in N} \frac{\|STn\|}{\|n + N(T|_N)\|}$$
$$\geq \inf_{n\in N} \frac{\gamma(S)\,\|Tn\|}{\|n + N(T|_N)\|} = \gamma(S)\gamma(T|_N) > 0$$

since $\gamma(S) = \|T|_E\|^{-1} > 0$ by the continuity of T. Therefore P is ubiquitously open. By Corollary V.1.12, since $\dim R(P) = \dim R(T) = \infty$, we have $\dim N(P) = \dim N(T) < \infty$. Therefore $T \in F_+$. Hence (ii) $\implies$ (i) in the case when T is continuous. Passing now to the general case when T is arbitrary, we note that for any subspace M of X,

$$\gamma(T|_M) > 0 \iff \gamma(TG|_{G^{-1}M}) = \gamma(T|_M \circ G_{T|_M}) > 0$$

(by IV.3.11) and hence T is ubiquitously open if and only if TG is ubiquitously open. Since TG is continuous, it follows from what has been proved that $TG \in F_+$. But clearly $TG \in F_+ \implies T \in F_+$ by IV.3.4. Therefore (ii) $\implies$ (i). □

V.12.5 Corollary. *The following properties are equivalent:*

(i) *T is ubiquitously open*
(ii) *$\tilde{T}(M)^- \subset \tilde{T}(\overline{M})$ for every linear subspace M of $D(\tilde{T})$*
(iii) *$\tilde{T}$ is ubiquitously open.*

PROOF. Combine Theorems V.9.22 and V.12.4. □

V.12.6 Exercise Let E be a closed linear subspace of X. Then Q_E^X is ubiquitously open if and only if E is either finite dimensional or finite codimensional.

V.12.7 Exercise Let P be a linear projection. Then P is ubiquitously open if and only if either its range or its kernel is finite dimensional.

V.12.8 Corollary. *The following properties are equivalent:*

(i) *T is ubiquitously open*
(ii) *$\tilde{T}$ is open, and either $\alpha(\tilde{T}) < \infty$ or $\dim R(T) < \infty$.*

PROOF. Assume (i). Then either $\dim R(T) < \infty$ or $T \in F_+$ (by V.12.4). In the latter case $\tilde{T} \in \phi_+$ (V.7.6) and hence $\tilde{T}$ is open (III.3.2) and $\alpha(\tilde{T}) < \infty$. In the former case, $\dim R(\tilde{T}) < \infty$ and so $\tilde{T}$ is open by II.3.2.

Conversely, assume that (ii) holds. If $\dim R(T) < \infty$ then T is ubiquitously open. Otherwise, if $\tilde{T}$ is open and $\alpha(\tilde{T}) < \infty$, then $\tilde{T} \in \phi_+$ (III.3.2). Hence $T \in F_+$ (V.7.6). Therefore T is ubiquitously open by Theorem V.12.4. □

V.12.9 Corollary. *Let $D(T)$ be infinite dimensional. Then the following properties of the minimum modulus of T are equivalent:*

(i) *For each $M \in \mathcal{I}(D(T))$ we have*

$$\sup_{N \in \mathcal{I}(M)} \gamma(T|_N) > 0.$$

(ii)
$$\inf_{M \in \mathcal{I}(D(T))} \sup_{N \in \mathcal{I}(M)} \gamma(T|_N) > 0.$$

PROOF. Assume (i). Then T is ubiquitously open (II.3.2). Hence by Theorem V.12.4, $T \in F_+$. Hence

$$\inf_{M \in \mathcal{I}(D(T))} \sup_{N \in \mathcal{I}(M)} \gamma(T|_N) \geq \inf_M \sup_N \inf_{n \in N} \frac{\|Tn\|}{\|n\|} = \inf_{M \in \mathcal{I}(D(T))} \tau(T|_M).$$

Let E be a finite codimensional subspace of $D(T)$ for which $QT|_E$ is injective and open. Then $M \cap E$ has finite codimension in M for $M \in \mathcal{I}(D(T))$ (by IV.1.2) and then $\tau(T|_M) = \tau(T|_{M \cap E})$ (by IV.1.3). Therefore

$$\inf_{M \in \mathcal{I}(D(T))} \tau(T|_M) = \inf_{M \in \mathcal{I}(E)} \tau(T|_M) = (\Delta(QT|_E)^{-1})^{-1}$$

by IV.3.7. But $(QT|_E)^{-1}$ is continuous and so $\Delta((QT|_E)^{-1}) \leq \|(QT|_E)^{-1}\| < \infty$. Hence (ii) follows. □

V.12.10 Exercise The sum of a ubiquitously open linear relation and a finite rank linear relation is ubiquitously open.

V.13 CANONICAL FACTORISATION

If T is a linear operator then it has a canonical factorisation $T = \hat{T}Q_{N(T)}$. We shall now describe the analogous factorisation of a linear relation.

V.13.1 Definition Let X and Y be vector spaces and $T \in LR(X,Y)$. Let us introduce the (standard) notation

(1)
$$[x] := Q^X_{N(T)}x.$$

The **injective component** $\hat{T}$ of T is the relation $\hat{T} \in LR(X/N(T), Y)$ given by

(2)
$$G(\hat{T}) := \{([x], y) : (x, y) \in G(T)\}.$$

We have

$$(3) \qquad Tx = \hat{T}[x] \qquad (x \in D(T))$$

and

$$(4) \qquad T = \hat{T}Q^X_{N(T)}$$

called the **canonical factorisation** of T.

If X is endowed with a norm, then $X/N(T)$ is normed if and only if $N(T)$ is closed; for example if T is closed or more generally, if T^{-1} is closable.

In the sequel we shall revert to our ongoing assumption that $X, Y, \ldots$ are normed spaces, except where stated otherwise, and $\hat{T}$ will only be used when $N(T)$ is closed in $D(T)$.

V.13.2 Proposition. *Let $N(T)$ be closed. Then*

(a) *T is closed if and only if $\hat{T}$ is closed*

(b) $\|T\| = \left\|\hat{T}\right\|$.

PROOF. (a) Let T be closed, and let $([x_n], y_n)$ be a sequence in $G(\hat{T})$ converging to $([x], y)$. Then there exists a sequence (z_n) in $N(T)$ such that $\lim_{n\to\infty}(x_n + z_n) = x$. We now have $(x_n + z_n, y_n) \to (x, y)$ and since T is closed, $x \in D(T)$, so $[x] \in D(\hat{T})$ and $y \in Tx = \hat{T}[x]$. Therefore $\hat{T}$ is closed. Conversely, if $\hat{T}$ is closed and (x_n, y_n) is a sequence in $G(T)$ converging to (x, y), then $([x_n], y_n) \to ([x], y)$ where $([x], y) \in G(\hat{T})$ since $\hat{T}$ is closed. Therefore $x \in D(T) = (Q^X_{N(T)})^{-1}D(\hat{T})$ and $y \in Tx = \hat{T}[x]$. Hence T is closed.

(b) We have $\|T\| = \left\|\hat{T}Q_{N(T)}\right\| \le \left\|\hat{T}\right\| \left\|Q_{N(T)}\right\| = \left\|\hat{T}\right\|$ by II.3.13. For the reverse inequality we may suppose $\|T\| < \infty$. In that case, for $x \in D(T)$ and $z \in [x]$ we have $\left\|\hat{T}[x]\right\| = \|Tz\| \le \|T\| \, \|z\|$ whence

$$\begin{aligned} \left\|\hat{T}[x]\right\| &\le \|T\| \inf\{\|z\| : z \in [x]\} \\ &= \|T\| \, \|[x]\| \,. \end{aligned}$$

Therefore $\left\|\hat{T}\right\| \le \|T\|$. □

V.13.3 Proposition. *Let $N(T)$ be closed. Then with the usual canonical identifications, we have*

$$G(T') = G(\hat{T}').$$

PROOF. We shall first verify that $D(T') = D(\hat{T}')$. For $y' \in D(\hat{T}')$ we have for $x \in D(T)$,

$$|y'Tx| = \left|y'\hat{T}[x]\right| \leq \left\|y'\hat{T}\right\| \|[x]\| \leq \left\|y'\hat{T}\right\| \|x\|,$$

so that $y' \in D(T')$ by Proposition III.1.2. For $y' \in D(T')$ we have for $x \in D(T)$, $\left\|y'\hat{T}[x]\right\| \leq \|y'T\| \|x\|$. Hence taking the infimum over z in $[x]$,

$$\left\|y'\hat{T}[x]\right\| \leq \|y'T\| \|[x]\|.$$

Therefore $y' \in D(\hat{T}')$. Hence $D(T') = D(\hat{T}')$.

Now let $y' \in D(T')$. Then for $x \in D(T)$,

$$\begin{aligned} T'y'(x) &= y'Tx = y'\hat{T}Q_{N(T)}x = \hat{T}'y'Q_{N(T)}x \\ &= Q'_{N(T)}\hat{T}'y'x. \end{aligned}$$

Hence

$$(1) \qquad T' = Q'_{N(T)}\hat{T}' = J_{N(T)^\perp}\hat{T}'.$$

Hence for $y' \in D(T')$ we have

$$T'y' = \hat{T}'y'$$

showing that $G(T') = G(\hat{T}')$. □

V.13.4 Remarks (a) The equality (1) in the proof of V.13.3 shows that

$$(\hat{T}Q^X_{N(T)})' = (Q^X_{N(T)})'\hat{T}'$$

despite the fact Theorem III.1.6 is not applicable.

(b) If $N(T)$ is closed in $D(T)$, one can instead proceed with the factorisation of $TJ_{D(T)}$ to obtain similar results. This has the added advantage that the adjoints are both single valued.

V.13.5 Proposition. *Let $N(T)$ be closed in $D(T)$. If T is partially open then $\hat{T}$ is partially open.*

PROOF. Let E be a finite codimensional subspace of $D(T)$ for which $T|_E$ is open. Write $Q := Q^X_{N(T)}$. We have

$$\begin{aligned}\gamma(\hat{T}|_{QE}) &= \sup\{\lambda : \left\|\hat{T}Qe\right\| \geq \lambda \left\|Qe\right\|, \quad e \in E\} \\ &\geq \sup\{\lambda : \left\|Te\right\| \geq \lambda \left\|e + N(T|_E)\right\|, \quad e \in E\} \\ &= \gamma(T|_E) > 0.\end{aligned}$$

Since QE is finite codimensional in $Q(D(T)) = D(\hat{T})$ (by I.6.1), $\hat{T}$ is partially open. □

V.13.6 Proposition. *Let X and Y be Banach spaces and let T be a closed linear relation. Then T is partially open if and only if it is open.*

PROOF. Let T be partially open. Then TG_T is partially open by Corollary IV.3.11. Furthermore X_T is a Banach space (Proposition IV.3.3). Consequently we may suppose that T is bounded. Moreover since T is open if and only if QT is open, we may suppose that T is single valued. We have $\hat{T}$ partially open by Proposition IV.13.5. Now $\hat{T}$ is an injective bounded linear operator with a partially continuous inverse. Hence $\hat{T}$ is an isomorphism (see Exercise V.4.12). Therefore T is open (by Theorem II.3.11). The converse statement is trivial. □

V.13.7 Exercise Give a short proof of V.13.2 by making use of the isometry $G(\hat{T}) \equiv G(T)/(N(T) \times \{0\})$.

V.14 COMPACTNESS OF THE ADJOINT

According to Corollary V.5.15, the adjoint T' is compact if and only if $\tilde{T}$ is compact. Here we examine some related properties.

V.14.1 Proposition. *The following properties are equivalent:*

(i) *T is precompact*

(ii) *T is continuous, and $Q_F T$ is precompact for every (alternatively, for some) finite dimensional subspace F of Y.*

PROOF. (i) $\Longrightarrow$ (ii): Assume (i). Let $\{x_n\}$ be a bounded sequence in $D(T)$. Then $\{x_n\}$ has a subsequence, assumed for simplicity to be itself, such that $\{Q_T T x_n\}$ is a Cauchy sequence. Write $x_{nm} := x_n - x_m$ $(n, m \in \mathbb{N})$. Then we have

$$\begin{aligned} & d(Tx_{nm}, T(0)) \to 0 \qquad (n, m \to \infty) \\ \Longrightarrow\ & d(Q_F T x_{nm}, Q_F T(0)) \to 0 \qquad (n, m \to \infty) \\ \Longrightarrow\ & Q_M(Q_F T x_{nm}) \to 0 \qquad (n, m \to \infty) \end{aligned}$$

where

$$M := \overline{Q_F T(0)}. \tag{1}$$

Therefore $Q_F T$ is precompact. Hence (i) $\Longrightarrow$ (ii).

Conversely, assume that (ii) holds. Let F be a finite dimensional subspace of Y, and let $\{x_n\}$ be a bounded sequence in $D(T)$. Then with M as in (1), $Q_M(Q_F T x_n)$ has a Cauchy subsequence, which we assume for simplicity to be itself. Select $y_n \in T x_n$ arbitrarily $(n \in \mathbb{N})$. Now select a subsequence of $\{Q_M Q_F T x_n\}$ (again for the sake of simplicity assumed to be $\{Q_M Q_F T x_n\}$ itself) such that

$$d(Q_F(y_n - y_{n+1}), Q_F T(0)) = \|Q_M Q_F T(x_n - x_{n+1})\| < 2^{-n} \tag{2}$$

for each $n \in \mathbb{N}$. We may therefore select $\{w_n\} \subset T(0)$ and $\{f_n\} \subset F$ so that $\|(y_n - f_n - w_n) - (y_{n+1} - f_{n+1} - w_{n+1})\| < 2^{-n}$ for each $n \in \mathbb{N}$. But then

$$\|Q_T(y_n - f_n) - Q_T(y_{n+1} - f_{n+1})\| < 2^{-n} \qquad n \in \mathbb{N} \tag{3}$$

The sequence $\{Q_T(y_n - f_n)\}$ is Cauchy since for every $p \in \mathbb{N}$,

$$\begin{aligned} \|Q_T(y_{n+p} - f_{n+p}) - Q_T(y_n - f_n)\| &\le \sum_{k=n}^{n+p-1} \|Q_T((y_{k+1} - f_{k+1}) - (y_k - f_k))\| \\ &\le \sum_{k=n}^{\infty} 2^{-k} = 2^{-n+1} \to 0 \text{ as } n \to \infty. \end{aligned} \tag{4}$$

Moreover since $Q_T f_n = Q_T y_n - Q_T(y_n - f_n) = Q_T T x_n - Q_T(y_n - f_n)$ with both $\{Q_T T x_n\}$ and $\{Q_T(y_n - f_n)\}$ now known to be bounded, the sequence $\{Q_T f_n\}$ is a bounded sequence in the finite dimensional space $Q_T F$. It must therefore have a convergent subsequence, for simplicity assumed to be $\{Q_T f_n\}$ itself. It now follows from (4) that the sequence $Q_T T x_n = Q_T y_n = Q_T(y_n - f_n) + Q_T f_n$, $n \in \mathbb{N}$, is Cauchy showing that T is precompact. $\square$

V.14.2 Corollary. *Let Y be complete. The following properties are equivalent:*

(i) *T is partially precompact*
(ii) *T is partially continuous and for some finite dimensional subspace F of Y, $Q_F T$ is precompact.*

PROOF. Assume (i). Then $T|_M$ is precompact for some finite codimensional subspace M of $D(T)$. By V.9.2 and V.9.3 we may select a finite dimensional $F \subset Y$ such that $Q_{Q_T F} Q_T T = Q_{\overline{T(0)}+F} T$ is continuous (to see this note that Q_T maps $\mathcal{F}(Y)$ onto $\mathcal{F}(Y)/\overline{T(0)}$). But then $Q_{\overline{T(0)}+F} T$ must be precompact since it is a finite dimensional extension of the precompact operator $Q_{\overline{T(0)}+F} T|_M$. If we now set $S = Q_F T$, (ii) follows on noticing that $Q_{\overline{T(0)}+F} T = Q_S S$.

Hence (ii) follows, by Proposition V.14.1.

Conversely, suppose that (ii) is satisfied. Then $T|_M$ is continuous for some finite codimensional subspace of $D(T)$, and $Q_F T|_M$ is precompact. But then $T|_M$ is precompact by Proposition V.14.1. Hence T is partially precompact. □

V.14.3 Corollary. *If T is partially precompact, then T' is compact.*

PROOF. We may clearly suppose Y to be complete. Let T be partially precompact. By Corollary V.14.2 there exists a finite dimensional subspace $F \subset Y$ for which $Q_F T$ is precompact. Write $M := \overline{Q_F T(0)}$. Then $Q_M Q_F T$ is a precompact linear operator and hence from standard results (see e.g. Goldberg [3; III.1.11], $(Q_M Q_F T)' = (Q_F T)' J_{M^\perp}$ is compact. Since $M^\perp \supset D((Q_F T)')$, it follows that $(Q_F T)'$ is compact. Hence $T' J_{F^\perp} (= (Q_F T)')$ is compact, where $F^\perp$ is a closed finite codimensional subspace of X'. Hence by Proposition IV.1.3 and Theorem V.2.2,

$$\overline{\Gamma}_0(T' J_{F^\perp}) = \overline{\Gamma}_0(T') = 0$$

and T' is compact. □

V.14.4 Example *The converse of Corollary V.14.3 is false.* For instance, if $D(T') = \{0\}$, then T' is compact. However, in such a case T will not even be partially continuous (except in the trivial case when $\dim Y < \infty$). Examples of differential operators with $D(T') = \{0\}$ were given in Section V.5

V.14.5 Theorem. *Let T be closable and Y complete. Then T' is compact if and only if T is compact.*

PROOF. Let T' be compact. Then $D(T')$ is $\sigma(Y',Y)$ closed by Corollary III.3.5, and so $D(T') = D(T')^{\top\perp} = T(0)^{\perp}$ since T is closable (Propositions II.5.5 and III.1.4(d)). Hence T is continuous by Proposition III.3.6(a). Now we have $QTJ_{D(T)}$ single valued and continuous, with $(QTJ_{D(T)})' = Q_{T'(0)}T'J_{D(T')}$ compact (see Theorem III.1.10(c)). Therefore QT is compact, by Corollary V.5.15. Hence T is compact.

Conversely, let T be compact. Then $\tilde{T}$ is compact by IV.4.9 and V.2.2. The compactness of T' now follows by another application of V.5.15. □

V.14.6 Corollary. *Let T be completely closable and partially precompact. Then T is precompact.*

PROOF. By Corollary V.14.3 and Theorem V.5.15, $\tilde{T}$ is compact. Hence T is precompact (see Exercise V.14.7). □

V.14.7 Exercise It T is completely closable and $\tilde{T}$ is compact, then T is precompact.

The following is an example of a discontinuous linear operator with a compact adjoint. It shows in particular that $\tilde{T}$ may be compact without T being precompact.

V.14.8 Example Let $T \in LR(X,\mathbb{R})$ be a discontinuous linear functional defined on X. It is easy to see that $\tilde{T}(0) = \mathbb{R}$; to see this, choose a sequence $\{x_n\} \subset X$ such that $x_n \to 0$ and $Tx_n \to y$ for $y \in \mathbb{R}$. Hence $\tilde{T}$ is compact. Since $D(T') = \{0\}$, T' is single valued and compact.

V.15 STABILITY OF THE INDEX

V.15.1 Definition If $\alpha(\tilde{T})$ and $\alpha(T')$ are not both infinite we define the quantity $\overline{\kappa}(T)$ as follows:

(1) $$\overline{\kappa}(T) := \alpha(\tilde{T}) - \alpha(T').$$

V.15.2 Remarks

(1) $\alpha(T') = \overline{\beta}(T) = \overline{\beta}(\tilde{T})$ (see III.7.2). Hence $\overline{\kappa}(T) \geq \kappa(\tilde{T})$ and $\overline{\kappa}(T) \geq \kappa(T)$ whenever these inequalities are defined.

(2) If T is upper semi-Fredholm, then $\alpha(T) \leq \alpha(\tilde{T}) < \infty$, as the Essential State Diagram V.8.4 shows. In this case $\alpha(T') = \beta(\tilde{T})$ (see V.7.6).

(3) Since T is lower semi-Fredholm if and only if $\tilde{T} \in \phi_-$, in this case also we have $\alpha(T') = \beta(\tilde{T})$.

V.15.3 Proposition. *The quantity $\overline{\kappa}(T)$ exists whenever $\kappa(T')$ exists. If indeed either $R(T')$ is closed or X is reflexive, then*

$$\overline{\kappa}(T) = -\kappa(T')$$

whenever either quantity exists.

PROOF. To see the first part observe that since

$$\overline{R(T')} \subset R(\tilde{T}')^{\top\perp} = N(\tilde{T})^{\perp} \tag{1}$$

by III.1.4, we clearly have that

$$N(T'') = R(T')^{\perp} = \overline{R(T')}^{\perp} \supset N(\tilde{T})^{\perp\perp}$$

with

$$\alpha(T'') \geq \dim N(\tilde{T})^{\perp\perp} = \dim(N(\tilde{T})'') = \alpha(\tilde{T})$$

by III.1.8.

If now either X is reflexive or $R(T')$ is closed, then the norm closure of $R(T')$ will coincide with its $\sigma(X', \tilde{X})$ closure (III.4.4) and we shall then conclude that $\alpha(T'') = \alpha(\tilde{T})$. □

V.15.4 Example The compact linear operator

$$T : \ell_1 \to \ell_1 : \{\alpha_k\} \to \{\alpha_k/k\}$$

is injective with dense range. Hence $\alpha(T) = 0 = \overline{\beta}(T) = \alpha(T')$, and so $\overline{\kappa}(T) = 0$. However

$$T' : \ell_\infty \to \ell_\infty : \{\alpha_k\} \to \{\alpha_k/k\}$$

has range contained in and norm dense in $\hat{c}_0$. Thus $\infty = \operatorname{codim} \overline{R(T')} = \overline{\beta}(T') = \alpha(T'')$. Consequently $\kappa(T') = -\infty$.

Observe that while $\overline{R(T')} = \hat{c}_0$, the bipolar theorem informs us that

$$R(T')^{\top\perp} = \overline{R(T')}^{\top\perp} = (\hat{c}_0)^{\top\perp} = \ell_\infty.$$

V.15.5 Lemma. (Finite dimensional extensions)

Let $S, T \in LR(X,Y)$ and let S be an extension of T such that $\dim D(S)/D(T) = n < \infty$. *Then*

(a) *If T is closed, then S is closed*
(b) *If $T(0)$ is closed and $R(T)$ is closed, then $R(S)$ is closed*
(c) *If T has an index, then $\kappa(S) = n + \kappa(T)$.*

PROOF. We have $D(S) = D(T) + N$ where $D(T) \cap N = \{0\}$ and $\dim N = n$.

(a) Suppose that T is closed. Then $G(QT)$ is closed (by Proposition II.5.3). Now

$$G(S) = G(T) + G(S|_N) \tag{1}$$

since if $(x + s, y) \in G(S)$ where $x \in D(T)$ and $s \in N$, then $y = y_1 + y_2$ where $y_1 \in Tx$ and $y_2 \in Ss$ and we have $(x + s, y_1 + y_2) = (x, y_2) + (s, y_2)$ where $(x, y_1) \in G(T)$ and $(s, y_2) \in G(S|_N)$. Hence

$$G(QS) = G(QT) + G(QS|_N) \tag{2}$$

(since $S(0) = T(0)$, where $Q = Q_T$). But QS is single valued and therefore $\dim(QS|_N) \leq \dim N < \infty$. It follows that $G(QS|_N)$ is finite dimensional. This implies, by (2), that $G(QS)$ is closed. Therefore S is closed by Proposition II.5.3.

(b) Let $R(T)$ be closed. Then $R(QT)$ is closed. From (2) we have

$$R(QS) = R(QT) + QS(N)$$

where $QS(N)$ is finite dimensional. Therefore $QR(S)$ is closed. Hence $R(S)$ is closed.

(c) Let q denote the quotient map defined on Y with kernel $T(0)$ (as in I.6.2). Then $\kappa(S) = \kappa(qS)$ and $\kappa(T) = \kappa(qT)$, by Proposition I.6.10. We may therefore assume that T and S are single valued. It is clearly enough to prove (c) for the case $n = 1$. Then $N = \text{sp}\{w\}$ for some non zero $w \in N$. There are two cases to consider: If $Sw \notin R(T)$, then $R(S) = R(T) + \text{sp}\{Sw\}$ and so $\beta(T) = \beta(S) + 1$. Furthermore it is clear that then $N(T) = N(S)$, whence $\alpha(T) = \alpha(S)$. Therefore $\kappa(S) = 1 + \kappa(T)$. On the other hand, if $Sw \in R(T)$, then $R(T) = R(S)$ and $\beta(T) = \beta(S)$. Also $Sw = Tx = Sx$ for some $x \in D(T)$, whence $S(w - x) = 0$, where $w - x \notin D(T)$. Therefore $\alpha(S) = \alpha(T) + 1$. It follows again that $\kappa(S) = 1 + \kappa(T)$. □

V.15.6 Theorem. *Let $S, T \in LR(X,Y)$ where S is single valued and $D(S) \supset D(T)$. If $\tilde{T}$ has an index, and if $\|S\| < \gamma(\tilde{T})$, then*

$$\overline{\kappa}(T+S) = \kappa(\tilde{T}).$$

PROOF. We may clearly suppose that $D(\tilde{S}) = X$. In that case S' will be bounded and single valued (see III.1.4 and III.3.6), and then so will S''. Since also $\|S'\| = \left\|\tilde{S}\right\| = \|S\|$ (by III.3.6(c)) and $\gamma(T') = \gamma(\tilde{T})$ (by III.3.6(d)), and furthermore, $(T+S)' = T' + S'$ (by III.1.5), it is sufficient to show that $\overline{\kappa}(T' + S') = \overline{\kappa}(T')$, by virtue of V.15.3, and II.3.2 and III.4.2 applied to T'.

In view of the foregoing, we may assume without loss of generality, that S is a bounded linear operator, X and Y are Banach spaces, T is closed and has an index, and $\|S\| < \gamma(T)$. Moreover it is sufficient to consider the case when $\alpha(T) < \infty$, since the other case, when $\alpha(T') < \infty$, is covered by passing to the adjoint.

Suppose that $\alpha(T) < \infty$. Then there is a relatively closed subspace M of $D(T)$ such that $D(T) = M + N(T)$, $M \cap N(T) = \{0\}$. Since $\gamma(T) > 0$, $R(T) = R(T|_M)$ is closed by the Open Mapping Theorem, so $\gamma(T|_M) > 0$. By the small perturbation theorem III.7.4, we have for $\|S\| < \gamma(T|_M)$

$$\alpha(T|_M + S) = 0. \tag{1}$$

Furthermore, by the perturbation theorem for F_+-relations V.3.2, we have

$$T|_M + S \in \phi_+$$

provided $\|S\| < \gamma(T|_M)$ and $\dim M = \infty$ (see V.5.1). In particular, $R(T|_M + S)$ is closed and therefore $\overline{\beta}(T|_M + S) = \beta(T|_M + S)$. It now follows from Corollary III.7.7 that

$$\beta(T|_M + S) = \beta(T|_M) \tag{2}$$

whenever $\|S\| < \gamma(T|_M)$.

It now follows from the finite dimensional extension lemma V.15.5 that

$$\kappa(T+S) = \kappa(T|_M + S) + \alpha(T) = \kappa(T|_M) + \alpha(T) = \kappa(T)$$

provided $\|S\| < \gamma(T|_M)$.

Passing to the general case $\|S\| < \gamma(T)$, the small perturbation theorem III.7.4 shows that $\alpha(T+S) < \infty$ (so $T+S$ has an index). Let I denote the closed unit interval $[0,1]$ and let $Z = \mathbb{Z} \cup \{-\infty, \infty\}$ (here $\mathbb{Z}$ denotes the set of integers). Define $\psi : I \to Z$ by $\psi(\lambda) = \kappa(T + \lambda S)$. Let I have the usual topology and Z the discrete topology. By what has been proved, ψ is a continuous map (indeed, constant in each sufficiently small interval). Consequently $\psi(I)$ is a connected set and therefore consists of just one point. Hence $\kappa(T) = \psi(0) = \psi(1) = \kappa(T+S)$ as required. □

V.15.7 Corollary. *Let X and Y be complete and T closed. Then for any linear operator S satisfying $D(S) \supset D(T)$ and $\|S\| < \gamma(T)$ we have*

$$\kappa(T + S) = \kappa(T).$$

PROOF. It is enough to observe that $T + S$ is closed and apply Theorem V.15.6. □

V.16 THE STABILITY OF COMPLEMENTED RANGES

This section is concerned with a particular aspect of the following general question: Given a linear relation T belonging to a certain class $\mathcal{C}$, such that $R(T)$ (or $\overline{R(T)}$) is topologically complemented, and given another relation S such that $D(S) \supset D(T)$, under what conditions on S do we have $R(T + S)$ (or $\overline{R(T+S)}$, respectively) topologically complemented for all $T \in \mathcal{C}$?

The class $\mathcal{C}$ considered in this chapter is the class of upper semi-Fredholm relations. An affirmative solution is obtained for single valued perturbations which include the precompact operators.

This complementation problem is, as we have seen in earlier sections, connected with the existence of continuous selections of a given linear relation (Section II.4), and the left invertibility of F_+-relations modulo the finite rank operators (Theorem V.10.2).

V.16.1 Lemma. *Let T be an F_+-relation and let $\overline{R(T)}$ be topologically complemented in Y. Let S be a continuous single valued relation with $D(S) \supset D(T)$ and $R(S)$ contained in a subspace which is topologically complementary to $\overline{R(T)}$. Then $T + S \in F_+$ and $\overline{R(T+S)}$ is topologically complemented in Y. Furthermore (i) if T is injective and open, then so is $T + S$ with in addition $R(T + S)$ closed whenever $R(T)$ is closed, and (ii) if X is complete and T is closed then $R(T)$ and $R(T + S)$ are closed.*

PROOF. Consider first the case when T is injective and open. We assume without loss of generality that $D(T) = X$. Let P be any bounded projection from Y onto $\overline{R(T)}$ with complementary projection Q, where $R(Q) \supset R(S)$. We shall verify that

$$R(T+S) \subset N(Q - ST^{-1}P) \subset \overline{R(T+S)}. \tag{1}$$

Let $x \in D(T)$. Then

$$\begin{aligned}(Q - ST^{-1}P)(T+S)x &= (Q - ST^{-1}P)(Tx + Sx)\\ &= (Q - ST^{-1}P)Tx + (Q - ST^{-1}P)Sx \qquad \text{(by I.3.1(c))}\\ &= -S(x + T^{-1}0) + Sx\\ &= -Sx + Sx \qquad \text{(since } T \text{ is injective)}\\ &= 0.\end{aligned}$$

Hence

(2) $$R(T+S) \subset N(Q - ST^{-1}P).$$

Now let $y \in N(Q - ST^{-1}P)$. Since Q and $ST^{-1}P$ are single valued, we have

(3) $$(Q - ST^{-1}P)y = 0.$$

Now

$$y = Py + Qy = Py + ST^{-1}Py.$$

Hence for $y' \in N(T' + S')$ we have

(4) $$y'y = y'(Py + ST^{-1}Py) = P'y'y + (T^{-1}P)'S'y'y$$

(since $y' \in D(S')$ and $T^{-1}P$ is continuous and single valued). But since $D(S') = Y'$,

(5) $$(T+S)' = T' + S'.$$

Hence as $(T' + S')(y') = (T' + S')(0)$,

$$\begin{aligned}(T^{-1}P)'S'y'y + (T^{-1}P)'T'y'y &= (T^{-1}P)'(T+S)'(0)y\\ &= (T+S)'(0)T^{-1}Py = 0((T+S)T^{-1}Py) = 0.\end{aligned}$$

Thus

$$(T^{-1}P)'S'y'y = -(T^{-1}P)'T'y'y$$

(since $y'T(0) = 0$ as $y' \in D(T')$).
Substituting in (4),

(6) $$y'y = P'y'y - (T^{-1}P)'T'y'y.$$

But

$$
\begin{aligned}
(T^{-1}P)'T'y'y &= (T^{-1}P)'T'y'(Py+Qy) \\
&= (T'y')(T^{-1}P)Py + (T'y')(T^{-1}P)Qy \\
&= y'TT^{-1}Py \\
&= y'(Py + T(0)) \\
&= y'Py \\
&= P'y'y. \qquad (7)
\end{aligned}
$$

Hence from (6) and (7), $y'y = 0$. But $(T+S)' = T' + S'$ since $D(S') = Y'$. Therefore $N(Q - ST^{-1}P) \subset N(T'+S')^{\top} = \overline{R(S+T)}$ and (1) is established.

It is easy to verify that the linear operator $P + ST^{-1}P$ is a bounded projection with range $N(Q - ST^{-1}P)$. Hence $N(Q - ST^{-1}P) = \overline{R(T+S)}$ is topologically complemented.

Since $R(T) \cap R(S) = \{0\}$ and T is injective, it is not difficult to show that then $T+S$ must be injective. Moreover observe that if $x \in N(P + ST^{-1}P)$, then $Px = -ST^{-1}Px \in R(S) \subset R(Q)$. This can only be true if $Px = 0$ and hence $N(P) \subset N(P + ST^{-1}P) \subset N(P)$. Thus *both* $\overline{R(T)}$ and $\overline{R(T+S)}$ are topologically complemented by $V := N(P)$ and both are then isomorphic to Y/V. Consequently since $R(S) \subset V$, T is open $\iff$ $QT = Q(T+S)$ is open $\iff$ $T+S$ is open, and $R(T)$ is closed $\iff$ $Q(T+S)$ is surjective $\iff$ $R(T+S)$ is closed. Hence (i) follows.

We now consider the general case where $T \in F_+$. Again we assume without loss of generality that $D(T) = X$.

There exist closed subspaces M and F, with F finite dimensional, such that $M + F + N(T) = X$, for which $T|_M$ is injective and open (Theorem V.1.6) and $M \cap F = F \cap N(T) = \{0\}$. Now $T(X) = TM + TF$, where $TM \cap TF = T(0)$ with

$$
T(F) = W + T(0), \quad W \cap T(0) = \{0\} \quad \text{and} \quad \dim W < \infty \qquad (8)
$$

for some subspace W (see I.6.4). Hence

$$
\overline{R(T)} = \overline{R(T|_M)} + W, \quad \overline{R(T|_M)} \cap W = \{0\}.
$$

It is now clear that $\overline{R(T|_M)}$ is topologically complemented and furthermore, that $R(S)$ is contained in a closed subspace complementary to $\overline{R(T|_M)}$. Hence by the first part, $\overline{R(T|_M + S)} = \overline{R(T|_M + S|_M)}$ is topologically complemented and $T|_M + S|_M$ is injective and open. Clearly $T + S \in F_+$ by V.1.6. Moreover

$$\begin{aligned} R(T+S) &\subset (T+S)(M) + T(F) + S(F+N(T)) \\ &= (T+S)(M) + W + S(F+N(T)). \end{aligned}$$

Now since W and $S(F + N(T))$ are finite dimensional,

$$\overline{(T+S)(M)} \subset \overline{R(T+S)} \subset \overline{(T+S)(M)} + W + S(F+N(T)). \tag{9}$$

Since $\overline{(T+S)(M)}$ is topologically complemented in Y, it is topologically complemented in $\overline{R(T+S)}$. It is also clear that

$$\overline{(T+S)(M)} + W + S(F+N(T))$$

is topologically complemented in Y. Therefore $\overline{R(T+S)}$ is topologically complemented in Y.

Suppose that X is complete and T is closed. Then $T \in \phi_+$ (see Exercise V.1.21, or use Theorem III.5.3), so $R(T)$ is closed. Let M and F be defined as before and let P and Q be complementary continuous projections of $M + F$ onto M and F respectively. Then TP is an F_+-relation. To check that TP is closed let (x_n, y_n) be a sequence in $G(TP)$ converging to (x, y). Then (Px_n, y_n) is a sequence in $G(T)$ converging to (Px, y), and since T is closed, $y \in TPx$. Therefore TP is closed. Since TP is a closed F_+-relation, it follows that $R(TP)$ is closed (see Exercise V.1.21). Since $T|_M$ is injective and open, $R(T|_M + S|_M)$ is closed by part (i). Since

$$R(T+S) = R(T|_M + S|_M) + R((T+S)|_{F+N(T)})$$

we have

$$R(T|_M + S|_M) \subset R(T+S) \subset R(T|_M + S|_M) + N,$$

where $\dim N < \infty$, and where the two subspaces on the left and right are closed. Therefore $R(T + S)$ is closed. □

V.16.2 Corollary. *Let X and Y be Banach spaces and let T be a ϕ_+-relation with complemented range. If S is a bounded linear operator with range contained in a closed subspace complementary to $R(T)$, then $T + S \in \phi_+$ and $R(T + S)$ is closed and complemented.*

V.16.3 Lemma. *Let M and F be subspaces of X, with M topologically complemented and F finite dimensional. Then $M+F$ is topologically complemented.*

PROOF. We may clearly suppose that $M \cap F = \{0\}$. Furthermore, it is enough to consider the case when $\dim F = 1$.

By the Hahn Banach theorem, there exists $x' \in X'$ which vanishes on M and is non zero at some point $f \in F$ (since M is a closed subspace, being topologically complemented).

Since M is topologically complemented in X, it is topologically complemented in $N(x')$ (as $N(x') \supset M$). Let U denote the bounded projection defined on X with range $N(x')$ and kernel F. Let V be a bounded projection with domain $N(x')$ and range M, and define the mapping P on X by

$$Px = VUx + (I-U)x \tag{1}$$

It only remains to verify that P is a bounded linear projection with range $M+F$. We have

$$\begin{aligned} P^2x &= P(VUx) + P(I-U)x \\ &= VUx + (I-U)VUx + (I-U)x \\ &= VUx + (I-U)x \\ &= Px. \end{aligned}$$

Hence P is idempotent. Clearly $R(P) \subset M+F$, while if $x = m+f$ $(m \in M, f \in F)$, then $Px = VUm + (I-U)(m+f) = Vm + (I-U)m + f = m+f$, and hence $R(P) = M+F$.

Finally to show that P is continuous, let $x_n \to 0$. Then $Ux_n \to 0$ and $(I-U)x_n \to 0$. Hence

$$Px_n = VUx_n + (I-U)x_n \to 0$$

as required. □

V.16.4 Proposition. *Let T be an upper semi-Fredholm relation with $\overline{\beta}(T) < \infty$ and let the relation S be continuous and single valued with $D(S) \supset D(T)$, having a strictly singular adjoint. Then $\overline{\beta}(T+S) < \infty$. If in addition $T+S$ is closed and X is complete, then $R(T+S)$ is closed.*

PROOF. Since $(SJ_{D(T)})' = Q_{D(T)^\perp}S'$ is strictly singular, we may suppose without loss of generality that $D(T) = X$, and consequently S' is single valued.

Now $T' \in F_+$ by Proposition V.5.13. Hence $T' + S' \in F_+$ (Theorem V.7.10) and since S is bounded and single valued, $(T+S)' = T' + S'$ (see III.1.5). Then

$$\overline{\beta}(T+S) = \dim Y/\overline{R(T+S)} = \dim R(T+S)^{\perp}$$
$$= \dim N((T+S)') = \dim N(T'+S') < \infty.$$

If $T + S$ is closed and X is complete then $R(T+S)$ is closed by V.1.21. □

V.16.5 Lemma. *Let T be an F_+-relation with $\overline{R(T)}$ topologically complemented in Y. Let S be a continuous and single valued strictly singular relation having a strictly singular adjoint, and let $D(S) \supset D(T)$. Then $\overline{R(T+S)}$ is topologically complemented.*

PROOF. As in the proof of Proposition V.16.4, we assume without loss of generality that $D(S) = D(T)$. In that case S' is single valued. Let P be a bounded projection with domain Y and range $\overline{R(T)}$. Since $(PS)' = S'P'$ is strictly singular, $\overline{R(T+PS)}$ has finite codimension in $\overline{R(T)}$ by Proposition V.16.4 (regarding T and PS as relations in $LR(X, \overline{R(T)})$). Therefore $\overline{R(T+PS)}$ is topologically complemented in Y. But $T + S = T + PS + (I-P)S$ where $(I-P)S$ is a continuous linear operator with range contained in $R(I-P)$ which is complementary to $\overline{R(T)}$. Hence by Lemma V.16.1, $\overline{R(T+S)}$ is topologically complemented. □

V.16.6 Theorem. *Let T be an F_+-relation with $\overline{R(T)}$ topologically complemented. Let S be a strictly singular operator having a strictly singular adjoint and let $D(S) \supset D(T)$. Then $\overline{R(T+S)}$ is topologically complemented.*

PROOF. Again we assume $D(T) = X$. Since S is strictly singular, it is partially continuous (Theorems V.2.6 and V.4.3). Thus we may select $M \in \mathcal{C}(X)$ so that $S|_M$ is continuous. Since both J_M^X and S are single valued, it follows that $(SJ_M)'$ is an extension of the strictly singular operator $Q_{M^\perp}S'$. But by V.9.4, $(SJ_M)'$ must indeed also be strictly singular. An application of Lemma V.16.3 combined with the fact that $\overline{R(T|_M)}$ and $\overline{R(T|_M + S|_M)}$ are finite codimensional in $\overline{R(T)}$ and $\overline{R(T+S)}$ respectively, yields the fact that $\overline{R(T)}$ (respectively $\overline{R(T+S)}$) is topologically complemented if and only if the same is true of $\overline{R(T|_M)}$ (respectively $\overline{R(T+S|_M)}$). This can also be seen by noting that by I.6.1 there exists $W \subset R(T)$ with $\dim W < \infty$ such that $R(T|_M) + W = R(T)$. But then $\overline{R(T|_M)} + W = \overline{R(T)}$. The result now follows on noticing that $\overline{R(T|_M + S|_M)}$ must be topologically complemented by Lemma V.16.5. □

V.16.7 Corollary. *Let X and Y be Banach spaces and T a closed upper semi-Fredholm relation with topologically complemented range. Let S be a strictly singular operator having a strictly singular adjoint, and let $D(S) \supset D(T)$. Then the range of $T+S$ is topologically complemented.*

V.16.8 Examples Examples where both S and S' are strictly singular, with S single valued, are listed below. The verifications are left as an exercise.

(1) S closed, Y complete and S' compact (see IV.4.9 and V.5.15).
(2) S semicontinuous and S' both strictly singular and strictly cosingular. For example, if S is a (bounded or unbounded) finite rank operator.
(3) X and Y Banach spaces with Y subprojective, S bounded and S' strictly singular. (For the definition of subprojective see Goldberg [3; p.92]. For further examples of this type, see R.J. Whitley [1]).
(4) S weakly compact and X and Y' both Dunford Pettis spaces (see Diestel and Uhl [1; 176]).
(5) X a Dunford Pettis space, Y a reflexive space, and S any bounded linear operator (see Goldberg [3; p.89]).
(6) S bounded and the pairs X, Y and X', Y' both totally incomparable (in the sense that all pairs of infinite dimensional subspaces are non isomorphic). For example, if X is reflexive and Y has no infinite dimensional reflexive subspace or quotient.

V.16.9 Theorem. *Let T be a closed F_+-relation with topologically complemented range, and let X be complete. Let S be a bounded strictly singular operator having a strictly singular adjoint. then $R(T+S)$ is topologically complemented.*

PROOF. Since $\overline{R(T+S)}$ is topologically complemented by Theorem V.16.6, it is sufficient to show that $R(T+S)$ is closed. But $T+S \in F_+$ (Theorem V.7.10), and $T+S$ is closed since S is bounded and single valued (see Exercise II.5.16).

The rest of the proof is now somewhat similar to that of Proposition V.16.4 Let A be the linear relation $T+S$ regarded as an element of $LR(X, \overline{R(T+S)})$. Then A is closed, and $A' \in \phi_-$, so $R(A')$ is closed. Moreover, A' is injective. Hence A is surjective. Therefore $R(T+S)$ is closed. □

V.17 UBIQUITOUS PROPERTIES OF THE QUANTITIES

The following characterisation of strict singularity can be inferred from earlier results of this chapter: T is strictly singular if and only if every infinite dimensional subspace M of $D(T)$ contains another such subspace N for which $\Gamma(T|_N) = 0$. This latter condition could be paraphrased as "$\Gamma(T)$ is ubiquitously zero". This and other ubiquitous properties of the quantities

$\Gamma.\Delta, \Gamma_0, \overline{\Gamma}_0, \tau, \tau_0, \overline{\tau}_0$, and the quantities they generate, will now be investigated. Some new characterisations of various classes of linear operators and relations will be derived (Theorem V.17.5). As a corollary, some uniform boundedness properties of the quantities follow (Corollary V.17.6).

V.17.1 Definitions

Given a quantity $f(T)$, the quantitites **generated** by f are quantities of the form $\inf_{M_1} f(T|_{M_1})$, $\sup_{M_1} f(T|_{M_1})$, $\sup_{M_1} \inf_{M_2} f(T|_{M_2})$, $\inf_{M_1} \sup_{M_2} f(T|_{M_2})$, $\inf_{M_1} \sup_{M_2} \inf_{M_3} f(T|_{M_3}), \ldots$ where $M_1 \in \mathcal{I}(D(T))$, $M_2 \in \mathcal{I}(M_1), M_3 \in \mathcal{I}(M_2) \ldots$.

Let $\mathbb{P}$ be a property of linear relations. Then the property $\mathbb{P}$ is called **hereditary** if for every T with property $\mathbb{P}$ and every infinite dimensional subspace M of $D(T)$, $T|_M$ has the property $\mathbb{P}$.

Let $\mathcal{A}$ denote the set of quantities $\{\Gamma, \Delta, \Gamma_0, \overline{\Gamma}_0, \tau, \tau_0, \overline{\tau}_0\}$ and $[\mathcal{A}]$ the quantities generated by $\mathcal{A}$. Clearly $[\mathcal{A}]$ contains $\mathcal{A}$. We have seen that $\overline{\tau}_0(T) \leq f(T) \leq \overline{\Gamma}_0(T)$ for each $f \in [\mathcal{A}]$ (see IV.1.1(1)).

Let $f \in \mathcal{A}$. We abbreviate here for convenience the phrase "for each $M \in \mathcal{I}(D(T))$ there exists $N \in \mathcal{I}(M)$ such that" simply as $(\forall M)(\exists N)$ and then define

$$[A]_f := \{T : \ (\forall M)(\exists N) f(T|_N) = 0\}$$
$$[B]_f := \{T : \ (\forall M)(\exists N) f(T|_N) > 0\}$$
$$[C]_f := \{T : \ (\forall M)(\exists N) f(T|_N) = \infty\}$$
$$[D]_f := \{T : \ (\forall M)(\exists N) f(T|_N) < \infty\}.$$

V.17.2 Proposition. *For each $f \in [\mathcal{A}]$ we have $\overline{\tau}_0(T) \leq f(T) \leq \overline{\Gamma}_0(T)$.*

PROOF. We have $\inf_{M\in\mathcal{I}(D(T))} \overline{\tau}_0(T|_M) \geq \overline{\tau}_0(T)$ by Proposition IV.2.7. Therefore $f(T) \geq \overline{\tau}_0(T)$ by (1) and (2) of IV.1.1. Similarly $f(T) \leq \overline{\Gamma}_0(T)$. □

V.17.3 Proposition. *Let $\mathbb{P}$ be a hereditary property for linear relations. Then the following properties are equivalent:*

(i) *T is nowhere $\mathbb{P}$*
(ii) *T is ubiquitously not $\mathbb{P}$.*

PROOF. Clearly (i) $\Longrightarrow$ (ii). Assume (ii). Let $M \in \mathcal{I}(D(T))$. Suppose $T|_M \in \mathbb{P}$. There exists $N \in \mathcal{I}(M)$ such that $T|_N \notin \mathbb{P}$, contradicting the hereditary property of $\mathbb{P}$. Hence (ii) $\Longrightarrow$ (i). □

V.17.4 Lemma. *The relation T is nowhere continuous if and only if $T \in [C]_{\overline{\tau}_0}$.*

PROOF. Assume that $T \in [C]_{\overline{\tau}_0}$. Let $M \in \mathcal{I}(D(T))$. Then there exists $N \in \mathcal{I}(M)$ such that $\overline{\tau}_0(QT|_N) = \infty$, and therefore $\|QT|_N\| = \infty$. Hence $QT|_M$ is discontinuous. Thus QT is nowhere continuous. Therefore T is nowhere continuous.

For the converse we first show

$$\text{(1)} \qquad T \text{ nowhere continuous} \implies T \in [C]_{\tau_0}.$$

Assume $T \notin [C]_{\tau_0}$. Then there exists $M \in \mathcal{I}(D(T))$ for which $\tau_0(T|_N) < \infty$ for all $N \in \mathcal{I}(M)$, and then $Q(T|_N)$ does not have a single valued precompact inverse (Proposition V.4.6). Now we can assume that $\dim N(QT|_M) < \infty$ otherwise (1) follows immediately. In that case there exists $E \in \mathcal{I}(M)$ with $N(QT|_M) + E = M$ and $N(QT|_M) \cap E = \{0\}$. Then for any $N \in \mathcal{I}(E)$ we have $QT(N) \in \mathcal{I}(QT(E))$ and $(QT|_N)^{-1} = (QT|_E)^{-1}|_{QT(N)}$ and therefore $(QT|E)^{-1}$ has no infinite dimensional precompact restriction. Hence $(QT|_E)^{-1}$ is an F_+–operator (Theorem V.1.6). Let W be a finite codimensional subspace of $QT(E)$ such that $((QT|_E)^{-1}|_W)^{-1}$ is continuous. We have

$$((QT|_E)^{-1}|_W)^{-1} = QT|_{((QT)^{-1}W)\cap E} \qquad \text{and}$$
$$((QT)^{-1}W) \cap E \in \mathcal{I}(D(T)).$$

Thus QT is somewhere continuous, whence T is somewhere continuous. The result now follows from the fact that $\tau_0 = \overline{\tau}_0$ (Theorem V.4.16). □

V.17.5 Theorem. *Let $T \in LR(X,Y)$. Then for each (alternatively, for some) $f \in [\mathcal{A}]$ we have*

(a) *T is strictly singular if and only if $T \in [A]_f$*
(b) *T is upper semi-Fredholm if and only if $T \in [B]_f$*
(c) *T is nowhere continuous if and only if $T \in [C]_f$*
(d) *T is partially continuous if and only if $T \in [D]_f$.*

PROOF. We may clearly assume that $\dim D(T) = \infty$.

(a) Let T be strictly singular. Then QT is ubiquitously precompact (Theorem V.4.5) and hence $QT \in [A]_{\overline{\Gamma}_0}$ (Theorem V.2.2). Hence $T \in [A]_f$ since $f \leq \overline{\Gamma}_0$ by Proposition V.17.2. Conversely let $T \in [A]_f$. Then $QT \in [A]_{\overline{\tau}_0}$ as $\overline{\tau}_0 \leq f$ (Proposition V.17.2). Let $M \in \mathcal{I}(D(T))$. Then there exists $N \in \mathcal{I}(M)$ for which $QT|_N$ has a nontrivial precompact restriction (Theorems V.2.4 and

V.1.6). Thus QT is ubiquitously precompact, hence strictly singular (Proposition V.4.5). Therefore T is strictly singular.

(b) Let $T \in [B]_f$. Then $T \in [B]_{\overline{\Gamma}_0}$ since $\overline{\Gamma}_0 \geq f$ (Proposition V.17.2). Consequently QT is ubiquitously not precompact (Theorem V.2.2), hence nowhere precompact by Proposition V.17.3. Therefore QT, and hence T, is upper semi-Fredholm (Theorem V.1.6). Conversely, if T is upper semi-Fredholm, then T is ubiquitously upper semi-Fredholm, whence $T \in [B]_{\overline{\tau}_0}$ (Theorem V.2.4). Therefore $T \in [B]_f$ since $f \geq \overline{\tau}_0$ (Proposition V.17.2).

(c) Let T be nowhere continuous. Then $T \in [C]_{\overline{\tau}_0}$ by Lemma V.17.4 and hence $T \in [C]_f$ for all $f \in [\mathcal{A}]$ by Proposition V.17.2. Conversely, if $T \in [C]_f$ then $T \in [C]_{\overline{\Gamma}_0}$ as $f \leq \overline{\Gamma}_0$ and so QT is ubiquitously not continuous by IV.2.14. Therefore T is nowhere continuous by Proposition V.17.3.

(d) Since $T \in [D]_f$ if and only if $QT \in [D]_f$, and T is partially continuous if and only if QT is partially continuous, it suffices to consider only the case when T is single valued. Accordingly, suppose that T is a linear operator. Evidently T is continuous if and only if G is an isomorphism. We have the following chain of implications: $T \in [D]_{\overline{\tau}_0} \Rightarrow T \in [D]_{\tau_0}$ (Theorem V.4.16) $\Rightarrow (\forall M)(\exists N)\tau_0(T|_N) < \infty \Rightarrow (\forall M)(\exists N)\Gamma_0(G|_{G^{-1}N}) > 0$ (Proposition IV.3.6) $\Rightarrow G \in [B]_{\Gamma_0} \Rightarrow G \in F_+$ (by part (b)) $\Rightarrow$ there exists $M \in \mathcal{C}(X_T)$ such that $(G|_M)^{-1}$ is continuous $\Rightarrow \exists M \in \mathcal{C}(X_T)$ such that $T|_{GM}$ is continuous $\Rightarrow \exists M \in \mathcal{C}(D(T))$ such that $T|_M$ is continuous (Proposition IV.1.2) $\Rightarrow T$ is partially continuous $\Rightarrow T$ is ubiquitously continuous $\Rightarrow (\forall M)(\exists N)\overline{\Gamma}_0(T|_N) < \infty$ (Corollary IV.2.14) $\Rightarrow T \in [D]_f$ (since $f \leq \overline{\Gamma}_0$) $\Rightarrow (\forall M)(\exists N)\overline{\tau}_0(T|_N) < \infty$ (since $\overline{\tau}_0 \leq f$) $\Rightarrow T \in [D]_{\overline{\tau}_0}$. Hence $T \in [D]_f$ if and only if T is partially continuous. □

From Theorem V.17.5 we easily deduce some uniform boundedness properties for the quantities in $[\mathcal{A}]$:

V.17.6 Corollary. *We have:*

(a) *For any f in $[\mathcal{A}]$, if $f(T|_M) < \infty$ for each $M \in \mathcal{I}(D(T))$ then* $\sup_M f(T|_M) < \infty$.

(b) *For any $f \in [\mathcal{A}]$, if $f(T|_M) > 0$ for each $M \in \mathcal{I}(D(T))$ then* $\inf_M f(T|_M) > 0$.

PROOF. (a) The case $f = \overline{\Gamma}_0$ is trivial. In the other cases if $f(T|_M) < \infty$ for each M then $T \in [D]_f$ and so T is partially continuous by Theorem V.17.5(b). Hence there exists $E \in \mathcal{C}(X)$ such that $T|_E$ is continuous. If $f \in \mathcal{A} \setminus \overline{\Gamma}_0$ then $f(T|_M) \leq \Gamma_0(T|_M) = \Gamma_0(T|_{M\cap E})$ (Proposition IV.1.3) $= \overline{\Gamma}_0(T|_{M\cap E}) \leq \overline{\Gamma}_0(T|_E) = \Gamma_0(T|E)$ by Proposition IV.2.7, while for $f = \overline{\Gamma}_0$, $f(T|_M) \leq \Gamma_0(T|_E)$

also (as T is continuous for this case). Now consider $f(T) = \sup_M g(T|_M)$ where $g \in \mathcal{A}$. Then $f(T|_M) \leq \Gamma_0(T|_E)$ by the preceding. Hence for $f \in [\mathcal{A}]$, $f(T|_M) \leq \Gamma_0(T|_E) = \Gamma_0(T) < \infty$.

(b) This may be verified by a suitable combination of IV.2.7(c), V.2.4, V.17.2 and IV.17.5(b). □

The uniform boundedness property of Γ was noted in Cross [5; 4.2] where it was shown that $\Gamma(T|_M) < \infty$ for each $M \implies \sup_M \Gamma(T|_M) = \Delta(T) < \infty$, and consequently that partial continuity of T is equivalent to $\Delta(T) < \infty$.

In fact a stronger form of Corollary V.17.6 holds which can be paraphrased as follows:

(1) *$f(T)$ ubiquit. finite for some $f \in [\mathcal{A}] \implies f(T) < \infty$ for every $f \in \mathcal{A} \setminus \overline{\Gamma}_0$*

(2) *$f(T)$ ubiquit. positive for some $f \in [\mathcal{A}] \implies f(T) > 0$ for every $f \in [\mathcal{A}]$.*

In view of Proposition V.17.2 we have the following stronger implication:

(3) $$\overline{\Gamma}_0(T) \textit{ ubiquitously positive} \implies \overline{\tau}_0(T) > 0.$$

V.17.7 Exercises

(a) Suppose that a scalar λ can be found such that $(T - \lambda I)^{-1}$ is a strictly singular operator. Then T is nowhere continuous.

(b) Let $X = \ell_1$, $Y = \ell_2$ and let A be any linear relation with $D(A) = \ell_1$ and $R(A) = \ell_2$. Let $T := \overline{A}$. Then *T is a continuous and strictly singular relation from ℓ_1 onto ℓ_2*. Continuity follows from the Closed Graph Theorem, while QT (and hence T) is clearly strictly singular. Moreover T is compact if and only if $Y/T(0)$ is finite dimensional.

(c) Let $X = C[0,1] = Y$ and let T be defined by $G(T) := \{(f,g) : g$ is an antiderivative of $f\}$. Then T is compact, $T(0)$ is one dimensional, and T^{-1} is a closed densely defined upper semi–Fredholm operator.

(d) Let T be defined by $G(T) = G(I_X) + G(O_X)$ where I_X and O_X are respectively the identity and the zero operator on X. Then $T(0) = X$ and thus $QT = O$. Here T is a precompact relation, yet T^{-1} is continuous since $\gamma(T) = \gamma(QT) = \gamma(O_X) = \infty$.

(e) Let X and Y be Banach spaces and T an everywhere defined discontinuous linear operator with one dimensional range. Then $\overline{\Gamma}_0(T) = \infty$. However $\overline{T}$ is continuous by the Closed Graph Theorem, so that $\overline{\Gamma}_0(\overline{T}) < \infty$. By contrast if T is a closable linear operator, then $f(T) = f(\overline{T})$ for all $f \in \{\Delta, \Gamma, \Gamma_0, \overline{\Gamma}_0, \tau, \tau_0, \overline{\tau}_0\}$.

V.18 SUMMARY OF EQUIVALENT FORMULATIONS OF THE UPPER SEMI-FREDHOLM PROPERTY

The following properties are equivalent to $T \in F_+$:

(1) There exists a closed finite codimensional subspace M of $D(T)$ such that $T|_M$ is injective and open. (V.1.6)
(2) $QT \in F_+$. (V.1.1(2))
(3) There exists $M \in \mathcal{C}(D(T))$ such that $T|_M$ is bounded below. (V.1.2)
(4) T has no infinite dimensional precompact restriction. (V.1.6)
(5) $\tilde{T} \in \phi_+$ (i.e., $R(\tilde{T})$ is closed and $\alpha(\tilde{T}) < \infty$). (V.7.6)
(6) T has no singular sequence. (V.1.11)
(7) For every bounded subset W of $D(T)$, if $QT(W)$ is precompact, then W is precompact. (V.1.11)
(8) $TG \in F_+$. (V.2.5)
(9) $\Gamma(T) > 0$, or $\dim D(T) < \infty$. (V.2.4)
(10) $\tau_0(T) > 0$, or $\dim D(T) < \infty$. (V.2.4)
(11) $\overline{\tau}_0(T) > 0$, or $\dim D(T) < \infty$. (V.2.4)
(12) $T' \in \phi_-$. (V.7.5)
(13) $Q_F T \in F_+$ for some finite dimensional subspace $F \subset Y$. (V.7.7)
(14) $T + F \in F_+$ for some (every) relation F with $\dim R(F) < \infty$. (V.7.7)
(15) $T + S \in F_+$ for some (every) strictly singular S with $\dim S(0) < \infty$. (V.7.10)
(16) $T + K \in F_+$ for some (every) precompact K with $\dim K(0) < \infty$. (V.7.11)
(17) $T + S - S \in F_+$ for some (every) S with $\dim S(0) < \infty$. (V.7.8)
(18) $T|_E \in F_+$ for some (every) core subspace of $D(T)$. (IV.4.8, V.2.4)
(19) $\tilde{T}(M)^- \subset \tilde{T}(\overline{M})$ for all linear subspaces $M \subset \tilde{X}$. (V.9.22, V.7.6)
(20) There exist linear operators A, F such that A is partially continuous, $\dim R(F) < \infty$ and $AT = I_{D(T)} + F$. (V.10.2)
(21) T^{-1} is partially continuous and $\dim T^{-1}(0) < \infty$. (V.10.4)
(22) T is ubiquitously open and $\dim R(T) = \infty$, or $\dim D(T) < \infty$. (V.12.4)
(23) $T \in [B]_f$ for some (every) $f \in [\mathcal{A}]$. (V.17.5)

In the case when T is single valued, each of the above is equivalent to:

(24) $\dim N(T + S) < \infty$ for all precompact linear operators S. (V.7.11)

V.19 NOTES AND REMARKS

Theorem V.1.6, which characterises upper semi-Fredholm relations, is the set valued version of Cross [5; 2.2]. The latter uses results of Kato [1] and Goldberg [3; III.1.9]. The operator version of Theorem V.1.11 is Cross [7; Theorem 21]. Singular sequences are used to characterise ϕ_+-operators in Banach spaces by Wolf [1] (also by E. Balslev and C.F. Schubert; see Lebow and Schechter [1]).

The terms **partially continuous** and **nowhere continuous** are due to the author (Cross [5]).

The characterisation of precompact linear relations of Theorem V.2.2 ($\overline{\Gamma}_0(T) = 0$) is the quantified version (suitably generalised) of Goldberg [3; III.2.3], due to Lacey [1]. Exercise V.2.17 is from Cross [11] (cf. Martin and Swart [1]).

Bounded strictly singular operators are defined in Kato [1] where the stability of the class of closed upper semi-Fredholm operators under bounded strictly singular perturbation is demonstrated. The notion has been generalised by van Dulst [1] and Cross [3, 4]. The book Kato [2] describes the historical development of the theory of closed upper semi-Fredholm operators in Banach spaces. The perturbation result Theorem V.3.2(b) is formally identical to Schechter [1; 2.14] (see Cross [5; 3.9]). Further similar formulas are derived in Schechter and Whitley [1].

The ubiquitous concept (Definition V.4.1) was introduced for linear operators in Cross [3] and for linear relations in Cross and Pillay [1]. The result of Exercise V.4.10 on the comparability of norms can be found, with an independent proof, in Drewnowski [2]. The equality $\tau_0 = \overline{\tau}_0$ (Theorem V.4.17) was derived jointly by L.E. Labuschagne and the author.,

Section V.5 covers the main results of Cross and Labuschagne [2] (see also Labuschagne [1, 6]).

The single-valued version of Corollary V.5.15 generalising Schauder's theorem can be stated as follows: Let $T : X \to Y$ be a densely defined linear operator. Then T' is compact if and only if the operator $Q_{\tilde{T}(0)}\tilde{T}$ is compact.

According to A. Pelczynski [1], a bounded linear operator $T : X \to Y$, where X and Y are Banach spaces, is called **strictly cosingular** if there does not exist an infinite dimensional subspace $M \subset X$ such that the operator $Q_M T$ is surjective. Definition V.5.18 can be shown to be equivalent to Pelczynski's. For further details, the reader is referred to Cross and Labuschagne [2] and Labuschagne [1, 4, 6].

Operator ranges (Section V.6) are historically associated with endomorphism ranges. A systematic study of endomorphism ranges in Hilbert space was carried out by Dixmier [2]. A survey of the main results was made by Fillmore and Williams [1]. Operator ranges in Banach spaces were studied in the papers Andrew and Patterson [1], Cross [1], Embry [1], Fonf [1], Plichko [2], Shevchik [1], and others, and more recently, by Cross, Ostrovskii and Shevchik [1] where further references may be found (see also Cross and Shevchik [1]). Calkin relations were introduced by Cross [17]. The Calkin property for operators has been studied in Edgar, Ernest and Lee [1], Goldberg and Thorpe [1], Harte and Shannon [1], Tarafdar [1], and Vladimirskii [2].

The Essential States and resulting state diagrams of Sections V.7 and V.8 are from Cross [18]. (See Cross [20] for some related properties).

Section V.16 is the multivalued version of Cross [8]. The latter contains a generalisation of a result on topologically complemented ranges of Holub [1]. Section V.17 is based on the paper Cross and Pillay [1].

CHAPTER VI
Spectral Theory

Throughout this chapter T will denote a linear relation in $LR(X)$, where X is a normed space over the complex field $\mathbb{C}$. We shall write $\lambda - T := \lambda I_{D(T)} - T$.

VI.1. THE SPECTRUM AND THE RESOLVENT OPERATOR

VI.1.1 Definitions Given $T \in LR(X)$, let us write for $\lambda \in \mathbb{C}$,

(1) $$R(\lambda, T) := (\lambda - T)^{-1}$$

called the **resolvent** of T (corresponding to λ) and

(2) $$T_\lambda := R(\lambda, T)^{\sim}$$

called the **complete resolvent** of T. It is simple to verify that T_λ is the resolvent of $\tilde{T}$, i.e.

(3) $$T_\lambda = (\lambda - \tilde{T})^{-1}.$$

The **resolvent set** of T is the set

(4) $$\rho(T) := \{\lambda \in \mathbb{C} : T_\lambda \text{ is everywhere defined and single valued }\}.$$

It is clear from the Closed Graph theorem for linear operators that T_λ is a bounded linear operator (defined on $\tilde{X}$) if and only if $\lambda \in \rho(T)$. In that case T_λ is called a **resolvent operator**.

The **spectrum** of T is the set $\sigma(T) := \mathbb{C} \setminus \rho(T)$.

Observe that

(5) $$\rho(T) = \rho(\overline{T})$$

(6) $$\sigma(T) = \sigma(\overline{T})$$

VI.1.2 Exercise Verify that $\rho(T)$ consists of all $\lambda \in \mathbb{C}$ for which $\lambda - T$ is injective, open and has dense range. In particular, $T(0)$ is closed in the range of $\lambda - T$ for each $\lambda \in \rho(T)$.

VI.1.3 Proposition. *$\rho(T)$ is an open set.*

PROOF. We may clearly suppose that X is complete and T is closed (see (3) of VI.1). Let $\lambda \in \rho(T)$. Then $\gamma(\lambda - T) > 0$. Let $|\mu - \lambda| < \gamma(\lambda - T)$. Then by Corollary III.7.7, the relation $\mu - T = (\mu - \lambda)I + (\lambda - T)$ is injective, surjective and open. Therefore $\mu \in \rho(T)$. Consequently $\rho(T)$ is open. □

VI.1.4 Proposition. (Resolvent Equation)
Let $\lambda, \mu \in \rho(T)$. Then

$$T_\mu - T_\lambda = (\lambda - \mu)T_\mu T_\lambda.$$

PROOF. Without loss of generality we shall suppose that X is complete and T closed. Let $x \in X$. We have

$$\begin{aligned}(\lambda - \mu)T_\mu T_\lambda x &= T_\mu(\lambda - \mu)T_\lambda x = T_\mu(\lambda - T - \mu + T)T_\lambda x \\ &= T_\mu T_\lambda^{-1} T_\lambda x - T_\mu T_\mu^{-1} T_\lambda x \\ &= T_\mu(x + T_\lambda^{-1}(0)) - (T_\lambda x + T_\mu(0)) \quad \text{(by I.3.1)} \\ &= T_\mu x + T_\mu T(0) - T_\lambda x \quad \text{(since } T_\mu(0) = \{0\}) \\ &= T_\mu x - T_\lambda x \quad \text{(since } T_\mu T(0) = T_\mu T_\mu^{-1}(0) = T_\mu(0) = \{0\}). \end{aligned}$$ □

VI.1.5 Corollary. *If $\lambda, \mu \in \rho(T)$, then $T_\mu T_\lambda = T_\lambda T_\mu$.*

We recall the following simple result.

VI.1.6 Lemma. *Let B be a bounded linear operator from a Banach space X into itself. If $\|B\| < 1$, then $I - B$ is invertible and*

$$\lim_{n\to\infty} \left\| (I - B)^{-1} - \sum_{k=0}^{n} B^k \right\| = 0.$$

VI.1.7 Theorem. *Let $\lambda \in \rho(T)$. Then*

$$\lim_{\mu \to \lambda} (\mu - \lambda)^{-1}(T_\mu - T_\lambda) = -T_\lambda^2.$$

PROOF. Again we shall assume without loss of generality that X is complete and T is closed. By Proposition VI.1.3, there is a neighbourhood of λ contained in $\rho(T)$. Let $\mu \neq \lambda$ in such a neighbourhood. Since T_λ and T_μ commute (by Corollary VI.1.5), we have from the Resolvent Equation VI.1.4,

$$(\mu - \lambda)^{-1}(T_\lambda - T_\lambda) + T_\lambda^2 = (T_\lambda - T_\mu)T_\lambda.$$

Hence

$$\left\|(\mu - \lambda)^{-1}(T_\mu - T_\lambda) + T_\lambda^2\right\| \leq \|T_\mu - T_\lambda\| \, \|T_\lambda\| . \tag{1}$$

Now let $|\mu - \lambda| < \|T_\lambda\|^{-1}$. Then $\sum_{n=0}^{\infty} |\mu - \lambda|^n \, \|T_\lambda\|^n < \infty$. Hence by Lemma VI.1.6, the linear operator $I - (\mu - \lambda)T_\lambda$ is invertible and

$$(I - (\mu - \lambda)T_\lambda)^{-1} = \sum_{n=0}^{\infty} (\mu - \lambda)^n T_\lambda^n \tag{2}$$

(where convergence is in the uniform operator topology). Now for $x \in D(T)$ we have

$$\begin{aligned} T_\lambda^{-1}(I - (\lambda - \mu)T_\lambda)x &= T_\lambda^{-1}(x - (\lambda - \mu)T_\lambda x) \\ &= T_\lambda^{-1}x - (\lambda - \mu)T_\lambda^{-1}T_\lambda x \\ &= T_\lambda^{-1}x - (\lambda - \mu)x + T_\lambda^{-1}(0) = (\mu - T)x. \end{aligned}$$

Therefore

$$T_\mu = (I - (\lambda - \mu)T_\lambda)^{-1}T_\lambda \tag{3}$$

Hence from (2) and (3)

$$T_\mu = \sum_{n=0}^{\infty} (\lambda - \mu)^n T_\lambda^{n+1}. \tag{4}$$

Consequently

$$T_\mu - T_\lambda = T_\lambda \sum_{n=1}^{\infty} (\lambda - \mu)^n T_\lambda^n.$$

Hence

$$\|T_\mu - T_\lambda\| \leq \|T_\lambda\| \sum_{n=1}^{\infty} |\mu - \lambda|^n \, \|T_\lambda\|^n . \tag{5}$$

Letting $\mu \to \lambda$ in (1) and using (5) gives the desired result. □

VI.1.8 Corollary. *The family $\{T_\lambda : \lambda \in \rho(T)\}$ of resolvent operators is holomorphic.*

VI.1.9 Corollary. *Let $\lambda \in \rho(T)$. Then*

$$|\lambda - \mu| \leq \|T_\lambda\|^{-1} \implies T_\mu = \sum_{n=0}^{\infty} (\lambda - \mu)^n T_\lambda^{n+1}.$$

VI.1.10 Example *A bounded linear relation T such that both $\rho(T)$ and $\rho(T^{-1})$ are empty.* Let T be the relation whose graph is $X \times X$. Then $T = T^{-1}$, $\|T\| = 0$ and $T_\lambda(0) = \tilde{X}$ $(\lambda \in \mathbb{C})$. Hence T has the required property, provided only that $X \neq \{0\}$.

VI.1.11 Proposition. $\sigma(T) = \sigma(T')$.

PROOF. Follows immediately from the State Diagram for closed linear relations III.6.5. □

VI.2 CLASSIFICATION OF THE SPECTRUM: EIGENVALUES

VI.2.1 Definitions A scalar λ such that $N(\lambda - T) \neq \{0\}$ is called an **eigenvalue** of T.

Let λ be an eigenvalue of T. Then the non zero subspace $N(\lambda - T)$ is called the **eigenspace** of T corresponding to λ, and the dimension of $N(\lambda - T)$ is called the **geometric multiplicity** of λ. The non zero vectors in $N(\lambda - T)$ are called **eigenvectors**.

Clearly, if λ is an eigenvalue of T, then $\lambda \in \sigma(T)$.

The set $\sigma(T)$ is decomposed into the following three disjoint sets:

$P\sigma(T)$, called the **point spectrum** of T, consisting of the eigenvalues of T.

$R\sigma(T)$, called the **residual spectrum** of T, consisting of the points $\lambda \in \mathbb{C}$ such that $\lambda - T$ is injective but does not have dense range.

$C\sigma(T)$, called the **continuous spectrum** of T, consisting of the points $\lambda \in \mathbb{C}$ such that $\lambda - T$ is injective and has dense range, but is not open.

In the notation of Section III.6 (The State Diagram), the corresponding states are as follows:

VI.2.2 Proposition. *Let $\lambda \in \sigma(T)$. Then*

$$(1)\qquad \lambda \in P\sigma(T) \iff \lambda - T \in 3$$

$$(2)\qquad \lambda \in R\sigma(T) \iff \lambda - T \in III \text{ and } \lambda - T \notin 3$$

$$(3)\qquad \lambda \in C\sigma(T) \iff \lambda - T \notin III \text{ and } \lambda - T \in 2$$

PROOF. It only remains to verify (3). Let $\lambda \in C\sigma(T)$. Then $\lambda - T \in I \cup II$ and $\lambda - T \in 1 \cup 2$. But $\lambda - T$ is not open, so $\lambda - T \notin 1$. Therefore $\lambda - T \in I_2 \cup II_2$ as required.

The reverse implication is immediate. □

VI.2.3 Proposition. *Let $\mu \in \rho(T)$ and let $\lambda \neq \mu$. Then $N(\lambda - T) = N((\mu - \lambda)^{-1} - (\mu - T)^{-1})$; in particular, λ is an eigenvalue of T if and only if $(\mu - \lambda)^{-1}$ is an eigenvalue of $(\mu - T)^{-1}$, and corresponding eigenvalues have equal geometric multiplicities.*

PROOF. Let $x \in D(T)$, $x \neq 0$. Then

$$\begin{aligned} & x \in N(\lambda - T) \\ \iff\ & (\lambda - T)x = (\lambda - T)(0) = T(0) \\ \iff\ & (\mu - T)x = (\mu - \lambda)x + T(0) \\ \iff\ & x + (\mu - T)^{-1}(0) = (\mu - \lambda)(\mu - T)^{-1}x + (\mu - T)^{-1}T(0) \\ \iff\ & x = (\mu - \lambda)(\mu - T)^{-1}x \text{ (since } \mu \in \rho(T)) \\ \iff\ & 0 = (I - (\mu - \lambda)(\mu - T)^{-1})x \\ \iff\ & 0 = ((\mu - \lambda)^{-1} - (\mu - T)^{-1})x \end{aligned}$$

Hence $N(\lambda - T) = N((\mu - \lambda)^{-1} - (\mu - T)^{-1})$. □

VI.2.4 Example In Example VI.1.10, $P\sigma(T) = \sigma(T) = \mathbb{C}$, since $(\lambda - T)x = (\lambda - T)(0) = X$ for all $\lambda \in \mathbb{C}$.

In the following example, the point spectrum is empty.

VI.2.5 Example Let $X = \ell_2$ and let K be the linear operator defined by

$$K(\{x_n\}) = (0, x_1, \frac{1}{2}x_2, \frac{1}{3}x_3, \ldots).$$

Then K *is an injective compact operator with* $\sigma(T) = \{0\}$.

VI.2.6 Example Let K be the operator of the previous example (VI.2.5) and let E be a non zero subspace of X such that $E \cap R(K) = \{0\}$. Define $T \in LR(X)$ by

$$G(T) = G(K) + \{0\} \times E.$$

Then T is a bounded compact linear relation. Since $N(T) = K^{-1}(E)$, T is injective as $E \cap R(K) = \{0\}$. Also T is closed if and only if E is closed (see II.5.1). Now suppose that E is closed and let $\lambda \in \mathbb{C}$, $\lambda \neq 0$. Since $\lambda - K$ is surjective by the index theorem (see Goldberg [3; V.2.2]), given $0 \neq e \in E$ there exists $x \in X$, $x \neq 0$, such that $(\lambda - K)x = e$. Then $(\lambda - T)x = E = (\lambda - T)(0)$. Hence λ is an eigenvalue of T. *We have constructed a bounded compact closed injective linear relation T such that $P\sigma(T) = \mathbb{C} \setminus \{0\}$, and $\sigma(T) = \mathbb{C}$.* Moveover, in this example we can choose E so that $\dim T(0) = \dim E < \infty$.

VI.2.7 Remark The inverse of the compact operator K of Example VI.2.6 is single valued and has empty spectrum. (See Theorem VI.4.2).

VI.2.8 Exercise Let $D(T) = \{0\}$, $\dim X = \infty$ and $D(T') = \{0\}$. Show that $\rho(T)$ is empty.

VI.3 EXISTENCE OF A NON EMPTY SPECTRUM

We saw in Example VI.2.6 that a non single valued bounded linear relation T may have a spectrum that coincides with the whole complex plane; in contrast, in the single valued case the spectrum is contained in a compact disc of radius $\|T\|$. A similar state of affairs holds even in the finite dimensional case. For example, let $X = \mathbb{C}^2$ and let T be defined in terms of its graph by $G(T) = G(I) + \{0\} \times E$ where E may be one or two dimensional. Then T is bounded, $T(0) = E$, and $\sigma(T) = \mathbb{C}$.

If the linear relation T is densely defined, but not single valued, then by Proposition VI.1.11 its spectrum coincides with the spectrum of its adjoint T', which is single valued. Thus the preceding example also shows that the spectrum of a continuous single valued relation may have spectrum equal to $\mathbb{C}$ (provided of course, that its domain is not the whole space).

The following proposition shows that the spectrum of a non single valued relation may be a proper subset of $\mathbb{C}$.

VI.3.1 Proposition. *Let the relation T be injective with dense range. Then*

$$\sigma(T) \subset \{\lambda \in \mathbb{C} : |\lambda| \geq \gamma(\tilde{T})\}.$$

PROOF. If $\gamma(\tilde{T}) = 0$ then $0 \in \sigma(\tilde{T}) = \sigma(T)$ and there is nothing to prove.

Now suppose that $\gamma(\tilde{T}) > 0$. Since T^{-1} is a continuous and densely defined linear operator, its completion $(T^{-1})^{\sim} = \tilde{T}^{-1}$ is a closed and bounded linear operator defined on the Banach space $\tilde{X}$. We may therefore assume without loss of generality that X is complete and T is closed, open, injective and surjective. Let $\lambda \in \mathbb{C}$, $|\lambda| < \gamma(T)$. Then by the perturbation result III.7.7 together with III.4.2(b), the relation $\lambda - T$ is injective and surjective, i.e. $\lambda \in \rho(T)$. Hence $\sigma(T) \subset \{\lambda \in \mathbb{C} : |\lambda| \geq \gamma(\tilde{T})\}$. □

VI.3.2 Proposition. *Let T be continuous, $\overline{D(T)} = X$ and $\rho(T)$ nonempty and unbounded. Then*

$$\lim_{\substack{|\lambda|\to\infty \\ \lambda\in\rho(T)}} \|T_\lambda\| = 0.$$

PROOF. We may clearly suppose that $T = \tilde{T}$. We have from the Resolvent Equation VI.1.4, for $\lambda, \mu \in \rho(T)$ and μ fixed,

$$(T_\lambda - T_\mu)(\mu - T) = (\mu - \lambda)T_\lambda T_\mu(\mu - T)$$

$$\text{(i.e.)} \qquad T_\lambda(\mu - T) - I_{D(T)} = (\mu - \lambda)T_\lambda I_{D(T)}.$$

Hence by II.3.13, $|\lambda - \mu|\,\|T_\lambda\| \leq 1 + \|\mu - T\|\,\|T_\lambda\|$. Since $\|\mu - T\| < \infty$, $(1 - \dfrac{\|\mu - T\|}{|\lambda - \mu|})\,\|T_\lambda\| \leq \dfrac{1}{|\lambda - \mu|}$. Letting $|\lambda| \to \infty$, $\lim\limits_{|\lambda|\to\infty} \|T_\lambda\| = 0$. □

We have noted that a closed densely defined linear operator may have empty spectrum (see Remark VI.2.7). This shortcoming falls away in the presence of continuity, as the following theorem shows. With all the ingredients in place, the proof is now entirely standard.

VI.3.3 Theorem. *Let X be a non zero normed space and let T be a continuous and densely defined linear relation in $LR(X)$. Then the spectrum of T is non empty.*

PROOF. Since $\sigma(\tilde{T}) = \sigma(T)$, and replacing T by $\tilde{T}$ if necessary, we may for simplicity assume that T is closed (so that $D(T) = X$) and X complete.

Suppose that $\rho(T) = \mathbb{C}$. By Theorem VI.1.7 we have for $x \in X$ and $x' \in X'$,

$$\lim_{\mu\to\lambda} \frac{x'T_\lambda x - x'T_\mu x}{\lambda - \mu} = -x'T_\lambda^2 x.$$

Thus the (single valued) function $f(\lambda) = x'T_\lambda x$ is an entire analytic function. Moreover

$$f(\lambda) \leq \|x'\|\,\|T_\lambda\|\,\|x\|$$

where $\lim_{|\lambda|\to\infty} \|T_\lambda\| = 0$ by Proposition VI.3.2. From Liouville's theorem it follows that $f(\lambda) = 0$ for all λ. Since x' is arbitrary, we have $T_\lambda x = 0$ for every $x \in X$. Hence

$$x + (\lambda - T)(0) = (\lambda - T)(0)$$
$$\text{i.e.} \qquad x + T(0) = T(0) \qquad (x \in X).$$

Hence $T(0) = X \neq \{0\}$. But then T is not bounded below, whence $0 \in \sigma(T)$, contradicting $\rho(T) = \mathbb{C}$. □

VI.4 COMPACTNESS OF THE AUGMENTED SPECTRUM

Let $\mathbb{C}_\infty$ denote the extended complex plane, $\mathbb{C}_\infty = \mathbb{C} \cup \{\infty\}$, endowed with the usual topology. Then $\mathbb{C}_\infty$ is a compact topological space.

The **Möbius** transformation

$$\eta(\lambda) = (\mu - \lambda)^{-1}$$

where μ is a fixed point of $\mathbb{C}$, is a topological homeomorphism from $\mathbb{C}_\infty$ onto itself.

VI.4.1 Definition The **augmented spectrum** of T is the set

$$\begin{aligned}\overline{\sigma}(T) &= \sigma(T) \cup \{\infty\} \text{ if } 0 \in \sigma(T^{-1}) \\ &= \sigma(T) \text{ otherwise.}\end{aligned}$$

Thus $\{\infty\} \not\subset \overline{\sigma}(T)$ if and only if $\tilde{T}$ is a bounded linear operator, if and only if $0 \notin \sigma(T^{-1})$.

The *augmented spectrum is non empty* (if $X \neq \{0\}$) since if $\{\infty\} \not\subset \overline{\sigma}(T)$ then $\overline{\sigma}(T) = \sigma(T) = \sigma(\tilde{T})$ is non empty by Theorem VI.3.3.

VI.4.2 Theorem. *Let $\mu \in \rho(T)$, $D(T) \neq \{0\}$. Then*

$$\eta(\overline{\sigma}(T)) = \sigma(T_\mu).$$

PROOF. We may clearly suppose that X is complete and T is closed. Write $S := (\mu - \lambda)((\mu - \lambda)^{-1} - T_\mu)$. We shall verify that

$$\lambda - T = S(\mu - T) \tag{1}$$

We have

$$\begin{aligned}\lambda - T &= \mu - T - (\mu - \lambda)I \\ &= (I - (\mu - \lambda)T_\mu)(\mu - T)\end{aligned}$$

(by I.4.2(e), since $D(T_\mu) = X$ and $T_\mu(0) = 0$)

$$= S(\mu - T)$$

as required.

Suppose that $\lambda \in \rho(T)$. Then S is injective since for $x \in X$,

$$\begin{aligned}Sx = 0(= S(0)) &\implies (\mu - \lambda)^{-1}x = T_\mu x \\ &\implies (\mu - \lambda)^{-1}(\mu - T)x = x + (\mu - T)(0) = x + T(0) \\ &\implies (\mu - T)x = (\mu - \lambda)x + T(0) \implies (\lambda - T)x = T(0) \\ &\implies x = 0\end{aligned}$$

as required, since $\lambda - T$ is both injective and open (hence bounded below).

Next we have from (1) that

$$X = R(\lambda - T) \subset R(S).$$

Hence S is surjective. Consequently, by the Open Mapping theorem III.4.2(b), S is open. It follows that $(\mu - \lambda)^{-1} \in \rho(T_\mu)$.

Conversely, let $(\mu - \lambda)^{-1} \in \rho(T_\mu)$. For $x \in D(T)$, we have from (1) that

$$\|(\lambda - T)x\| = \|S(\mu - T)x\| \geq \gamma(S(\mu - T))d(x, T_\mu S^{-1}(0)).$$

But $T_\mu S^{-1}(0) = T_\mu(0) = 0$ (since S is injective). Therefore

$$\|(\lambda - T)x\| \geq \gamma(S(\mu - T))\,\|x\| \tag{2}$$

where $\gamma(S(\mu - T)) \geq \gamma(S)\gamma(\mu - T)$ (by II.3.11, as S is injective). But $\gamma(S) > 0$ from the hypothesis, and $\gamma(\mu - T) > 0$ since $\mu \in \rho(T)$. Hence from (2), $\lambda - T$ is injective and open. Since $\lambda - T$ is surjective by (1), it follows that $\lambda \in \rho(T)$. □

VI.4.3 Corollary. *Let T have a non empty resolvent set.*

(a) *If $\tilde{T}$ is a bounded linear operator, then $\sigma(T)$ is a compact subset of $\mathbb{C}$.*

(b) *If $\tilde{T}$ is not a bounded linear operator, then $\{\infty\} \subset \overline{\sigma}(T)$, and $\overline{\sigma}(T)$ is a compact subset of $\mathbb{C}_\infty$.*

VI.5 SPECTRAL MAPPING THEOREM

In this section we shall establish the equality $p\sigma(T) = \sigma(p(\tilde{T}))$ for any complex polynomial p. (See Theorem VI.5.4).

VI.5.1 Proposition. *The relations $\lambda - T$ and $\mu - T$ commute ($\lambda, \mu \in \mathbb{C}$).*

PROOF. It is clear that the domains of $(\mu - T)(\lambda - T)$ and $(\lambda - T)(\mu - T)$ are each equal to $D(T^2)$. We now have for $x \in D(T^2)$

$$\begin{aligned}(\lambda - T)(\mu - T)x &= (\lambda - T)(\mu x - Tx)\\ &= (\lambda - T)\{\mu x - y : y \in Tx, y \in D(T)\}\\ &= \{\lambda\mu x - \lambda y - \mu Tx + Ty : y \in Tx, y \in D(T)\}\\ &= \lambda\mu x - (\lambda + \mu)Tx + T^2 x\\ &= (\mu - T)(\lambda - T)x\end{aligned}$$

as required. □

VI.5.2 Proposition. *Let the relations S and T in $LR(X)$ be completely closed, bounded below and surjective. Then ST has the same properties.*

PROOF. Clearly ST is injective. Furthermore, $\gamma(ST) \geq \gamma(S)\gamma(T)$ by Theorem II.3.11. Therefore ST is bounded below.

Next, $ST(X) = S(T(X)) = S(X) = X$. Therefore ST is surjective.

Finally, to show that ST is closed, let (x_k, y_k) be a sequence in $G(ST)$ converging to (x, y) in $\tilde{X} \times \tilde{X}$.

For each k there exists z_k such that $(x_k, z_k) \in G(T)$ and $(z_k, y_k) \in G(S)$. By the continuity and single valuedness of S^{-1}, $z_k = S^{-1}y_k$ converges in $\tilde{X}$. Since S is completely closed, it follows that $\lim z_k = z$ for some $z \in D(S)$, and that $(z, y) \in G(S)$. Now $(x_k, z_k) \to (x, z) \in X \times X \quad (k \to \infty)$ and since T is closed, we have $(x, z) \in G(T)$. Therefore $(x, y) \in G(ST)$. The proof is now complete. □

VI.5.3 Remark The proof of Proposition VI.5.2 shows that it is enough to assume that T is closed, rather than completely closed, for the stated conclusion to follow.

VI.5.4 Theorem. *Let X be a normed space and let $T \in LR(X)$. Then for any complex polynomial p we have $\sigma(p(\tilde{T})) = p(\sigma(T))$.*

PROOF. Fix $\lambda \in \mathbb{C}$ and let $p(\lambda) - \mu = c\prod_{j=1}^{n}(\lambda - \alpha_j)$. Then

$$p(T) - \mu I = c\prod_{j=1}^{n}(T - \alpha_i I) \tag{1}$$

which is well defined by Proposition VI.5.1.

Since $\sigma(T) = \sigma(\tilde{T})$, we shall assume without loss of generality that X is complete and T closed.

Let $\mu \in \sigma(p(T))$. If $\alpha_j \in \rho(T)$ for all $j = 1, \ldots, n$, then $p(T) - \mu$ would be bounded below and surjective, by Proposition VI.5.2. But then $\mu \in \rho(p(T))$, a contradiction. Hence for some j, $1 \leq j \leq n$, we have $\alpha_j \in \sigma(T)$. Since $p(\alpha_j) = \mu$, it follows that $\mu \in p(\sigma(T))$.

For the reverse inclusion, let $\mu \in p(\sigma(T))$, so that $p(\lambda) = \mu$ for some $\lambda \in \sigma(T)$. Then $\lambda = \alpha_j$ for some j, say $j = 1$. If $T - \alpha_1 I$ is bounded below, then it is not surjective (as $\alpha_1 \in \sigma(T)$). Hence from (1), $p(T) - \mu$ is not surjective. In that case $\mu \in \sigma(p(T))$.

Let us assume that $\mu \notin \sigma(p(T))$. Then from the preceding, $T - \alpha_j I$ is not bounded below for any j, $1 \leq j \leq n$. But $T - \alpha_j I$ is surjective for each j (since $p(T) - \mu I$ is surjective). Hence each $T - \alpha_j I$ is open, by the Open Mapping theorem III.4.2(b).

Writing $p := p(T) - \mu$, we have

$$p^{-1}(0) = (T - \alpha_n I)^{-1} \ldots (T - \alpha_1 I)^{-1}(0) = 0.$$

Hence $(T - \alpha_n I)^{-1}(0) = 0$. Therefore $T - \alpha_n I$ is bounded below, a contradiction. The proof of the theorem is now complete. □

VI.6 NOTES AND REMARKS

Favini and Yagi [1] define $\rho(T)$ to be the set $\{\lambda \in \mathbb{C} : (\lambda - T)^{-1}$ is a bounded linear operator on $X\}$ where X is a Banach space. They then obtain VI.1.3 and VI.1.8 under this definition (which coincides with ours when T is closed.).

The resolvent of T is defined by Arens [1] to be the set $\{\lambda \in \mathbb{C} : (\lambda - T)^{-1}$ is everywhere defined and single valued$\}$. This definition coincides with our Definition VI.1.1(4) if $T = \tilde{T}$ (i.e., if T is completely closed). In the case when

T is a closed linear operator and X is a Banach space, both these definitions coincide with the classical one.

Some properties of the spectra of the linear relations in Hilbert space are discussed in the book of Azizov and Iokhvidov [1; p.142] (see also Ritsner [1]).

The paper of P. Sorjonnen [1] investigates properties of symmetric linear relations in Hilbert and Pontrjagin spaces; in particular, eigenvalues are studied therein (ibid, Section 4.1). The paper also addresses the question: Can the theory of linear relations in Hilbert space be extended to a corresponding theory in an indefinite inner product space?

CHAPTER VII

The Essential Spectrum

Throughout this chapter, except where stated otherwise, X will denote a complex normed linear space and T a relation in $LR(X)$.

VII.1. PRELIMINARY PERTURBATION RESULTS

VII.1.1 Definitions The *essential resolvent set* of T is the set

$$\rho_e(T) := \{\lambda \in \mathbb{C} : \alpha(\lambda - \tilde{T}) < \infty \text{ and } \beta(\lambda - \tilde{T}) < \infty\}.$$

The *essential spectrum* of T is the complementary set

$$\sigma_e(T) := \mathbb{C} \setminus \rho_e(T).$$

We clearly have

(1) $$\rho_e(T) = \rho_e(\tilde{T}) = \rho_e(T').$$

VII.1.2 Proposition. *The following properties are equivalent for a given $\lambda \in \mathbb{C}$:*

(i) $\lambda \in \rho_e(T)$
(ii) $\lambda - T \in F_+ \cap F_-$
(iii) *The complete closure of $\lambda - T$ is a Fredholm relation.*

PROOF. (i) $\iff$ (iii). Let $\lambda \in \rho_e(T)$. Then $\lambda - \tilde{T}$ has closed range by Corollary V.6.5 and hence $\lambda - \tilde{T} \in \phi_+ \cap \phi_-$. The converse is trivial.

(ii) $\iff$ (iii). Combine Proposition V.1.9 and Corollary V.7.6. □

VII.1.3 Proposition. *$\rho_e(T)$ is an open set.*

PROOF. Clearly $\rho_e(T) = \emptyset$ or $\mathbb{C}$ if $D(T)$ is finite dimensional. Hence assume that $D(T)$ is infinite dimensional. Let $\lambda \in \rho_e(T)$. Then $\lambda - T \in F_+ \cap F_-$. Let $|\mu| < \Gamma(\lambda - T)$. Then $(\lambda + \mu) - T \in F_+$ by Theorem V.3.2. Also, since $\lambda - T \in F_-$, therefore $\lambda - T' \in \phi_+$ and so, excluding the trivial case when $\dim D(T') < \infty$, we have

$$|\mu| < \Gamma(\lambda - T') \implies (\lambda + \mu) - T' \in \phi_+ \implies (\lambda + \mu) - T \in F_-.$$

Hence $\lambda + \mu \in \rho_e(T)$ whenever $|\mu| < \min\{\Gamma(\lambda - T), \Gamma(\lambda - T')\}$. Therefore $\rho_e(T)$ is an open subset of $\mathbb{C}$. □

We have stability under finite rank perturbation:

VII.1.4 Lemma. *Let the relation F satisfy $D(F) \supset D(T)$ and $\dim R(F) < \infty$. Then*

$$\sigma_e(T + F) = \sigma_e(T).$$

PROOF. Let $\lambda \in \rho_e(T)$. Then $\lambda - T \in F_+$. Hence $\lambda - (T + F) \in F_+$ by V.7.10, since F is strictly singular. Likewise, $\lambda - (T+F) \in F_-$ by V.5.20. Hence $\rho_e(T) \subset \rho_e(T + F)$.

Now let $\lambda \in \rho_e(T + F)$. Then $\lambda - (T + F - F) \in F_+ \cap F_-$ by the preceding. Therefore $\lambda - T \in F_+$ by Lemma V.7.8. Again, from Proposition V.5.11, $Q_F(\lambda - T) = Q_F(\lambda - T + F - F) \in F_-$, and thus $\lambda - T \in F_-$. Therefore $\lambda \in \rho_e(T)$. □

(Lemma VII.1.4 becomes a special case of Theorem VII.3.2 when F is continuous.)

VII.1.5 Exercise Let T be densely defined. Show that $\lambda \in \rho_e(T)$ if and only if $\lambda I - T'$ is a (closed) Fredholm operator.

VII.2 RELATIVE BOUNDEDNESS AND RELATIVE COMPACTNESS

VII.2.1 Definitions A relation $S \in LR(X, Z)$ is said to be **T-bounded** if $D(S) \supset D(T)$ and there exist real numbers a, b for which the inequality

$$\|Sx\| \leq a\|x\| + b\|Tx\|$$

holds for all $x \in D(T)$. In that case the infimum of all the numbers b is called the **T-bound** of S.

The relation S is called **T-precompact** (or **precompact relative to** T) if $D(S) \supset D(T)$ and SG_T is precompact. Similarly, S is called **T-compact** (or **compact relative to** T) if SG_T is compact.

It is sometimes convenient to renorm the space X_T with some other equivalent norm. The following proposition is useful:

VII.2.2 Proposition. *The norms $\| \; \|_T$ and $\| \; \|_{\lambda - T}$ are equivalent.*

PROOF. We have for $x \in D(T)$,

$$\begin{aligned} \|x\|_{\lambda-T} &= \|x\| + \|(\lambda - T)x\| \leq \|x\| + |\lambda| \, \|x\| + \|Tx\| \\ &= (1 + |\lambda|) \, \|x\| + \|Tx\| \leq (1 + |\lambda|)(\|x\| + \|Tx\|) \\ &= (1 + |\lambda|) \, \|x\|_T \end{aligned}$$

and

$$\|x\|_T = \|x\| + \|Tx\| = \|x\| + \|\lambda x - (\lambda x - Tx)\| \le (1 + |\lambda|)\,\|x\| + \|(\lambda - T)x\|$$
$$\le (1 + |\lambda|)(\|x\| + \|(\lambda - T)x\|) = (1 + |\lambda|)\,\|x\|_{\lambda - T} \cdot$$

Therefore $\|\ \|_T$ and $\|\ \|_{\lambda I - T}$ are equivalent. □

VII.2.3 Proposition. *Let $S \in LR(X)$ be such that $S(0) \subset \overline{T(0)}$ and T-bounded with T-bound < 1. Then the norms $\|\ \|_T$ and $\|\ \|_{T+S}$ are equivalent.*

PROOF. Write $R := T + S$. Then for $x \in D(T)$ we have

$$\|Sx\| \le a\,\|x\| + b\,\|Tx\|$$

where $a \ge 0$, $0 \le b < 1$. Hence

$$\begin{aligned} -a\,\|x\| + (1-b)\,\|Tx\| &\le -\|Sx\| + \|Tx\| \\ &\le \|Rx\| \qquad (\text{since } S(0) \subset \overline{T(0)}) \\ &\le a\,\|x\| + (1+b)\,\|Tx\|\,. \end{aligned}$$

Choose any number $c > a$. Then

$$\begin{aligned} (c-a)\,\|x\| + (1-b)\,\|Tx\| &\le \|(S+T)x\| + c\,\|x\| \\ &\le (a+c)\,\|x\| + (1+b)\,\|Tx\| \end{aligned}$$

showing that the norms $\|\ \|_T$ and $\|\ \|_{T+S}$ are equivalent. □

VII.2.4 Exercise Show that if K is T-precompact, then $\tilde{K}$ is TJ^{-1}-compact, where $J := J_X^{\tilde{X}}$.

VII.2.5 Theorem. *Let S and K be linear relations in $LR(X,Y)$ with S T-bounded and K T-compact (resp., T-precompact) and single valued. Let $S - K$ have T-bound $< \alpha$ and let $\{a_n\}$ and $\{b_n\}$ be sequences such that $a_n \uparrow \infty$ and $b_n \downarrow \delta \ge \alpha$. If either*

(a) *K is closable (resp., completely closable), or*

(b) *X and Y are reflexive and T is closable,*

then the inequality

(1) $$\|Sx\| \le a_n \|x\| + b_n \|Tx\| \qquad (x \in D(T))$$

holds for all sufficiently large $n \in \mathbb{N}$. *In particular,* S *has* T*-bound* $< \alpha$.

PROOF. Suppose (1) does not hold for all sufficiently large $n \in \mathbb{N}$. Then there is a sequence x_n in $D(T)$ such that

(2) $$\|Tx_n\| = 1, \text{ and } \|Sx_n\| \ge a_n \|x_n\| + b_n \qquad (n \in \mathbb{N}).$$

But S is T-bounded, so $a\|x\| + b\|Tx\| \ge \|Sx\|$ for some $a, b > 0$ and all $x \in D(T)$. Hence for sufficiently large n,

(3) $$\|x_n\| \le \frac{b - b_n}{a_n - a}$$

showing that $\lim_{n\to\infty} x_n = 0$. Now $\|x_n\|_T = \|x_n\| + \|Tx_n\| \to 1 \quad (n \to \infty)$. Consequently the sequence $\{\|x_n\|_T\}$ is bounded.

(a) By the T-compactness (T-precompactness) of K, $\{Kx_n\}$ has a convergent (resp., Cauchy) subsequence, for simplicity assumed to be itself. Now $(x_n, Kx_n) \to (0, \lim Kx_n)$ (resp., $(0, \tilde{Y}-\lim Kx_n)$), so $Kx_n \to 0$ as K is closable (resp., completely closable).

By hypothesis,

$$\|(S-K)x\| \le c\|x\| + d\|Tx\| \quad \text{where } c, d > 0$$

and $d < \alpha$. Hence by the single valuedness of K,

(4) $$\|Sx_n\| \le \|(S-K)x_n\| + \|Kx_n\| \le c\|x_n\| + d + \|Kx_n\|$$

whence

(5) $$\limsup \|Sx_n\| \le d < \alpha.$$

But from (2), since $\alpha \le b_n$ we have

(6) $$\alpha \le \limsup b_n \le \limsup(b_n + a_n\|x_n\|) \le \limsup \|Sx_n\|$$

contradicting (5). Hence (1) follows.

(b) Since $X_{\tilde{T}}$ is reflexive (being isomorphic to a closed subspace of $X \times Y$), it follows that $\{x_n\}$ has a subsequence, again assumed to be itself, converging in the weak topology of $X_{\tilde{T}}$ (and hence also of X) to some point $x \in D(\tilde{T})$. Thus $x = 0$ by (3). Since K is T-compact, it is T-completely continuous and so $\|Kx_n\| \to 0$. The rest of the proof now proceeds exactly as in (a).

Finally, we verify that S has T-bound $< \alpha$. Let h be a number strictly between the T-bound of $S - K$ and α. Choose $b_n = h \ (n \in \mathbb{N})$. Then (1) with h in place of b_n gives

(7) $$\|Sx\| \le a_n\|x\| + h\|Tx\| \qquad (x \in D(T))$$

for all sufficiently large n. Thus S has T-bound $< \alpha$. □

VII.2.6 Corollary. *Let K be T-compact (resp., T- precompact) and single valued.*

If either
(a) K is closable (resp., completely closable), or
(b) X and Y are reflexive and T is closable,
then K has T-bound 0. Hence the norms $\| \ \|_T$ and $\| \ \|_{T+K}$ are equivalent.

PROOF. Put $S = K$ in Theorem VII.2.5. The second assertion follows immediately from Proposition VII.2.3. □

Theorem VII.2.9 below provides examples of T-precompact and T-compact relations. First we note the following property:

VII.2.7 Proposition. *Let $\lambda \in \rho(T)$ and suppose that T_λ is compact. Then T_μ is compact for every $\mu \in \rho(T)$.*

PROOF. Let $\lambda, \mu \in \rho(T)$ and suppose that T_λ is compact. We have from the Resolvent Equation (VI.1.4),

$$T_\mu = ((\lambda - \mu)T_\mu + I)T_\lambda$$

where $(\lambda - \mu)T_\mu + I$ is a bounded linear operator defined on $\tilde{X}$. Therefore T_μ is compact. □

VII.2.8 Proposition. *Let $\lambda \in \rho(T)$. Then T_λ is compact if and only if $R(\lambda, T)$ is precompact.*

PROOF. Combine V.2.2 with the density invariance result IV.4.9, noting that $R(\lambda, T)$ is completely closable by II.5.5. □

VII.2.9 Theorem. *Let the relation $T \in LR(X)$ have a compact resolvent operator. Then every continuous relation S such that $D(S) \supset D(T)$ is T-precompact.*

PROOF. We may clearly suppose that $D(T)$ is infinite dimensional. Let $\lambda \in \rho(T)$, let $(\lambda - T)^{-1}$ be precompact and let S be a continuous linear relation in $LR(X)$ such that $D(S) \supset D(T)$. The continuity of S implies that SG_T is precompact whenever G_T is precompact. By Proposition VII.2.2, G_T is precompact if and only if $G_{\lambda-T}$ is precompact. Furthermore $\overline{\Gamma}_0(G_{\lambda-T}) = \Gamma_0(G_{\lambda-T})$ by the continuity of the graph operator $G_{\lambda-T}$. Hence it is sufficient to show that $\Gamma_0(G_{\lambda-T}) = 0$, by Theorem V.2.2.

Write $R := (\lambda - T)^{-1}$. Considered as a map from its domain onto its range. Since R is precompact, then clearly $\hat{R}$ is a precompact linear operator. Thus

$$(1) \qquad \Gamma_0(\hat{R}) = 0.$$

Write $\Pi := Q_{N(R)}^{D(R)}$. Clearly $\tau_0(\Pi^{-1}) = 1$ (by direct computation). Hence by Proposition IV.3.6, IV.3.7 and IV.2.11,

$$\begin{aligned}\Gamma_0(G_{R^{-1}}) &= (1+\tau_0(R^{-1}))^{-1} \leq (1+\tau_0(\Pi^{-1})\tau_0(\hat{R}^{-1}))^{-1} \\ &= (1+\Gamma_0(\hat{R})^{-1})^{-1} = 0, \quad \text{by (1)}.\end{aligned}$$

Hence $G_{R^{-1}}$ is precompact and the theorem is proved. □

A condition will now be found for T to have a precompact resolvent (see Corollary VII.2.13).

VII.2.10 Proposition. *If G_T is precompact then the linear relation $(\lambda - T)^{-1}$ is precompact whenever $\lambda - T$ is open and $T(0)$ is relatively closed in $R(\lambda - T)$ (in particular, whenever $\lambda \in \rho(T)$).*

PROOF. Since the norms $\| \, \|_T$ and $\| \, \|_{\lambda-T}$ are equivalent (Proposition VII.2.2), it is enough to consider the case $\lambda = 0$.

We first observe that the proof can be reduced to the single valued case. Indeed, suppose that the Proposition is true for single valued relations. Let G_T be precompact. Since $G_T = G_{QT}$ (see Exercise IV.3.12), we have $T^{-1}Q^{-1}$ precompact. Now let $\{y_n\}$ be a bounded sequence in $R(T)$. We have for $y \in R(T)$, since $T(0)$ is relatively closed

$$(1) \qquad d(T^{-1}y, T^{-1}(0)) = d(T^{-1}y, T^{-1}T(0)) = d(T^{-1}Q^{-1}Qy, 0).$$

Since $T^{-1}Q^{-1}$ is precompact and $\{Qy_n\}$ is bounded, there exists a subsequence of $\{y_n\}$, for simplicity assumed to be itself, such that

$$(2) \qquad d(T^{-1}Q^{-1}Q(y_n - y_k), 0) \to 0 \qquad (n, k \to \infty).$$

From (1) and (2) it follows that

$$d(T^{-1}(y_n - y_k), 0) \to 0 \qquad (n, k \to \infty).$$

Therefore T^{-1} is precompact.

Now let T be a single valued relation such that G_T is precompact. If $D(T)$ is finite dimensional, then clearly T^{-1} is precompact. Hence assume that $\dim D(T) = \infty$. By Proposition IV.3.6 and Theorem V.2.2,

$$(3) \qquad (1+\tau_0(T))^{-1} = \Gamma_0(G_T) = \overline{\Gamma}_0(G_T) = 0.$$

Therefore $\tau_0(T) = \infty$. Consequently, $T \in F_+$ (Theorem V.2.4), and so there exists $M \in \mathcal{P}(D(T))$ making $T|_M$ open and injective. We also have $G_{T|_M}$ precompact by Proposition V.1.20.

Since $T|_M$ is single valued and injective, we have from Proposition IV.3.7 and Theorem V.2.2,

$$\begin{aligned} 0 = \overline{\Gamma}_0(G_{T|_M}) &= \Gamma_0(G_{T|_M}) = (1+\tau_0(T|_M))^{-1} \\ &= (1+\Gamma_0((T|_M)^{-1})^{-1})^{-1}. \end{aligned}$$

Hence $\Gamma_0((T|_M)^{-1}) = 0 = \overline{\Gamma}_0((T|_M)^{-1})$ (since $(T|_M)^{-1}$ is continuous). Therefore $(T|_M)^{-1}$ is precompact. Consequently T^{-1} is precompact, by Proposition V.1.19. □

As a partial converse to Proposition VII.2.10, we have the following:

VII.2.11 Proposition. *Let $N(T)$ and $T(0)$ be relatively closed in $D(T)$ and $R(T)$ respectively. If T^{-1} is precompact, then $G_{\hat{T}}$ is precompact. If in addition $N(T)$ is finite dimensional, then G_T is precompact.*

PROOF. We may clearly suppose that $D(\hat{T})$ is infinite dimensional and furthermore, that $D(T) = X$, $R(T) = Y$ and $N(T)$ is closed. From the canonical factorisation

$$T = \hat{T}Q_{N(T)}$$

and the inequalities

$$\begin{aligned} \overline{\Gamma}_0(Q_{N(T)}^{-1}\hat{T}^{-1}) &\le \overline{\Gamma}_0(Q_{N(T)}^{-1})\overline{\Gamma}_0(\hat{T}^{-1}) \\ \overline{\Gamma}_0(Q_{N(T)}^{-1}\hat{T}^{-1}) &\ge \tau_0(Q_{N(T)}^{-1})\overline{\Gamma}_0(\hat{T}^{-1}) \end{aligned}$$

(see Proposition IV.2.15 and IV.2.16) we see that T^{-1} is precompact if and only if $\hat{T}^{-1}$ is precompact (Theorem V.2.2), since $\tau_0(Q_{N(T)}^{-1}) = 1$ (as $D(\hat{T})$ is infinite dimensional) and $\hat{T}^{-1}$ is single valued.

Suppose that T is single valued. Then by Proposition IV.3.7,

$$\Gamma_0(\hat{T}^{-1}) = \tau_0(\hat{T})^{-1}.$$

Hence if T^{-1} is precompact, then $\hat{T}^{-1}$ is precompact and

$$0 = \overline{\Gamma}_0(\hat{T}^{-1}) = \Gamma_0(\hat{T}^{-1}) = \tau_0(\hat{T})^{-1},$$

so $\tau_0(\hat{T}) = \infty$, and consequently $\Gamma_0(G_{\hat{T}}) = (1+\tau_0(\hat{T}))^{-1} = 0$ (IV.3.6). It follows that $G_{\hat{T}}$ is precompact whenever T is single valued.

If T is not single valued, then since $T(0)$ is relatively closed, with the same argument used to obtain (1) of VII.2.10

$$d(T^{-1}y, T^{-1}(0)) = d(T^{-1}Q^{-1}Qy, 0)$$

and hence T^{-1} is precompact if and only if $(QT)^{-1}$ is precompact. Consequently if T^{-1} is precompact, then $G_{(QT)\hat{}}$ is precompact. But

$$QT = (QT)\hat{}\,Q_{N(QT)} = (QT)\hat{}\,Q_{N(T)}.$$

Hence for $x \in D(T) = D(QT)$, we have

$$\begin{aligned}\|[x]\|_{(QT)\hat{}} &= \|[x]\| + \|(QT)\hat{}[x]\| = \|[x]\| + \|Tx\| \\ &= \|[x]\| + \|\hat{T}[x]\| = \|[x]\|_{\hat{T}}\,.\end{aligned}$$

Therefore $G_{\hat{T}} = G_{(QT)\hat{}}$. Hence $G_{\hat{T}}$ is precompact.

Now suppose that $N(T)$ is finite dimensional and let P be a bounded projection defined on $D(T)$ with kernel $N(T)$. Then for any $x \in R(P)$ we have

$$\|P\|^{-1}\|x\| = \|P\|^{-1}\|Px\| \leq \|x + N(P)\| = \|[x]\| \leq \|x\|$$

and so

$$\|P\|^{-1}(\|x\| + \|Tx\|) \leq \|[x]\| + \left\|\hat{T}[x]\right\| \leq \|x\| + \|Tx\|$$

since $\|P\| \geq 1$. Consequently, $G_{\hat{T}}$ is precompact if and only if $G_T|_{G^{-1}(R(P))}$ is precompact, where $R(P) \in \mathcal{C}(D(T))$. But $\overline{\Gamma}_0(G_T) = \Gamma_0(G_T) = \Gamma_0(G_T|_{G^{-1}R(P)})$. Hence G_T is precompact. □

VII.2.12 Corollary. *Let $T(0)$ be closed and let $\{\lambda \in \mathbb{C} : \lambda - T \text{ is open}\} \neq \emptyset$. Then the following properties are equivalent:*

(i) *G_T is precompact*
(ii) *$(\lambda I - T)^{-1}$ is precompact and $N(\lambda - T)$ is finite dimensional whenever $\lambda - T$ is open.*

PROOF. By VII.2.2, since G_T is an isometry on $N(T)$, it is an isomorphism on $N(\lambda - T)$. Hence if G_T is precompact, then $N(\lambda - T)$ is finite dimensional. Now combine VII.2.10 and VII.2.11. □

VII.2.13 Corollary. *Let $\rho(T) \neq \emptyset$. Then G_T is precompact if and only if T has a compact resolvent operator.*

VII.2.14 Exercise Generalise Proposition VII.2.10 as follows: Let the relation $T \in LR(X,Y)$ and let G_T be precompact. Then $(A+T)^{-1}$ is precompact for any continuous single valued relation A such that $D(A) \supset D(T)$ whenever $(A+T)^{-1}$ is continuous and $T(0)$ is relatively closed in $R(A+T)$.

VII.2.15 Proposition. *Let X and Y be reflexive Banach spaces and let $S, T \in LR(X,Y)$. If T is closable, and if S is single valued and T-compact, then $T+S$ is closable.*

PROOF. Let $y \in (\overline{T+S})(0)$, i.e. $(0,y) \in G(\overline{T+S})$. Then there exist sequences $\{x_n\} \subset X$, and $\{t_n\}, \{s_n\} \subset Y$ such that $(x_n, t_n + s_n) \to (0,y)$ where $(x_n, t_n+s_n) \in G(T+S)$, $t_n \in Tx_n$ and $s_n = Sx_n$ $(n \in \mathbb{N})$.

Since S has T-bound 0 by Corollary VII.2.6, the norms $\| \ \|_T$ and $\| \ \|_{T+S}$ are equivalent (Proposition VII.2.3). Hence for some constant $c > 0$,

$$c(\|x_n\| + \|Tx_n\|) \leq \|x_n\| + \|(T+S)x_n\|$$
$$= \|x_n\| + d(Tx_n + Sx_n, T(0)) \leq \|x_n\| + \|s_n + t_n\|$$

where $\|x_n\| + \|s_n + t_n\| \to \|y\|$ $(n \to \infty)$. Therefore $\{\|x_n\|_T\}$ is bounded. Consequently, as S is T-compact, the sequence $\{Sx_n\}$ has a subsequence, for simplicity assumed to be itself, which is Cauchy. Write $s := \lim Sx_n$ (where $s \in Y$).

It follows that $\lim t_n = y - s \in Y$. We now have $(x_n, t_n) \to (0, y-s)$ in $G(\overline{T})$. Since T is closable, $y - s \in T(0)$. Hence $(x_n, QTx_n) = (x_n, Qt_n) \to (0, Q(y-s)) = (0,0)$. Therefore $\|x_n\| + \|QTx_n\| = \|x_n\| + \|Tx_n\| \to 0$, i.e. $\|x_n\|_T \to 0$. By the equivalence of the norms $\| \ \|_T$ and $\| \ \|_{T+S}$ we thus have

$$\lim \|x_n\|_{T+S} = 0.$$

Therefore $\|(T+S)x_n\| = \|Q(t_n + s_n)\| \to 0 = Qy$. Hence $y \in T(0) = (T+S)(0)$. □

VII.2.16 Exercise Let X be infinite dimensional, $x_0 \in X$ a unit vector and f a discontinuous linear functional on X. Define $T \in LR(X)$ by $Tx := x_0 f(x)$ $(x \in X)$. Then $T(0) = \{0\}$, T^{-1} is compact (in fact $Q_{T^{-1}}T^{-1} = 0$) while G_T is not precompact.

VII.3 STABILITY OF THE ESSENTIAL SPECTRUM

Let T be a linear relation in $LR(X)$ and suppose that the complex number λ is in the resolvent set of T. Then if S is any precompact relation satisfying $D(S) \supset D(T)$ and $\dim S(0) < \infty$, we have

$$(1) \qquad \rho_e(T+S) = \rho_e(T).$$

This is an immediate consequence of the perturbation results V.5.20 and V.7.10.

Our aim in this section is to show that (1) holds more generally for completely closable T-precompact perturbations (Theorem VII.3.2). In particular, for relations having a compact resolvent operator (see Corollary VII.2.13 and Remark VII.3.4), the equality (1) will be valid for all bounded linear operators S (Corollary VII.3.3).

As in the rest of this chapter, T denotes an arbitrary linear relation in $LR(X)$.

VII.3.1 Lemma. *Let $T \in F_+(X,Y)$. Then*

$$T \in F_- \iff TG \in F_-.$$

PROOF. Let $T \in F_-$. Then

$$(1) \qquad \alpha((TG)') = \overline{\beta}(TG) = \overline{\beta}(T) < \infty.$$

Now since $T \in F_+$, we have $TG \in F_+$ by Corollary V.2.5. Hence $(TG)' \in \phi_-$ (Corollary V.7.5) and so $(TG)'$ has closed range. Hence from (1), $(TG)' \in \phi_+$. Therefore $TG \in F_-$.

The converse is contained in Proposition V.5.24. □

VII.3.2 Theorem. *Let $T \in LR(X)$ be an arbitrary linear relation, and let S be any T-compact (resp., T-precompact) relation satisfying* $\dim S(0) < \infty$. *If either S is closable (resp., completely closable), or X is reflexive and T is closable, then*

$$\sigma_e(T+S) = \sigma_e(T).$$

PROOF. First suppose that S is single valued. Let $\lambda \in \mathbb{C}$. Suppose that $\lambda - T \in F_+$. Then $(\lambda - T)G_T \in F_+$ (Corollary V.2.5 and Proposition VII.2.2). Since SG_T is compact (resp., precompact), therefore $(\lambda - (T+S))G_T \in F_+$ by V.7.10. Hence by the equivalence of the norms $\| \ \|_T$ and $\| \ \|_{T+S}$ (Corollary VII.2.6)

$$(1) \qquad (\lambda - (T+S))G_{T+S} \in F_+$$

and by the equivalence of $\| \ \|_{T+S}$ and $\| \ \|_{\lambda-(T+S)}$ (Proposition VII.2.2),

$$(\lambda - (T+S))G_{\lambda-(T+S)} \in F_+.$$

Hence, again by Corollary V.2.5,

$$\lambda - (T+S) \in F_+.$$

Observing that each step in the above argument is reversible, we obtain the equivalence

$$\lambda - T \in F_+ \iff \lambda - (T+S) \in F_+. \tag{2}$$

Now suppose that $\lambda \in \rho_e(T)$. Then by Lemma VII.3.1,

$$(\lambda - T)G_{\lambda-T} \in F_-$$

and hence by Proposition VII.2.2,

$$(\lambda - T)G_T \in F_-.$$

Therefore by the perturbation theorem V.5.12,

$$(\lambda - T)G_T - SG_T \in F_-$$

$$\text{i.e.} \qquad (\lambda - (T+S))G_T \in F_-$$

whence

$$(\lambda - (T+S))G_{\lambda-(T+S)} \in F_- \qquad \text{(Proposition VII.2.2 and VII.2.3).}$$

Hence

$$(\lambda - (T+S)) \in F_- \qquad \text{(Lemma VII.3.1).}$$

Consequently by (2), $\lambda \in \rho_e(T+S)$. We have thus shown that

$$\rho_e(T) \subset \rho_e(T+S). \tag{3}$$

(since the inclusion holds trivially if $\rho_e(T)$ is empty).

Now let $\lambda \in \rho_e(T+S)$. From the equivalence of the norms $\| \ \|_T$ and $\| \ \|_{T+S}$, S is $(S+T)$-compact (resp., $(S+T)$-precompact). Hence from (3), we conclude that

$$\rho_e(T+S) \subset \rho_e(T+S-S) = \rho(T).$$

Therefore $\sigma_e(T+S) = \sigma_e(T)$.

Now suppose that $\dim S(0) < \infty$. Then there exists a continuous projection P defined on $R(S)$ with kernel $S(0)$, and

$$S = PS + S - S. \tag{5}$$

Moreover PSG_T is compact (resp., precompact) (Corollary V.2.11) and so from what has been proved

$$\sigma_e(T) = \sigma_e(T + PS). \tag{6}$$

But $R(S - S) = S(0)$ is finite dimensional, and $\|S - S\| = 0$. Therefore by the finite rank perturbation result VII.1.4,

$$\sigma_e(T + PS) = \sigma_e(T + PS + S - S) = \sigma_e(T + S). \tag{7}$$

Hence from (6) and (7),

$$\sigma_e(T) = \sigma_e(T + S). \qquad \square$$

VII.3.3 Corollary. *Let the relation T have a compact resolvent operator and let S be any bounded linear operator. Then*

$$\sigma_e(T + S) = \sigma_e(T).$$

PROOF. Combine Theorems VII.2.9 and VII.3.2. □

VII.3.4 Remark It is observed in Edmunds and Evans [1; p.427] that if Ω is a nonempty bounded set in $\mathbb{R}^n$ having smooth boundary, and T is a nth-order differential operator in $L^2(\Omega)$ with smooth enough coefficients, then X_T is isomorphic to a closed subspace of the Sobolev space $W^{n,2}(\Omega)$ which is compactly embedded in $L^2(\Omega)$. Such an operator T has a compact resolvent operator.

VII.4 NOTES AND REMARKS

The definition of essential spectrum VII.1.1 coincides with the standard definition for bounded or closed operators in Banach spaces. There are many different but related definitions of essential spectra in operator theory. Five of these are studied in Edmunds and Evans [1; Chapter IX]. We have limited our discussion to one of these. Theorem VII.3.2 is a generalization of the stability result of H. Weyl [1] (see Edmunds and Evans [1; IX.2.1]). The requirement that all relations be closed is unsuitably restrictive for perturbation theory since the sum of two closed relations (or operators) is not in general closed. Our aim in this chapter has been to retain the standard properties without imposing closability restrictions. We do not yet know to what extent these restrictions are necessary in Theorem VII.3.2.

CHAPTER VIII
The Second Adjoint of a Linear Relation

VIII.1 PRELIMINARY RESULTS

VIII.1.1 Notation Given $T \in LR(X,Y)$, we shall denote Q_T, $Q_{T'}$ and $Q_{T''}$ by Q, Q_1 and Q_2 respectively when T is understood.

We clearly have from Proposition III.1.4 and Theorem III.1.10,

(1) $D(T'')^{\top\perp} = D(T)''$ (since $D(T)'' = D(T)^{\perp\perp}$)

(2) $T''J_{D(T'')^{\top\perp}} = (Q_1T')' = (TJ_{D(T)})''$

(3) $Q_2T'' = (T'J_{D(T')})'$

Furthermore, since $(T'J_{T(0)^{\perp}})' = Q_{T(0)^{\perp\perp}}T''$ by Theorem III.1.6(a),

(4) $(QT)'' = Q_{T(0)^{\perp\perp}}T'' = Q''T''$.

If J denotes the canonical injection map of X into X'', and M is a subset of X, we write $\hat{M} := JM$; in particular, $\hat{x} := Jx \quad (x \in X)$.

VIII.1.2 Proposition. *We have*

(a) $D(T)^\frown \subset D(T'')$, *and*

(b) $Q_2T''\hat{x} = Q_2(Tx)^\frown \quad (x \in D(T))$.

PROOF. Let $y' \in D(T')$. Then for all $x' \in T'y'$ and $x \in D(T)$ we have $x'x = y'Tx$ (see III.1.1(4)). Hence

$$(1) \qquad T'y'x = y'Tx.$$

Clearly

$$(2) \qquad T'y'x = \hat{x}(T'y').$$

Now for $y'' \in Y''$ we have $(Q_2y'')(y') = (y''|_{D(T')})(y')$. Hence

$$(3) \qquad Q_2y'' = y''J^{Y'}_{D(T')}.$$

It follows from (1), (2) and (3) that

$$(4) \qquad \hat{x}(T'y') = y'Tx = (Tx)^\frown(y') = Q_2(Tx)^\frown(y').$$

Therefore $\hat{x}T'$ is continuous and single valued on $D(T')$, whence $\hat{x} \in D(T'')$ and (a) follows. Then by (4) and VIII.1.1(3),

$$(Q_2T''\hat{x})y' = ((T'J_{D(T')})'\hat{x})y' = \hat{x}T'y' = Q_2(Tx)^\frown(y')$$

and (b) follows. $\square$

VIII.1.3 Proposition. *Let T' be continuous. Then*

$$(Q_{\tilde{T}}\tilde{T})'' = (T' J_{D(T')})' = Q_2 T''.$$

PROOF. Since T' is continuous, $D(T') = \tilde{T}(0)^{\perp}$ (by III.4.6(a) and III.5.3). Hence by Proposition III.1.4 and Theorem III.1.10,

$$Q_2\tilde{T}'' = (\tilde{T}' J_{D(\tilde{T}')})' = (\tilde{T}' J_{\tilde{T}(0)^{\perp}})' = (Q_{\tilde{T}})''. \quad \square$$

VIII.1.4 Remark It is easy to see that *in general $(QT)''$ and Q_2T'' need not be equal even if T' is continuous.* Indeed, suppose that T is single valued and everywhere defined with $D(T') = \{0\}$ (see [Goldberg [3; p53]). Then $G(T') = \{0\} \times \{0\}$, $G((QT)'') = G(T'') = X'' \times Y''$ and $G(Q_2T'') = X'' \times \{0\}$.

VIII.1.5 Proposition. *The following properties are equivalent:*

(i) *T'' is bounded*

(ii) *T' is continuous and single valued*

(iii) *$\tilde{T}$ is bounded.*

PROOF. In the notation of the Inverse State Diagram we have from III.6.7 and the Closed Graph Theorem III.4.2, T'' bounded $\iff T'' \in I \iff T' \in 1 \iff \tilde{T} \in I \iff \tilde{T}$ is bounded. $\square$

VIII.1.6 Proposition. *Consider the properties:*

(i) *T' is open*

(ii) *The norm and $\sigma(X', \tilde{X})$ closures of $R(\tilde{T}')$ coincide (for example, if $R(T')$ is dense in X')*

(iii) *$N(T'') = N(\tilde{T})''$.*

Then (i) $\implies$ (ii) $\implies$ (iii).

In particular, (iii) holds whenever $\tilde{X}$ is reflexive.

PROOF. Let T' be open. Then $R(T') = \overline{R(T')} = R(\tilde{T}')^{\top\perp}$ by the Open Mapping and Closed Range theorems. Hence (i) $\implies$ (ii). Now suppose (ii) holds. Then $N(T'') = R(T')^{\perp} = R(\tilde{T}')^{\top\perp\perp} = N(\tilde{T})^{\perp\perp} = N(\tilde{T})''$. Therefore (ii) $\implies$ (iii). $\square$

VIII.1.7 Remarks (a) By the Open Mapping and Closed Range Theorems

(1) *T'' is open $\iff$ T' is open $\iff$ $\tilde{T}'$ is open.*

(b) Let us also note the inverse duals of VIII.1.5 and VIII.1.6. We have the following implications:

(2) *T'' is open* $\iff$ *T' is open* $\iff$ $R(T'') = R(T)^{\perp\perp}$

(3) *T' is continuous* $\implies$ *the norm and $\sigma(Y', \tilde{Y})$ closures of $D(\tilde{T}')$ coincide* $\implies T''(0) = \tilde{T}(0)^{\perp\perp} \implies (Q_{\tilde{T}}\tilde{T})'' = Q_{T''}T''$

(4) *$D(T')$ dense in Y'* $\implies T''(0) = \tilde{T}(0)''$

(5) *If $\tilde{Y}$ is reflexive, then $T''(0) = \tilde{T}(0)^{\perp\perp}$.*

(6) *If T is single valued and continuous, then T'' is single valued and continuous, and conversely.*

Since in general $T''(0) \supset \tilde{T}(0)^{\perp\perp}$, we may also note that

(7) *T'' single valued* $\implies$ *$\tilde{T}$ single valued* $\implies$ *T single valued.*

(8) *T'' injective* $\implies$ *$\tilde{T}$ injective* $\implies$ *T injective.*

VIII.1.8 Proposition. *Q_1T' is $\sigma(Y', Y)$–$\sigma(Q_1X', D(T))$ continuous. Hence, if $\{y'_\alpha\}$ is a net in $D(T')$ which is $\sigma(Y', Y)$ convergent to $y' \in D(T')$, then $\lim T'y'_\alpha x = T'y'x$ for all $x \in D(T)$.*

PROOF. Let $\{y'_\alpha\}$ be a net in $D(T') \subset Y'$ which is weak*-convergent to 0. Then for $x \in D(T)$ and $y \in Tx$ we have

$$(Q_1T'y'_\alpha)x = y'_\alpha Tx = y'_\alpha y \to 0.$$

Hence the result follows. □

VIII.1.9 Proposition. *$T''(B_{D(T'')})$ is contained in the $\sigma(Y'', D(T'))$-closure of $(TB_{D(T)})\hat{}$.*

PROOF. Let $x'' \in B_{D(T'')}$. By Goldstine s theorem there is a net $\{x_\alpha\}$ in $B_{D(T)}$ such that $\sigma(D(T)'', D(T)') - \lim \hat{x}_\alpha = x''$. By Proposition VIII.1.8, and Proposition VIII.1.2(b) we have for all $y' \in D(T')$

$$(Tx_\alpha)\hat{}(y') = Q_2(Tx_\alpha)\hat{}(y') \to Q_2T''x''y' = T''x''y'.$$

Hence if $y_\alpha \in Tx_\alpha$ and $y'' \in T''x''$, then

$$\sigma(Y'', D(T')) - \lim \hat{y}_\alpha = y''$$

as required. □

VIII.1.10 Proposition. *$G(T')$ is $\sigma(Y',Y)\times\sigma(X',D(T))$ closed in $Y'\times X'$.*

PROOF. Write $\sigma_1 := \sigma(Y',Y)$, $\sigma_2 := \sigma(X',D(T))$ and let (y'_α, x'_α) be a net in $G(T')$ which is $\sigma_1 \times \sigma_2$-convergent to (y',x'). For $(x,y) \in G(T)$ we have

$$y'y = \lim y'_\alpha y = \lim y'_\alpha Tx = \lim T'y'_\alpha x = \lim x'_\alpha x = x'x.$$

Hence $(y',x') \in G(T')$. □

VIII.1.11 Corollary. *$T'B_{D(T')}$ is $\sigma(X',D(T))$ closed (hence norm closed).*

PROOF. Let $\{y'_\alpha\}$ be a net in $B_{D(T')}$, let $x'_\alpha \in Ty'_\alpha$, and suppose that $\{x'_\alpha\}$ is $\sigma(X',D(T))$ convergent to $x' \in X'$. By the Banach-Alaoglu Theorem, $\{y'_\alpha\}$ has a subnet, for simplicity assumed to be itself, which is $\sigma(Y',Y)$ convergent to some point $y' \in B_{Y'}$. Now we have a net (y'_α, x'_α) in $G(T')$ which is $\sigma(Y',Y) \times \sigma(X',D(T))$ convergent to (y',x'). Hence by Proposition VIII.1.10, $(y',x') \in G(T')$. Thus $x' \in T'y' \subset T'B_{D(T')}$. □

VIII.1.12 Exercise Let $\tilde{X}$ and $\tilde{Y}$ be reflexive. Show that $G(T'') = G(\tilde{T})^\frown$. In particular, $D(T'') = D(\tilde{T})^\frown$ and $T''\hat{x} = (\tilde{T}x)^\frown$ for $x \in D(\tilde{T})$.

VIII.1.13 Proposition. *Let $\hat{y} \in T''B_{'''}$. Then y belongs to the $\sigma(Y, D(T'))$ closure of TB'.*

PROOF. Let $\hat{y} \in T''x''$ where $x'' \in D(T'') \subset D(T)^{\perp\perp} = D(T)''$ and $\|x''\| \leq 1$. By Goldstine sTheorem there is a net $\{x_\alpha\}$ in $B_{D(T)}$ such that $x'x_\alpha \to x''x'$ for every $x' \in D(T)'$. But $D(T)' = X'/D(T)^\perp$. Hence $\sigma(X'',X') - \lim \hat{x}_\alpha = x''$. For each index α, choose $y_\alpha \in Tx_\alpha$ arbitrarily. By Proposition VIII.1.8 we have for $y' \in D(T')$,

$$y'y_\alpha = y'Tx_\alpha = (T''\hat{x}_\alpha)y' \to (T''x'')y' = \hat{y}y' = y'y$$

whence $y' \in \sigma(Y, D(T')) - \mathrm{cl}(TB_{DT})$. □

VIII.1.14 Lemma. $R(T'') \cap \hat{Y} \subset \overline{R(T)}\,\hat{}$.

PROOF. We have by Proposition III.1.4,

$$R(T'') \subset R(T'')^{\top\perp} = N(T')^{\perp} = R(T)^{\perp\perp} = (R(T)^{\wedge})^{\top\perp}.$$

Hence

$$R(T'') \cap \hat{Y} \subset (R(T)^{\wedge})^{\top\perp} \cap \hat{Y} = (R(T)^{\perp\top})^{\wedge}$$
$$= \overline{R(T)}\,\hat{}. \quad \square$$

VIII.2 WEAKLY COMPACT LINEAR RELATIONS

Let us first recall some facts about bounded linear operators in Banach spaces. Suppose that X and Y are Banach spaces and $T : X \to Y$ a bounded linear operator. Then T is said to be *weakly compact* if $\overline{T(B_X)}$ is weakly compact (equivalently, if T transforms each bounded sequence into a sequence having a weakly convergent subsequence). For some pairs (X, Y) of Banach spaces, all bounded linear operators are weakly compact; for example, if X is arbitrary and Y is reflexive, or if X is a $C(K)$ space (where K is a compact Hausdorff space) and Y is separable. The simplest example of the latter is $X = \ell_\infty$ and $Y = c_0$ or ℓ_p $(1 \leq p < \infty)$. For some other pairs (X, Y) of Banach spaces, the classes of weakly compact operators and compact operators coincide; this is the case when $X = c_0$, for instance.

We have the following standard characterization:

VIII.2.1 Theorem. *Let $T : X \to Y$ be a bounded linear operator where X and Y are Banach spaces. Then the following properties are equivalent:*

(i) *T is weakly compact*
(ii) *T' is $\sigma(Y', Y) - \sigma(X', X'')$ continuous*
(iii) $T''(X'') \subset \hat{Y}$ (See Swartz [1; p.423]).

The corresponding properties for linear relations will now be investigated.

VIII.2.2 Definition Let $T \in LR(X, Y)$. Then T is said to be **weakly compact** if TB_X is relatively $\sigma(Y, D(T'))$-compact.

Notice that

(1) *T is weakly compact if and only if the single valued relation QT is weakly compact,* since the domains of T' and $(QT)'$ coincide, and $D(T') \subset T(0)^{\perp}$.
(2) If T is continuous, then $(QY)' = T(0)^{\perp} = D(T')$, and so the definition is the standard one for the continuous linear operator QT.

The definition VIII.2.2 is found to admit discontinuous weakly compact linear operators or relations. However, Theorem VIII.2.4 below shows in particular that

(3) $\tilde{T}$ *weakly compact* $\implies$ $\tilde{T}$ *continuous*

so there are no closed unbounded weakly compact linear relations in Banach spaces.

VIII.2.3 Proposition. *Let T be weakly compact and let $S \in LR(Y, Z)$ be continuous with $D(S) \supset R(T) + \overline{T(0)}$. Then ST is weakly compact.*

PROOF. First assume that T is single valued. Then $ST(0) = S(0)$, so $Q_{ST}ST = (Q_S S)T$. Since ST is weakly compact if and only if $Q_{ST}ST$ is weakly compact, we may assume that S is single valued. Let $\{x_\alpha\}$ be a bounded net in $D(T)$. Then $\{Tx_\alpha\}$ has a $\sigma(Y, D(T'))$ convergent subnet, assumed for simplicity to be itself. Let $y \in Y$ be a $\sigma(Y, D(T'))$ limit point of the $\sigma(Y, D(T'))$ convergent net $\{Tx_\alpha\}$. We have for $z' \in D((ST)') \subset D(S') = Z'$,

$$\lim z'STx_\alpha = z'Sy \quad (\text{since } z'S \in D(T') \text{ - see III.1.6}).$$

Therefore ST is weakly compact.

For the general case, consider $SQ^{-1}QT$. Then QT is weakly compact, and SQ^{-1} is continuous since $Q^{-1}(0) = \overline{T(0)} \subset D(S)$ (see II.3.13). Hence $SQ^{-1}QT$ is weakly compact. Therefore by Lemma V.2.9, $Q_{ST}ST$ is weakly compact (since $\overline{ST(0)} = \overline{SQ^{-1}QT(0)}$ by the continuity of S. Consequently ST is weakly compact. □

VIII.2.4 Theorem. *Let T be closable and weakly compact, and let Y be complete. Then T is continuous.*

PROOF. Since T is closable, QT is closable (II.5.7). Moreover, QT is weakly compact. We shall therefore suppose without loss of generality that T is single valued, and furthermore that X is complete, with T densely defined and non zero. Since T is closable, $D(T')$ is a total subset of Y' (see II.5.5 and III.1.4(d)). Hence the functional

$$(1) \qquad \|y\|_1 := \sup_{y' \in D(T')} \frac{|y'y|}{\|y'\| + \|T'y'\|} \qquad (y \in Y)$$

defines a renorming of Y. Write $Y_1 := (Y, \|\cdot\|_1)$ and let J_1 denote the identity from Y into Y_1. Write $T_1 := J_1 T$. Then T_1 is a continuous operator. By Proposition VIII.2.3, T_1 is weakly compact. Write

$$(2) \qquad W := J_{Y_1}^{\tilde{Y}_1} J_1.$$

We shall verify that

$$(3) \qquad R(\overline{T}_1) \subset WY.$$

Indeed, let $x \in X = D(\overline{T}_1)$ and select a sequence $\{x_n\}$ in $D(T)$ such that $\lim_{n\to\infty} \|x - x_n\| = 0$. By continuity we have $\lim T_1 x_n = \overline{T}_1 x \in \tilde{Y}_1$. By the weak compactness of T_1, there exists $y_1 \in Y_1$ which is a weak limit point of $\{T_1 x_n\}$ (since $D(T_1') = Y_1'$). Hence $y_1 = \overline{T}_1 x \in Y_1$ and (3) is established. The proof is completed by showing that $D(W^{-1}\overline{T}_1) = X$, for then, since $W^{-1}\overline{T}_1$ is evidently a closed linear operator defined everywhere on the Banach space X, it will be bounded by the Closed Graph theorem. Since $T = W^{-1}\overline{T}_1|_{D(T)}$, the continuity of T follows. We have by (3),

$$D(W^{-1}\overline{T}_1) = D(\overline{T}_1) = X$$

as required. □

Example V.9.12 shows that completeness of Y is essential in Theorem VIII.2.4 (see VIII.3.1).

Recall the definition of the operator $H := H_T$ (IV.3.18). The following lemma reduces the weak compactness property to the continuous single valued case. Note that HT is single valued, by Theorem IV.3.17.

VIII.2.5 Lemma. *T is weakly compact if and only if HT is weakly compact.*

PROOF. Let HT be weakly compact and let $\{x_\alpha\}$ be a net in $B_{D(T)}$. Then there exists $y \in Y$ and a subnet of $\{x_\alpha\}$, assumed to be itself, such that for every $z' \in HT(0)^\perp \subset Z'$

$$z'(Q_{HT}HTx_\alpha) \to z'Q_{HT}Hy \text{ for some } y \in Y.$$

Since $(Q_{HT})' = J_{HT(0)^\perp}$, this is equivalent to

$$(HT)'z'x_\alpha = T'H'z'x_\alpha \to z'Q_{HT}Hy = H'J_{HT(0)^\perp}z'y.$$

Let us recall Theorem IV.3.17. We have $(Q_{HT})' = J_{HT(0)^\perp}$ where $HT(0)^\perp = D((HT)') = Y'_{T'}$ by the continuity of HT, and $H' = G_{T'}$. We thus have

$$(1) \qquad T'G_{T'}z'x_\alpha \to G_{T'}z'y \qquad (z' \in Y'_{T'})$$

But $G_{T'}$ maps $Y'_{T'}$ onto $D(T')$. Hence from (1) we have for every $y' \in D(T')$,

$$T'y'x_\alpha \to y'y,$$

i.e.,

$$y'Tx_\alpha \to y'y \qquad (y' \in D(T')).$$

Therefore T is weakly compact.

The converse is contained in Proposition VIII.2.3. □

VIII.2.6 Theorem. *The following properties are equivalent:*

(i) $J_Y^{\tilde{Y}}T$ *is weakly compact*
(ii) $\tilde{T}$ *is weakly compact*
(iii) T' *is weakly compact.*

PROOF. (i) $\Longrightarrow$ (iii). Let $J_Y^{\tilde{Y}}T$ be weakly compact. Consider the regular contraction $S := Q_E JT$ where $E := D(T')^{\top}$ and $J := J_Y^{\tilde{Y}}$. Then S is weakly compact by Proposition VIII.2.3. Now S is closable and single valued, by Corollary III.4.11. Since $S' = T' J_{D(T')^{\top\perp}}$, where $D(T')^{\top\perp} \supset D(T')$, we may assume without loss of generality that $S = T$ and hence that Y is complete and T closable and single valued. Now T is a continuous linear operator (Theorem VIII.2.4) and therefore T' is weakly compact by Gantmacher s Theorem (see Swartz [1; p.241]).

(iii) $\Longrightarrow$ (ii). Let T' be weakly compact. Then clearly $T'G_{T'}$ is weakly compact (since a bounded net in $Y'_{T'}$ transforms under $G_{T'}$ into a bounded net in $D(T')$). But $T'G_{T'} = (HT)'$ by Theorem IV.3.17. Since $T' = \tilde{T}'$ it follows from Gantmacher sTheorem Swartz [1; loc.cit.] that $H_{\tilde{T}}\tilde{T}$ is weakly compact (see VIII.2.2(1)). Hence $\tilde{T}$ is weakly compact, by Lemma VIII.2.5.

(ii) $\Longrightarrow$ (i). Let $\tilde{T}$ be weakly compact and let $\{x_\alpha\}$ be a bounded net in $D(T)$. Since $\tilde{T}$ coincides with the complete closure of $J_Y^{\tilde{Y}}T$, we may as well suppose that Y is complete, i.e., $J_Y^{\tilde{Y}}T = T$. By hypothesis there exists $y \in Y$ and a subnet of $\{x_\alpha\}$, for simplicity assumed to be itself, such that for $y' \in D(T')$,

$$y'\tilde{T}x_\alpha \to y'y.$$

But $y'\tilde{T}x = T'y'x = y'Tx$ for $y' \in D(T')$ and $x \in D(T)$. Therefore

$$y'Tx_\alpha \to y'y \qquad (y' \in D(T'))$$

showing that T is weakly compact. □

VIII.2.7 Corollary.

(a) *If T is weakly compact, then T' is weakly compact.*
(b) *If Y is a Banach space and T' is weakly compact, then T is weakly compact.*

VIII.2.8 Proposition. *Let T' be continuous and let $T''(X'') \subset \hat{Y} + T''(0)$. Then T is weakly compact.*

PROOF. Let $\{x_\alpha\}$ be a net in $B_{D(T)}$. Then $\{\hat{x}_\alpha\}$ is a net in $B_{D(T)''}$. Since $B_{D(T)''}$ is weak*-compact, there exists a subnet of $\{\hat{x}_\alpha\}$, assumed for simplicity to be itself, which is weak*-convergent to some point $x'' \in B_{D(T)''}$. Since T' is continuous, $D(T'') = D(T)''$ by III.8.10. Hence we may assume $D(T) = X$. By the weak*-weak* continuity of T'' (see VIII.1.8) we have

$$Q_2 T'' \hat{x}_\alpha \to Q_2 T'' x'' \text{ in } \sigma(Q_2 Y'', D(T')).$$

By hypothesis, $Q_2 T'' x'' = Q_2 \hat{y}$ for some $y \in Y$. Hence for $y' \in D(T')$ we have

$$(Q_2 T'' \hat{x}_\alpha) y' = Q_2 (T x_\alpha)\hat{\ } y' \to Q_2 \hat{y} y'$$

i.e.,

$$y'(Q_1 T x_\alpha) \to y'(Q_1 y)$$

for all $y' \in D(T')$. This shows that QTx_α is $\sigma(QY, D(T'))$-convergent to Qy. Hence T is weakly compact. □

VIII.2.9 Lemma. *Let M and E be subspaces of $D(T'')$ and Y'' respectively, and let $\hat{Y} \subset E \subset Y''$. Then the following properties are equivalent:*

(i) $T''(M) \subset E + T''(0)$

(ii) *T' is $\sigma(D(T'), E) - \sigma(X', M)$ continuous.*

PROOF. Denote $\sigma_1 := \sigma(D(T'), E)$, $\sigma_2 := \sigma(X', D(T''))$.

(i) $\Longrightarrow$ (ii). Assume (i) and let $\{y'_\alpha\}$ be a net in $D(T')$ such that $\sigma_1 - \lim y'_\alpha = 0$. Let $x'' \in M$. By assumption there exists $y'' \in E$ such that $Q_2 T'' x'' = Q_2 y''$. Now for $y' \in D(T')$ we have

$$Q_2 T'' x'' y' = T'' x'' y' = x'' T' y' = x'' Q_1 T' y'$$

Hence

$$x'' Q_1 T' y'_\alpha = Q_2 T'' x'' y'_\alpha = Q_2 y'' y'_\alpha \to 0.$$

Therefore $\sigma_2 - \lim T' y'_\alpha = 0$, as required.

(ii) $\Longrightarrow$ (i). Assume that (ii) holds. Let $x'' \in B_M$. By assumption there exists a finite subset $F \subset E$ such that $F_0 \cap D(T') \subset (Q_1 T')^{-1}(\{x''\}_0)$. We claim that

$$\bigcap_{y'' \in F} N(Q_2 y'') \subset N(Q T'' x''). \tag{1}$$

Let $y' \in D(T') \backslash N(Q_2T''x'')$. Put $w' = 2(Q_2T''x''y')^{-1}y'$. Then $|x''Q_1T'w'| = |T''x''w'| = |Q_2T''x''w'| = 2 > 1$ and so $Q_1T'w' \notin \{x''\}_0$. Thus

$$w' \notin (Q_1T')^{-1}(\{x''\}_0) \supset F_0 \cap D(T') \supset \bigcap_{y'' \in F} N(y'') \cap D(T').$$

Hence

$$y' \notin \bigcap_{y'' \in F} N(y'').$$

Consequently there exists $y'' \in F$ for which $y''y' \neq 0$. Since $y' \in D(T') = D(Q_1T')$, this means that $Q_2y''y' \neq 0$, proving (1). It now follows from (1) (see Dunford and Schwartz [1; p.421]) that

$$Q_2T''x'' \in \text{sp}\{Q_2F\} \subset Q_2E. \quad \square$$

VIII.2.10 Corollary. *Let $\hat{Y} \subset E \subset Y''$. Then the following properties are equivalent:*

(i) *$T''(X'') \subset E + T''(0)$ and T' is continuous.*
(ii) *T' is $\sigma(D(T'), E) - \sigma(X', D(T)'')$ continuous.*

PROOF. Since $D(T'') = D(T)''$ if and only if T' is continuous by Proposition VIII.1.5, it only remains to verify (in view of Lemma VIII.2.9) that (ii) implies that $D(T'') = D(T)''$.

Let $x'' \in D(T)''$ and let $\{y'_n\}$ be a sequence in $D(T')$ such that $\lim_{n\to\infty} \|y'_n\| = 0$. Suppose that (ii) holds. Then $Q_1T'y'_n \to 0$ $(n \to \infty)$ in $\sigma(Q_1X', D(T)'')$. Hence $x''Q_1T'y'_n \to 0$. It follows that the linear functional $x''Q_1T'$ is norm continuous on $D(T')$. Therefore $x'' \in D((Q_1T')') = D(T'')$ (see Proposition III.1.2). $\square$

VIII.2.11 Theorem. *The following properties are equivalent:*

(i) *T is weakly compact.*
(ii) *T' is continuous and $T''(X'') \subset \hat{Y} + T''(0)$*
(iii) *T' is $\sigma(Y', \hat{Y}) - \sigma(X', D(T)'')$ continuous.*

PROOF. (i) $\Longrightarrow$ (ii). Let T be weakly compact. Let $x'' \in B_{D(T'')}$. By Proposition VIII.1.9 there is a net $\{x_\alpha\}$ in $B_{D(T)}$ such that

(1) $$Q_2(Tx_\alpha)\widehat{\ } \to Q_2T''x'' \text{ in } \sigma(Q_2Y'', D(T')).$$

But $Q(TB_{D(T')})$ is relatively $\sigma(QY, D(T'))$ compact. Hence $\{x_\alpha\}$ has a subnet, assumed for simplicity to be itself, such that

$$QTx_\alpha \to Qy \text{ in } \sigma(QY, D(T')). \tag{2}$$

Now for $x \in D(T)$ and $y' \in D(T')$ we have

$$\begin{aligned} y'QTx &= (QT)'y'x = T'y'x = \hat{x}T'y' \text{ (as } \hat{x} \in D(T'')) \\ &= (T''\hat{x})y' = (Q_2T''\hat{x})y' = Q_2(Tx)\hat{\ }(y') \text{ (by VIII.1.2(b)).} \end{aligned}$$

Since (2) is equivalent to

$$y'(QTx_\alpha) \to y'y \qquad (y' \in D(T'))$$

we have

$$Q_2(Tx_\alpha)\hat{\ }(y') \to Q_2\hat{y}y' \qquad (y' \in D(T')). \tag{3}$$

Comparing (1) and (3) and noting that the topology $\sigma(Q_2Y'', D(T'))$ is Hausdorff (indeed, Q_2Y'' is the norm dual of $D(T')$), we conclude that $Q_2T''x'' = Q_2\hat{y}$. Hence $Q_2T''(X'') \subset Q_2(\hat{Y})$. Since T' is continuous by Corollary VIII.2.7 the proof of the implication (i) $\Longrightarrow$ (ii) is complete.

(ii) $\Longrightarrow$ (iii). This follows immediately from Corollary VIII.2.10 upon substituting $E := \hat{Y}$.

The remaining implication (iii) $\Longrightarrow$ (i) is covered by Proposition VIII.2.8 and Corollary VIII.2.10. □

VIII.2.12 Example Let T be everywhere defined and single valued, with $D(T') = \{0\}$ (see Goldberg [1; p.53]). Then $G(T') = \{0\} \times \{0\}$ and $G(T'') = X'' \times Y''$. In particular, $T''(0) = Y''$. Then T *is weakly compact, while* $T''(X'') = Y''$.

VIII.2.13 Definition On Y'' we define the $\| \ \|_{D(T')}$ seminorm by

$$\|y''\|_{D(T')} := \sup\{|y''y'| : y' \in B_{D(T')}\} \qquad (y'' \in Y'').$$

Note that $\|y''\|_{D(T')} \le \|y''\| \qquad (y'' \in Y'')$.

VIII.2.14 Notation Let M be a subspace of Y'' and N a subspace of Y'.

(1) $D(T') - \mathrm{cl}(M)$ will denote the $\| \ \|_{D(T')}$ seminorm closure of M.
(2) $\hat{N}|M$ will denote the set of functionals $\{\hat{g}|_M : g \in N\}$
(3) $M_{D(T')}$ will denote M equipped with the $D(T')$ seminorm restricted to M.

VIII.2.15 Proposition. *Let E be a subspace of Y''. Then*

$$\overline{D(T')}\hat{}\,|\, E \subset (E_{D(T')})'.$$

PROOF. Let $y' \in \overline{D(T')}$, $y' \neq 0$. Let $\{y'_n\}$ be a sequence in $D(T')$ such that $y'_n \neq 0$ and $y'_n \to y'$. For each $y'' \in Y''$ we have

$$|((y')\hat{}/\, \|y'\|)y''| = \lim |y''(y'_n/\, \|y'_n\|)| \leq \|y''\|_{D(T')} \, .$$

Thus for all $y'' \in E$, $|(y')\hat{}(y'')| \leq \|y'\|\, \|y''\|_{D(T')}$, so $(y')\hat{}\,|\, E \in (E_{D(T')})'$. □

VIII.2.16 Proposition. *(a) Let $\hat{Y} \subset E \subset Y''$. If $D(T')\hat{}\,|\, E = (E_{D(T')})'$, then T' is continuous.*

(b) Let $\hat{Y} \subset E \subset D(T') - \mathrm{cl}(\hat{Y})$. Then $D(T')\hat{}\,|\, E = (E_{D(T')})'$ if and only if T' is continuous.

PROOF. (a) Suppose $D(T')\hat{}\,|\, E = (E_{D(T')})'$. By the previous proposition (VIII.2.15), we have $D(T')\hat{}\,|\, E = \overline{D(T')}\,\hat{}\,|\, E$. Since $\hat{Y} \subset E$, hence $D(T') = \overline{D(T')}$, and therefore T' is continuous by the Closed Graph Theorem (III.4.2).

(b) Necessity follows from (a). For the converse, assume that T' is continuous. Let $z' \in (E_{D(T')})'$ and define y'_0 by $y'_0 y := z'(\hat{y})$ $(y \in Y)$. We have for $y \in Y$,

$$|y'_0 y| = |z'(\hat{y})| \leq \|z'\|\, \|\hat{y}\|_{D(T')} \leq \|z'\|\, \|y\|$$

so $y'_0 \in Y'$. For each $x \in D(T)$,

$$|y'_0 Tx| \leq \|z'\|\, \|(Tx)\hat{}\|_{D(T')} = \|z'\| \sup\{|y'Tx| : y' \in B_{D(T')}\} \leq \|z'\|\, \|T'\|\, \|x\|\, ,$$

showing that $y'_0 \in D(T')$. Now let $y'' \in E$. Since $y'' \in D(T') - \mathrm{cl}(\hat{Y})$, there is a sequence $\{y_n\}$ in Y such that $\|\hat{y}_n - y''\|_{D(T')} \to 0$. Now

$$y''y'_0 = \lim y'_0 y_n = \lim z'\hat{y}_n = z'y''.$$

Therefore $z' = (y')\hat{}\,|\, E$. Hence $(E_{D(T')})' \subset D(T')\hat{}\,|\, E$. The reverse inclusion is clear. □

The following proposition is a generalization of the Banach-Alaoglu Theorem.

VIII.2.17 Proposition. *Let $\hat{Y} \subset E \subset D(T') - \mathrm{cl}(\hat{Y})$ and let T' be continuous. Then $B_{D(T')}$ is $\sigma(D(T'), E)$-compact.*

PROOF. By the preceding proposition (VIII.2.16), $(E_{D(T')})' = D(T')^\frown | E$. We shall first show that $B_{(E_{D(T')})'} = (B_{D(T')})^\frown | E$. Let $z' \in B_{(E_{D(T')})'}$. Then $\|z'\| \leq 1$ and there exists $y' \in D(T')$ such that $z' = (y')^\frown|_E$. Let $y \in B_Y$. Then $\|\hat{y}_{D(T')}\| \leq 1$, so $|y'y| = |z'(\hat{y})| \leq 1$. Thus $y' \in B_{D(T')}$, showing that $B_{(E_{D(T')})'} \subset (B_{D(T')})^\frown | E$. Now let $z' \in (B_{D(T')})^\frown | E$. Then there exists $y' \in B_{D(T')}$ such that $z' = (y')^\frown|_E$, considered as a functional in $(E_{D(T')})'$. Let $y'' \in B_{E_{D(T')}}$. Then $|z'y''| = |y''y'| \leq 1$, so $\|z'\| \leq 1$. Thus $(B_{D(T')})^\frown | E \subset B_{(E_{D(T')})'}$ and equality follows.

By the Banach-Alaoglu Theorem, $B_{(E_{D(T')})'}$ is $\sigma((E_{D(T')})', E_{D(T')})$-compact. From the preceding discussion, $B_{D(T')}$ is $\sigma(D(T'), E)$ compact. □

VIII.2.18 Corollary. *Let $\hat{Y} \subset E \subset D(T') - \mathrm{cl}(\hat{Y})$. If T' is $\sigma(D(T'), E) - \sigma(X', D(T)'')$ continuous, then T' is weakly compact.*

PROOF. By Corollary VIII.2.10, T' is continuous. Hence $B_{D(T')}$ is $\sigma(D(T'), E)$ compact by Proposition VIII.2.17. Therefore $Q_1 T' B_{D(T')}$ is $\sigma(Q_1 X', D(T)'')$ compact. Hence T' is weakly compact. □

VIII.2.19 Exercise Let X be complete and $D(T) = X$. Show that following properties are equivalent:

(i) T' is weak*-weak continuous
(ii) $\tilde{T}$ is weakly compact.

VIII.2.20 Proposition. *Let $\hat{Y} \subset E \subset Y''$. Consider the following properties:*

(i) *$(TB_X)^\frown$ is relatively $\sigma(E, D(T'))$ compact*
(ii) *$T''(X'') \subset E + T''(0)$.*

Then (i) $\implies$ (ii), and if T' is continuous then (i) and (ii) are equivalent.

PROOF. Suppose $Q_2(TB_X)^\frown$ is relatively $\sigma(Q_2E, D(T'))$ compact. Let $x'' \in B_{D(T'')}$. By Proposition VIII.1.9, there exists a net $\{x_\alpha\}$ in $B_{D(T)}$ such that $Q_2(Tx_\alpha)^\frown \to Q_2T''x''$ in the $\sigma(Q_2Y'', D(T'))$ topology. By assumption $\{x_\alpha\}$ has a subnet, assumed for simplicity to be itself, such that $Q_2(Tx_\alpha)^\frown \to Q_2y''$ with respect to the $\sigma(Q_2E, D(T')$ topology (hence in the $\sigma(Q_2Y'', D(T'))$ topology) for some $y'' \in E$. But since $Q_2Y'' = D(T')'$, the topology $\sigma(Q_2Y'', D(T'))$ is Hausdorff. Consequently $Q_2T''x'' = Q_2y''$. Hence (i) $\implies$ (ii).

Now suppose that T' is continuous (so that $D(T'') = D(T)'')$ and that $Q_2R(T'') \subset Q_2E$. Let $\{x_\alpha\}$ be a net in $B_{D(T)}$. Then $\{\hat{x}_\alpha\}$ is a net in $B_{D(T'')}$ and since $B_{D(T)''}$ is $\sigma(D(T)'', D(T)')$ compact by the Banach-Alaoglu Theorem, $\{x_\alpha\}$ has a subnet, assumed for simplicity to be itself, such that $\hat{x}_\alpha \to x''$ with respect to $\sigma(D(T)'', D(T)')$, for some $x'' \in X''$. By hypothesis there exists $y'' \in E$ such that $Q_2T''x'' = Q_2y''$. Now $Q_2(Tx_\alpha)\hat{}\to Q_2T''x''$ in the $\sigma(D(T')', D(T'))$ topology (Proposition VIII.1.8). Hence for all $y' \in D(T')$ we have

$$y''y' = Q_2y''y' = \lim Q_2(Tx_\alpha)\hat{}(y')$$

which shows that $Q(Tx_\alpha)\hat{}\to y''$ in the $\sigma(Q_2E, D(T'))$ topology, as required. □

VIII.2.21 Theorem. *Let E denote the subspace $D(T') - \mathrm{cl}(\hat{Y})$. Then the following properties are equivalent:*

(i) *$(TB')\hat{}$ is relatively $\sigma(E, D(T'))$ compact and T' is continuous.*
(ii) *$T''(X'') \subset E + T''(0)$ and T' is continuous*
(iii) *T' is $\sigma(D(T'), E) - \sigma(X', D(T)'')$ continuous*
(iv) *T' is weakly compact.*

PROOF. We have (i) $\Longrightarrow$ (ii) (Proposition VIII.2.20) $\Longrightarrow$ (iii) (Corollary VIII.2.10) $\Longrightarrow$ (iv) (Corollary VIII.2.18). It only remains to prove that (iv) $\Longrightarrow$ (i).

Assume that (iv) holds. Write $S := T'$ and $E_1 := (X')\hat{}$. Since (iv) implies that S is continuous by Theorem VIII.2.4, and then by Theorem VIII.2.11, $Q_{S''}R(S'') \subset Q_{S''}E_1$, hence by Lemma VIII.2.9, $Q_{S'}S'$ is $\sigma(D(S'), E_1) - \sigma(Q_{S'}Y'', D(S)'')$ continuous. By Corollary VIII.2.18, the subset $Q_{S'}S'B_{D(S')}$ is $\sigma(Q_{S'}Y'', D(S)'')$ compact, i.e. $Q_2T''B_{D(T'')}$ is $\sigma(Q_2Y'', D(T')'')$ compact. Let $\{x_\alpha\}$ be a net in $B_{D(T)}$. Then there exists a subnet of $\{x_\alpha\}$, assumed for simplicity to be itself, and a point $x'' \in B_{D(T'')}$ such that $Q_2T''\hat{x}_\alpha \to Q_2T''x''$ with respect to $\sigma(Q_2Y'', D(T')'')$. Thus $Q_2T''x''$ is in the $\sigma(Q_2(Y''), D(T')'')$ closure of $Q_2\hat{Y}$, which coincides with the norm closure. Hence there is a sequence $\{y_n\}$ in Y such that $\|Q_2\hat{y}_n - Q_2T''x''\|_{D(T')''} \to 0$. Let $y'' \in T''x''$. Noting that

$$D(T')\hat{}\subset D(T')^{\perp\perp} = D(T')'' = (Q_2Y'')'$$

it follows that

$$\sup_{y' \in B_{D(T')}} \|(Q_2\hat{y}_n - Q_2y'')y'\| = \sup_{y' \in B_{D(T')}} \|(\hat{y}_n - y'')y'\| \to 0,$$

whence $y'' \in D(T') - \mathrm{cl}(\hat{Y}) = E$. Hence $(TB_{D(T)})\hat{}$ is relatively $\sigma(Q_2E, D(T'))$ compact. Therefore (iv) $\Longrightarrow$ (i). □

VIII.3 EXAMPLES OF WEAKLY COMPACT RELATIONS

VIII.3.1 Example *If $D(T)$ is reflexive, then T is weakly compact.* Indeed, T is weakly compact if and only if $TJ_{D(T)}$ is weakly compact, so we may suppose that $D(T) = X$. Let X be reflexive. Then T' is continuous (Corollary III.4.3). Hence T is weakly compact by Proposition VIII.2.8.

VIII.3.2 Definition A normed space X is called *prereflexive* if its completion is reflexive.

VIII.3.3 Example *Let T' be continuous and let Y be a Banach space. If $D(T)$ is prereflexive, then T is weakly compact.* To see this, suppose that $D(T)$ is prereflexive. We have $D(T'') = D(T)''$ by the continuity of T'. Hence $D(T'') = (D(T)^{\sim})^{\frown}$. Consequently

$$R(Q_2T'') = Q_2T''(D(T)^{\sim})^{\frown} \subset (Q_2T''D(T)^{\frown})^{\sim}$$

(by the continuity of T'') $= (Q_2R(T)^{\frown})^{\sim}$ (by VIII.1.2(b)) $\subset (Q_2(\overline{R(T)})^{\frown})^{\sim} = Q_2\overline{R(T)}^{\frown} \subset Q_2\hat{Y}$. Therefore T is weakly compact by Theorem VIII.2.11.

VIII.3.4 Example *On every separable reflexive Banach space X there exists an everywhere defined weakly compact and nowhere continuous linear operator from X into ℓ_1.* Choose a countable dense subset $\{x_n : n \in \mathbb{N}\}$ of B_X. Define $S : \ell_1 \to X$ by

$$S(\alpha) = \sum_{n=1}^{\infty} \alpha_n s_n \text{ for } \alpha = (\alpha_1, \alpha_2, \dots) \in \ell_1 .$$

Clearly, S is bounded with $||S|| \leq 1$. Moreover, $S(B_{\ell_1})$ is dense in B_X. Consequently S is open by Proposition III.5.2, and so $S(\ell_1) = X$ by the Open Mapping theorem. Let M be an algebraic complement of $N(S)$ in ℓ_1. Then $S|_M$ is a continuous injective map of M onto X. Put $T := (S|_M)^{-1}$. Then T is an everywhere defined linear operator from X into ℓ_1, which is weakly compact by VIII.3.1. Clearly, T is nowhere continuous, since no infinite dimensional subspace of ℓ_1 is isomorphic to a subspace of a reflexive Banach spaces.

In contrast to Example VIII.3.4 we should note that there may not exist nowhere continuous linear maps defined on a given Banach space. We note the following result in Bogachev, Kircheim and Schachermeyer [1]:

VIII.3.5 Proposition. *Let T be an everywhere defined linear operator from ℓ_1 into a normed space Y. Then there exists a dense subspace Z of ℓ_1 such that the restriction $T|_Z$ is continuous.*

PROOF. We have $R(T) = \bigcup_{n=1}^{\infty} nB_{R(T)}$. Hence

$$\ell_1 = D(T) = \bigcup_{n=1}^{\infty} nT^{-1}(B_{R(T)}).$$

By Baire sCategory theorem, there is an integer m for which $mT^{-1}(B_Y)$ is dense in B_{ℓ_1}. Let $\{e_n\}$ be the canonical basis in ℓ_1. For each $n \geq 1$, select $x_n \in \ell_1$ satisfying $\|x_n - e_n\| < 2^{-n-1}$ and $\|Tx_n\| \leq m$. Then

$$\sum_{n=1}^{\infty} \|x_n - e_n\| \leq 1/2$$

and by the perturbation theorem Lindenstrauss and Tzafriri [2; Prop. 1.a.9] there exists $c > 0$ such that

$$\left\| \sum_{j=1}^{n} \lambda_j x_j \right\| \geq c \sum_{j=1}^{n} |\lambda_j|$$

for any finite set of real numbers $\lambda_1, \ldots, \lambda_n$. But this implies that

$$\|T|_{\text{sp}\{x_n\}}\| \leq m/c < \infty. \quad \square$$

VIII.3.6 Example *Let Y be reflexive and let $T \in LR(\ell_1, Y)$ with $D(T) = \ell_1$. Then T is weakly compact, and a dense subspace E of ℓ_1 exists for which $T|_E$ is continuous.* Indeed, since $D(T)$ is a Banach space, T' will be continuous, so (ii) of Theorem VIII.2.11 is satisfied. Therefore T is weakly compact. The remaining assertions follow immediately from Proposition VIII.3.5 upon substituting QT for T.

Unfortunately not all the properties enjoyed by bounded weakly compact operators in Banach spaces are shared by weakly compact relations. For example, if T is weakly compact, and S is a bounded linear operator, then TS may not be weakly compact. The situation is worse than this, as we see from the next example.

VIII.3.7 Example *A weakly compact operator between Banach spaces having a non weakly compact restriction.*

Let $X = L_1[0,1]$, $Y = L_q[0,1]$, $1 \leq q \leq \infty$, and define T by $D(T) = \{f \in L_1[0,1] : f'$ exists almost everywhere and $f' \in L_q[0,1]\}$, $Tf = f'$ $(f \in D(T))$, where f' is the derivative of f. Then $D(T') = (0)$ (see Example V.5.10). Hence T is weakly compact. Now let M be the dense subspace of $L_1[0,1]$ consisting of the absolutely continuous functions. Then $T|_M$ is a closed unbounded Fredholm operator (see e.g., Goldberg [3; Chap.VI]). Hence $T|_M$ is not weakly compact, by Theorem VIII.2.4.

Our next question is: Given a weakly compact relation T, for which subspaces E is the restriction $T|_E$ weakly compact? Such a class of subspaces is provided by Corollary VIII.3.9 below.

VIII.3.8 Proposition. *Let T be weakly compact and let M be a linear subspace of $D(T)$. The following implication holds:*

$$D((T|_M)') \subset D(T') \implies T|_M \text{ is weakly compact.}$$

PROOF. Follows immediately from the definition of weak compactness. □

VIII.3.9 Corollary. *Let T be weakly compact and let M be a linear subspace of $D(T)$. Suppose that either*

(i) *M is closed and of finite codimension in $D(T)$, or*
(ii) *M is a core of T.*

Then $T|_M$ is weakly compact.

PROOF. Apply Proposition VIII.3.8 (with Theorem III.1.6(b)) or Proposition IV.4.4, as appropriate. □

VIII.3.7 Example Let T be weakly compact. If T is either bounded below, or nowhere continuous, then $D(T)$ is reflexive.

VIII.4 WEAKLY COMPLETELY CONTINUOUS AND UNCONDITIONALLY CONVERGING LINEAR RELATIONS

If T is a bounded linear operator from a Banach Space X into a Banach Space Y, then T is said to be *weakly completely continuous* if it transforms weak Cauchy sequences into weakly convergent sequences, and *unconditionally converging* if it transforms weakly unconditionally convergent series into unconditionally convergent series. (See Diestel [1]).

In order to characterize these classes of operators, one introduces for a given normed space X, the following subsets of X'':

$$K(X) := \{x'' \in X'' : \text{there exists a sequence } \{x_n\} \text{ in } X \text{ such that } x'' = \sigma(X'', X') - \lim \hat{x}_n\}$$

$$N(X) := \Big\{x'' \in X'' : \text{there exists a weakly unconditionally Cauchy series } \Sigma x_n \text{ in } X \text{ such that } x'' = \sigma(X'', X') - \lim \sum_{k=1}^{n} \hat{x}_k \Big\}.$$

Clearly $\hat{X} \subset N(X) \subset K(X) \subset X''$.

The two classes of bounded linear operators in question may now be characterized as follows:

VIII.4.1 Theorem. *Let $T : X \to Y$ be a bounded linear operator, where X and Y are Banach spaces, and let M denote $K(X)$ (resp. $N(X)$). The following properties are equivalent:*

(i) *T is weakly completely continuous (resp. unconditionally converging)*
(ii) $T''(M) \subset \hat{Y}$
(iii) *T' is $\sigma(Y', Y) - \sigma(X', M)$ continuous.*

Theorem VIII.4.1 is due to J. Howard and K. Melendez [1]. The theorem reveals properties of the second adjoint T''.

Our aim in this section is to discover new properties of the second adjoint of a linear relation by studying the appropriate generalizations of the above classes of operators.

VIII.4.2 Definitions The linear relation T will be called

(a) **weakly completely continuous** if QT transforms weakly Cauchy sequences in $D(T)$ into $\sigma(QY, D(T'))$-convergent sequences,
(b) **unconditionally converging** if whenever Σx_n is weakly unconditionally convergent in $D(T)$, each subseries ΣQTx_{k_n} of ΣQTx_n is $\sigma(QY, D(T'))$ convergent.

The corresponding classes of linear relations will be abbreviated $WCC(X,Y)$ and $UC(X,Y)$ (or simply WCC and UC). The class of weakly compact relations in $LR(X,Y)$ will be denoted by $WC(X,Y)$ (or simply WC).

We have the inclusions

$$WC \subset WCC \subset UC \tag{1}$$

(see Corollary VIII.4.8. The second inclusion is obvious from the definitions.)

VIII.4.3 Proposition. *Let T' be continuous and let $T''(M) \subset \hat{Y} + T''(0)$, where $M = K(D(T))$ (resp. $N(D(T))$). Then $T \in WCC$ (resp. UC).*

PROOF. Let $T''(K) \subset \hat{Y} + T''(0)$ (where $K := K(D(T))$). Let $\{x_n\}$ be a $\sigma(X, X')$-Cauchy sequence in $D(T)$. Then by the uniform boundedness principle, the sequence of norms $\{\|x_n\|\}$ is bounded. Hence by the Banach-Alaoglu Theorem, the sequence $\{\hat{x}_n\}$ has a subsequence $\{\hat{x}_{k_n}\}$ which is $\sigma(X'', X')$ convergent to some point $x'' \in X''$. Then by the weak*-weak* continuity of Q_2T'' (Proposition VIII.1.8),

$$Q_2T''x'' = \sigma(Q_2Y'', D(T')) - \lim Q_2T''\hat{x}_{k_n}.$$

By hypothesis, since $x'' \in K(D(T))$, there exists $y \in Y$ for which $Q_2\hat{y} = Q_2T''x''$. Hence for $y' \in Y'$,

$$(Q_2T''\hat{x}_{k_n})y' = Q_2(Tx_{k_n})\hat{}(y') \to Q_2\hat{y}y' , \qquad \text{or equivalently}$$
$$y'(QTx_{k_n}) \to y'(Qy).$$

This shows that QTx_{k_n} is $\sigma(QY, D(T'))$-convergent to Qy. Now for $x' \in X'$, there exists $n_0 \in \mathbb{N}$ such that $|(\hat{x}_n - \hat{x}_{k_n})x'| < 1/n$ and $|(\hat{x}_{k_n} - x'')x'| < 1/n$ for all $n \geq n_0$. Hence $|(\hat{x}_n - x_k'')x'| < 2/n$ for $n \geq n_0$. Therefore QTx_n is $\sigma(QY, D(T'))$-convergent to Qy, showing that $T \in WCC$.

Now let $T''(N) \subset \hat{Y} + T''(0)$ (where $N := N(D(T))$). Let Σx_n be a weakly unconditionally convergent series in $D(T)$. Then $\Sigma\,|\hat{x}_n(x')| < \infty$ for each $x' \in D(T')$. It follows that the sequence of partial sums Σx_n is norm bounded. The proof now proceeds as in the previous case. □

VIII.4.4 Proposition. *Let $S \in LR(Y, Z)$ be continuous, and let $T \in WCC$ (resp. UC), with $D(S) \supset R(T) + \overline{T(0)}$. Then $ST \in WCC$ (resp. UC).*

PROOF. The proof is along the lines of the proof of Proposition VIII.2.3, with the appropriate modifications.

Let $T \in WCC$ and assume in the first instance that T is single valued. Then $ST(0) = S(0)$, so $Q_{ST}ST = (Q_SS)T$. Since $ST \in WCC$ if and only if $Q_{ST}ST \in WCC$, if and only if Q_SST is weakly completely continuous, we may assume that S is single valued. In that case $D(S') = Z'$ and hence by Theorem III.1.6(a), $(ST)' = T'S'$. Now let $\{x_n\}$ be a $\sigma(X, X')$-Cauchy sequence in $D(ST) = D(T)$. Then $\{Tx_n\}$ is $\sigma(Y, D(T'))$-convergent to some point $y \in Y$. We now have for $z' \in D((ST)') = D(T'S') \subset D(S') = Z'$, that $z'S = S'z' \in D(T')$ and consequently

$$\lim z'STx_n = z'Sy.$$

Thus $\{STx_n\}$ is $\sigma(Z, D((ST)'))$-convergent to $Sy \in Z$. Therefore $ST \in WCC$.

For the general case, consider $SQ^{-1}QT$. The $QT \in WCC$, and SQ^{-1} is continuous since $Q^{-1}(0) \subset \overline{T(0)} \subset D(S)$ (see II.3.13). Hence $SQ^{-1}QT \in WCC$. Therefore by Lemma V.2.9, $Q_{ST}ST \in WCC$. Consequently $ST \in WCC$.

Now let $T \in UC$. Let Σx_n be a weakly unconditionally convergent series in $D(ST)$, hence in $D(T)$. Then for each subseries ΣQTx_{k_n} there exists $y \in Y$ such that $y = \sigma(QY, D(T')) - \lim \sum_{i=1}^{n} QTx_{k_i}$. The rest of the proof now follows as in the previous case. □

The following Theorem is a generalization of Theorem VIII.2.4.

VIII.4.5 Theorem. *Let T be a closable unconditionally converging relation and let Y be complete. Then T is continuous.*

PROOF. The proof is similar to that of Theorem VIII.2.4, with only minor changes. As before, we may assume that T is single valued, densely defined, and nonzero, and that X is complete. We shall adopt the notation of the proof of VIII.2.4. Then T_1 is an unconditionally converging linear operator, by Proposition VIII.4.4. We shall verify that $R(\tilde{T}_1) \subset WY$. Let $x \in D(\tilde{T}_1)$ and select a sequence $\{x_n\}$ in $D(T)$ such that $\|x - \sum_{k=1}^{n} x_k\| \to 0$ and $\sum_{n=1}^{\infty} \|x_n\| < \infty$. Then Σx_n is weakly unconditionally convergent. Since $T_1 \in UC$ there exists $y_1 \in Y_1$ such that $y_1'(\sum_{k=1}^{n} T_1 x_k) \to y_1' y_1$ for every $y_1' \in Y_1'$. But $\tilde{T}_1(\Sigma x_n) = \tilde{T}_1 x$. Therefore $J_{Y_1}(y_1) = \tilde{T}_1 x$ as claimed. The rest of the proof proceeds exactly as in VIII.2.3. □

VIII.4.6 Corollary. *If T is an unconditionally converging relation, then T' is continuous.*

PROOF. Let $T \in UC$. Then $J_Y^{\tilde{Y}} T \in UC$, by Proposition VIII.4.4. We may therefore suppose that Y is complete. Let S be the regular contraction of T, i.e. $S := Q_{D(T')^\top} T$. Then $S \in UC$, again by Proposition VIII.4.4, and S is closable (Corollary III.4.11). Hence $S' = T'$ is continuous, by Theorem VIII.4.5. □

VIII.4.7 Theorem. *Let $M := K(D(T))$ (resp. $N(D(T))$). Then the following properties are equivalent:*

(i) *$T \in WCC$ (resp. UC)*
(ii) *T' is continuous and $T''(M) \subset \hat{Y} + T''(0)$*
(iii) *T' is both norm-norm continuous and $\sigma(D(T'), Y) - \sigma(X', M)$ continuous.*

PROOF. (i) $\Longrightarrow$ (ii). Case 1: $M = K(D(T)))$. Let $T \in WCC$ and write $K := K(D(T))$. Let $x'' \in K$, and let $\{x_n\}$ be a sequence in $D(T)$ such that weak*-lim $\hat{x}_n = x''$. Since $\{x_n\}$ is weak*-Cauchy, $\{QTx_n\}$ is $\sigma(QY, D(T'))$-convergent. Hence there exists $y \in Y$ such that $Qy = \sigma(QY, D(T')) - \lim QTx_n$. Now T' is continuous by Corollary VIII.4.6. Hence $D(T'') = D(T)''$ (Theorem III.8.10), and since $Q_2 T''$ is $\sigma(X'', X') - \sigma(Q_2 Y'', D(T'))$ continuous, we have

$$(1) \qquad (Q_2 T'' \hat{x}_n) y' = Q_2 (Tx_n)\hat{}\, y' \to (Q_2 T'' x'') y' \qquad (y' \in D(T')).$$

But since $y'(QTx_n) \to y'Qy$ $(y' \in D(T'))$ and noting that $Q_2(Tx)\hat{\ }y' = y'Tx$ for $x \in D(T)$, $y' \in D(T')$, if follows from (1) that

$$(Q_2T''x'')y' = (Q_2\hat{y})y' \qquad (y' \in D(T')).$$

Hence $Q_2T''x'' = Q_2\hat{y}$, showing that

$$T''(K) \subset \hat{Y} + T''(0).$$

Therefore (i) $\implies$ (ii).

Case 2: $M = N(D(T)) := N$. Assume $T \in UC$ and let $x'' \in N$. Then there exists a weakly unconditionally convergent series Σx_n in $D(T)$ such that $x'' = \sigma(X'', X') - \lim \sum_{k=1}^{n} \hat{x}_k$. Since $T \in UC$, for each subseries ΣTx_{k_n}, there exists $y \in Y$ such that $Qy = \sigma(QY, D(T')) - \lim \Sigma Tx_{k_n}$. The rest of the proof now proceeds as for the previous case.

The implication (ii) $\implies$ (iii) is covered by Corollary VIII.2.9.

(iii) $\implies$ (i). Assume that (iii) holds. Then $D(T'') = D(T)''$ and so $K, N \subset D(T'')$. Hence by Lemma VIII.2.9 (with $E = \hat{Y}$), (ii) holds. Hence (i) follows, by Proposition VIII.4.3. □

VIII.4.8 Corollary. $WC \subset WCC \subset UC$.

A Banach space X is weakly sequentially complete if and only if $K(X) = \hat{X}$, and contains no isomorphic copy of c_0 if and only if $N(X) = \hat{X}$ (see J. Howard and K. Melendez [1]). Combining these observations with Corollary III.4.3 and Proposition VIII.4.3 yields immediately the following example.

VIII.4.9 Proposition.

(a) *If $D(T)$ is weakly sequentially complete, then $T \in WCC$.*

(b) *If $D(T)$ is a Banach space containing no isomorphic copy of c_0, then $T \in UC$.*

We may relax the completeness condition on $D(T)$ in Proposition VIII.4.9, substituting an alternative assumption:

VIII.4.10 Proposition. *Let T' be continuous and let Y be a Banach space.*

(a) *If $D(T)^{\sim}$ is weakly complete, then $T \in WCC$.*

(b) *If $D(T)^{\sim}$ contains no isomorphic copy of c_0, then $T \in UC$.*

PROOF. Since T' is continuous, we have $D(T'') = D(T)''$. The result therefore follows from Proposition VIII.4.3, observing that in case (a), $K(D(T)^{\sim}) = (D(T)^{\sim})^{\widehat{}}$, and in case (b), $N(D(T)^{\sim}) = (D(T)^{\sim})^{\widehat{}}$. □

Proposition VIII.3.8 for weakly compact relations can be extended to WCC and UC relations. We have:

VIII.4.11 Proposition. *Let T be WCC (resp. UC), and let M be a linear subspace of $D(T)$. The following implication holds:*

$$D((T|_M)') \subset D(T') \implies T|_M \in WCC \quad (resp.\ UC).$$

PROOF. Follows immediately from the definitions. □

VIII.4.12 Corollary. *Let T be WCC (resp. UC) and let M be a linear subspace of $D(T)$. Suppose that either*

(i) *M is closed and finite codimensional in $D(T)$ or*

(ii) *M is a core of T.*

Then $T|_M$ is WCC (resp. UC).

In order to obtain a partial converse to Proposition VIII.4.9(a) we recall the following standard definition:

VIII.4.13 Definition We say that a subclass $\mathcal{A}$ of the class of all Banach spaces has the **three-space property** if it satisfies the following conditions: If M is a closed subspace of a Banach space X such that $X/M \in \mathcal{A}$ and $M \in \mathcal{A}$, then $X \in \mathcal{A}$.

For example, let the class $\mathcal{A}$ have the three space property and suppose that $\mathcal{A}$ contains the finite dimensional subspaces. Then if a Banach space X contains a closed finite codimensional subspace belonging to $\mathcal{A}$, then X belongs to $\mathcal{A}$.

A unified proof of the following results was given by T. Alvarez and V.M. Onieva [1]:

VIII.4.14 Proposition. *The classes of reflexive spaces, weakly sequentially complete spaces and Banach spaces containing no isomorphic copy of c_0, are spaces with the three space property.*

VIII.4.15 Proposition. *Let T be a weakly completely continuous F_+-relation. Then $D(T)$ is weakly sequentially complete.*

PROOF. Substituting QT for T if necessary, we may suppose that T is single valued. There exists a closed finite codimensional subspace M of $D(T)$ such that $T|_M$ is open and injective. We have $I_M = (T|_M)^{-1}(T|_M)$.

By Proposition VIII.4.12, $T|_M \in WCC$. Hence $I_M \in WCC$ by Proposition VIII.4.4. Consequently, every weak Cauchy sequence in M is $\sigma(M, M')$-convergent. Therefore M is weakly sequentially complete. It follows that $D(T)$ is complete, and hence by the three space property, weakly sequentially complete. □

VIII.4.16 Examples We construct a closed unbounded operator defined everywhere on a Banach space of each of the following types:

(i) weakly compact,
(ii) weakly completely continuous but not weakly compact,
(iii) unconditionally converging but not weakly completely continuous.

Let X be an infinite dimensional Banach space. Let T_0 be a restriction of the identity on X to a dense subspace of codimension one. Put $Y = R(T_0)$ and select a point $x_0 \in X \backslash D(T_0)$. Define $T \in L(X, Y)$ by $Tx = T_0 x$ $(x \in D(T_0))$ and $Tx_0 = 0$. Then T is an everywhere defined unbounded F_+-operator with one-dimensional null space. We shall verify that T is closable, hence closed. Let P, Q be complementary linear projections defined on X such that $R(P) = \text{sp}\{x_0\}$, $R(Q) = D(T_0)$. Then $Tx = Qx$. Let $(x_n, Tx_n) \to (0, y)$ $(y \in Y = QX)$. Then $(x_n, x_n - Px_n) \to (0, y)$, whence $Px_n \to -y \in PX$. But $PX \cap QX = \{0\}$. Hence $y = 0$, whence T is closable as claimed. By Example VIII.3.1, Proposition VIII.4.9 and suitable applications of Proposition VIII.4.14 (one such application is Proposition VIII.4.15), T has the property (i), (ii) or (iii) stated above for the following respective choices of X:

(i) X is reflexive space,
(ii) X a weakly sequentially complete non-reflexive space (e.g. ℓ_1 or L_1),
(iii) X a Banach space containing no isomorphic copy of c_0 which is not weakly sequentially complete (e.g. the James space; see Lindenstrauss and Tzafriri [1; p.25]).

We end this section with the analogue of Lemma VIII.2.5 for WCC and UC relations.

VIII.4.17 Proposition. *T is $WCC(UC)$ if and only if HT is $WCC(UC)$.*

PROOF. Let T be $WCC(UC)$ then so is HT, by Proposition VIII.4.4.

For the converse, we consider each of the two cases. From the definitions, it is enough to consider the case when T is single valued.

Case 1: $HT \in WCC$. Let $\{x_n\}$ be a weakly Cauchy sequence in $D(T)$. Then for some $y \in Y$ and every $z' \in D((HT)') = Y'_{T'}$ (see Theorem IV.3.17) we have

$$z'HTx_n \to z'Hy$$

i.e.

$$(HT)'z'x_n \to H'z'y' \tag{1}$$

But $(HT)' = T'G_{T'}$ (IV.3.17) and $D(G_{T'}) = Y'_T$. Hence

$$T'G_{T'}z'x_n \to H'z'y \qquad (z' \in Y'_{T'})$$

or equivalently, for every $y' \in D(T')$,

$$y'Tx_n = T'y'x_n \to y'y.$$

Hence $\{Tx_n\}$ is $\sigma(Y, D(T'))$ convergent to y. Therefore $T \in WCC$.

Case 2: $HT \in UC$. Let $HT \in UC$ and let the series Σx_n be weakly unconditionally convergent in $D(T)$. Consider a subseries ΣTx_{k_n}. By hypothesis, ΣHTx_{k_n} is $\sigma(HY, D(Y'_{T'}))$ convergent (since $Y'_{T'} = D((HT)')$, as noted above). Hence there exists $y \in Y$ such that for every $z' \in D(Y'_{T'})$, we have

$$z'(\sum_{j=1}^{n} HTx_{k_j}) \to z'Hy.$$

Thus

$$z'HT(\sum_{j=1}^{n} x_{k_j}) \to z'Hy.$$

i.e.

$$(HT)'z'(\sum_{j=1}^{n} x_{k_j}) \to H'z'y.$$

As before, this is equivalent to

$$T'G_{T'}z'(\sum_{j=1}^{n} x_{k_j}) \to G_{T'}z'y \qquad (z' \in Y'_{T'})$$

i.e.

$$T'y'(\sum_{j=1}^{n} x_{k_j}) \to y'y \qquad (y' \in D(T'))$$

i.e.

$$y'(\sum_{j=1}^{n} Tx_{k_j}) \to y'y \qquad (y' \in D(T'))$$

which shows that $T \in UC$. □

VIII.5 TAUBERIAN RELATIONS

Tauberian linear operators in Banach spaces were defined by N.J.Kalton and A. Wilansky in [1] as the class of bounded linear operators $T : X \to Y$ (where X and Y are Banach spaces) with the property $(T'')^{-1}(\hat{Y}) \subset \hat{X}$. We shall denote $(D(T))^{\sim}$ by $\tilde{D}(T)$.

VIII.5.1 Definitions The linear relation T is called **Tauberian** if

(1) $$(T'')^{-1}(\hat{Y}) \subset \tilde{D}(T)\hat{}.$$

More generally, let E be a subspace of $\tilde{X}$ satisfying $D(T) \subset E \subset \tilde{D}(T)$. Then T is called E-**Tauberian** if

(2) $$(T'')^{-1}(\hat{Y}) \subset \hat{E}.$$

In particular, $D(T)$-Tauberian relations are termed **strongly Tauberian**.

From (2) it follows immediately that

(3) $$T \text{ is } E\text{-Tauberian} \implies R(T'') \cap \hat{Y} \subset T''(\hat{E})$$

and

(4) $$T \text{ is strongly Tauberian} \implies R(T'') \cap \hat{Y} \subset R(T)\hat{} + T''(0).$$

VIII.5.2 Proposition. *For $y \in Y$ we have*

$$((QT)'')^{-1}((Qy)\hat{}) = (T'')^{-1}(\hat{y}).$$

PROOF. We have $(QT)'' = Q''T''$ by VIII.1.1(4). Therefore

$$((QT)'')^{-1}(Qy)\hat{} = (T'')^{-1}(Q'')^{-1}(Qy)\hat{}, \text{ where}$$
$$(Q'')^{-1}(Qy)\hat{} = (Q'')^{-1}Q''\hat{y} = \hat{y} + T(0)^{\perp\perp}.$$

But $T(0)^{\perp\perp} \subset \overline{T}(0)^{\perp\perp} = D(T')^{\top\perp\perp} \subset D(T')^{\perp} = T''(0)$.

Hence

$$((QT)'')^{-1}(Qy)\hat{} = (T'')^{-1}(\hat{y} + T''(0)) = (T'')^{-1}(\hat{y}). \quad \square$$

VIII.5.3 Corollary. *Let $D(T) \subset E \subset \tilde{D}(T)$. Then T is E-Tauberian if and only if QT is E-Tauberian.*

We note the following stability property of E-Tauberian relations:

VIII.5.4 Theorem. *Let the relation T be E-Tauberian, where $D(T) \subset E \subset \tilde{D}(T)$, and let S be a continuous weakly compact relation with $S(0) \subset \overline{T(0)}$ and $D(S) \supset D(T)$. Then $T + S$ is E-Tauberian.*

PROOF. Since $S(0) \subset \overline{T}(0)$ we have $\overline{(T+S)(0)} = \overline{T(0)}$, and hence by Corollary VIII.5.3, $T + S$ is E-Tauberian if and only if $Q_T(T+S)$ is E-Tauberian. Since now $Q_T S$ is a continuous weakly compact operator, we may thus suppose that T is single valued. Let $y \in Y$. We have

(1) $$(T'')^{-1}(\hat{y}) \subset \hat{E}.$$

Since S is continuous, $D(S') = S(0)^{\perp} \supset T(0)^{\perp} \supset D(T')$, and hence by III.1.5 $(T+S)'' = T'' + S''$. Let $x'' \in (T'' + S'')^{-1}\hat{y}$. Then

$$(T'' + S'')x'' \in \hat{y} + (T'' + S'')(0) = \hat{y} + T''(0)$$

where $S''x'' \in \hat{Y} + S''(0)$ by VIII.2.11. Write $S''x'' = \hat{y}_0 + S''(0)$. Since S is continuous and $S(0) \subset T(0)$, we get $S''(0) = D(S')^{\perp} = S(0)^{\perp\perp} \subset T(0)^{\perp\perp} \subset T''(0)$. We therefore have

$$(T'' + S'')x'' = T''x'' + \hat{y}_0 \in \hat{y} + T''(0)$$

whence

(2) $$T''x'' \in (y - y_0)\hat{} + T''(0).$$

Hence from (2),

$$x'' \in (T'')^{-1}(\hat{Y})$$

where from (1), $(T'')^{-1}(\hat{Y}) \subset \hat{E}$. Consequently $x'' \in \hat{E}$. Therefore $T + S$ is E-Tauberian. □

We shall now introduce a seminorm on the space Y that will make T continuous.

VIII.5.5 Notation For $y \in Y$ define

$$\|y\|_0 := \sup\{|y'y| : y' \in D(T'), \|y'\| + \|T'y'\| \leq 1\}.$$

Let $Y_{D(T')}$ and Y_0 denote the seminormed spaces $(Y, \| \ \|_{D(T')})$ and $(Y, \| \ \|_0)$. Evidently

(1) $$\|y\|_0 \leq \|y\|_{D(T')} \leq \|y\| .$$

Write

$$T_0 := J_Y^{Y_0} T.$$

VIII.5.6 Proposition. *The relation $T_0 \in LR(X, Y_0)$ is continuous and*

$$Y_0' \subset (Y_{D(T')})' \subset Y'$$

where the spaces are regarded as linear subspaces of the algebraic dual of Y.

PROOF. We have for $x \in D(T)$,

$$\|T_0 x\|_0 = \sup_{y' \in D(T')} \frac{|y'Tx|}{\|y'\| + \|T'y'\|} \leq \sup_{y' \in D(T')} \frac{\|T'y'\|\,\|x\|}{\|y'\| + \|T'y'\|} \leq \|x\|\,.$$

Hence $\|T_0\| \leq 1$; in particular, T_0 is continuous.

The inclusions $Y_0' \subset (Y_{D(T')})' \subset Y'$ are an immediate consequence of VIII.5.5(1). □

VIII.5.7 Theorem. *Let E be a linear subspace of $\tilde{D}(T)$ containing $D(T)$. The following properties are equivalent:*

(i) *$N(T'') \subset \hat{E}$.*

(ii) *Every bounded sequence $\{x_n\}$ in $D(T)$ for which $\sigma(Y, D(T'))-\lim Tx_n = 0$ has a $\sigma(E, X')$ convergent subsequence.*

If T is continuous, then (i) and (ii) are equivalent to:

(iii) *Every bounded sequence $\{x_n\}$ in $D(T)$ for which $\|Tx_n\|_{D(T')} \to 0$ has a $\sigma(E, X')$ convergent subsequence.*

PROOF. Since $N(T'') = N((QT)'')$ by Proposition VIII.5.2, $\sigma(Y, D(T')) - \lim Tx_n = 0$ if and only if $\sigma(QY, D((QT)')) - \lim QTx_n = 0$, and $\|Tx\|_{D(T')} = \|QTx\|_{D((QT)')}$, we may suppose (substituting QT for T if necessary) that T is single valued.

(i) $\Longrightarrow$ (ii). Assume (i) and let $\{x_n\}$ be a bounded sequence in $D(T)$ such that $\sigma(Y, D(T')) - \lim Tx_n = 0$, i.e. the numerical sequence $\{y'Tx_n\}$ converges to 0 for each $y' \in D(T')$. The set $\{\hat{x}_n\}$ is relative $\sigma(X'', X')$-compact in X'' (by the Banach-Alaoglu Theorem) and so $\{\hat{x}_n\}$ has a subnet $\{\hat{x}_{n'}\}$ which is $\sigma(X'', X')$-convergent to some point $x'' \in X''$. But for $y' \in D(T')$,

$$y'Tx_n = (T'y')x_n = \hat{x}_n T'y' = (T''\hat{x}_n)y' \to 0.$$

Thus if $y_n'' \in T''\hat{x}_n$ $(n \in \mathbb{N})$, then $(\hat{x}_{n'}, y_{n'}'')$ is a net in $G(T'')$ which is $\sigma(X'', X') \times \sigma(Y'', D(T'))$ convergent to $(x'', 0)$. Hence $(x'', 0) \in G(T'')$, by Proposition VIII.1.10. Therefore $x'' \in N(T'') \subset \hat{E}$. Let $x'' = \hat{x}$, where $x \in E$. Then we have $\sigma(E, X') - \lim x_{n'} = x$. This shows that the set $\{x_n\}$ is relatively $\sigma(E, X')$ countably compact, hence relatively $\sigma(E, X')$ sequentially compact, i.e. $\{x_n\}$

has a $\sigma(E, X')$-convergent subsequence. Therefore (i) $\Longrightarrow$ (ii). The second assertion follows immediately.

(ii) $\Longrightarrow$ (i). Assume that (ii) holds, and let $x'' \in N(T'')$ where $\|x''\| = 1$. Suppose in the first instance that T is continuous; in particular we have $D(T') = Y'$, $D(T'') = D(T)^{\perp\perp}$ and T'' is single valued. Choose a net (x_α) in $B_{D(T)}$ such that $\sigma(X'', X') - \lim \hat{x}_\alpha = x''$. Then we have $\sigma(Y'', Y') - \lim T''\hat{x}_\alpha = T''x'' = 0$, and thus $\sigma(Y, Y') - \lim Tx_\alpha = 0$. Let $C_\alpha = \text{co}\{x_\gamma : \gamma \geq \alpha\}$. Since the $\sigma(Y, Y')$ and norm closures of the convex set TC_α coincide and contain 0, there exists a sequence $\{c_\alpha^n\}$ in C_α for which $\|Tc_\alpha^n\| \to 0$. By (ii) $\{c_\alpha^n\}$ has a subsequence, which we assume for simplicity to be itself, which is $\sigma(E, X')$ convergent to some point $c_\alpha \in E$; moreover, since C_α is convex, $c_\alpha \in \overline{C}_\alpha$ (norm closure). By Proposition VIII.1.10, $\hat{c}_\alpha \in D(T'')$ and $T''\hat{c}_\alpha = 0$. Now let (v_n) be an arbitrary sequence in the set $\{c_\alpha\}$. Since $v_n \in E \subset \tilde{D}(T)$ there exists a sequence $\{u_n\}$ in $D(T)$ such that $\|u_n - v_n\| \leq 1/n$. Then

$$\|Tu_n\| = \|T''(u_n - v_n)\hat{}\| \leq \|T''\| / n \to 0.$$

Hence by (ii), (u_n) has a subsequence, which for simplicity we assume to be itself, $\sigma(E, X')$ convergent to $u \in E$, say. Now for $x' \in X'$ we have

$$|x'(u - v_n)| \leq |x'(u_n - u)| + |x'(v_n - u_n)| \leq |x'(u_n - u)| + \|x'\| / n \to 0.$$

Thus $\sigma(E, X') - \lim v_n = u \in E$. This shows that $\{c_\alpha\}$ is relatively $\sigma(E, X')$ sequentially compact and hence (by Eberlein s Theorem, and the angelic property Floret [1; Ch3]) relatively $\sigma(E, X')$ compact. Therefore $\{c_\alpha\}$ has a $\sigma(E, X')$ convergent subnet, again assumed to be itself. Let $\sigma(E, X') - \lim c_\alpha = c$. We claim that $\hat{c} = x''$. Indeed if W is a closed convex $\sigma(X'', X')$ neighbourhood of x'' we can determine α_0 such that $\hat{x}_\alpha \in W$ for $\alpha \geq \alpha_0$, and since $W \supset (\overline{C}_\alpha)\hat{}$ (norm closure) for $\alpha \geq \alpha_0$, we have $\hat{c}_\alpha \in W$ for $\alpha \geq \alpha_0$. Consequently $x'' = \hat{c}$ as claimed. Since $c \in E$, (i) follows. This completes the proof of (ii) $\Longrightarrow$ (i) for the case when T is continuous.

Now let T be an arbitrary single valued relation. Consider the statement (iii):

It is clear that (ii) implies (iii). In the case when T is continuous, the preceding proof shows that (iii) implies (i) (since the stated subsequence was derived from the norm convergence $\|Tc_\alpha^n\| \to 0$). It follows that (i), (ii) and (iii) are equivalent for continuous linear relations.

In order to show that (ii) implies (i) it is enough to consider the case when T is closable. To see this, put $S := Q_{\overline{T}(0)}T$. Then S is closable (Corollary III.4.11), with $D(S') = D(T')$, and $y'Sx = y'Tx$ $(y' \in D(T'),\ x \in D(T))$. Hence if T satisfies (ii), then so does S. Then if S satisfies (i), i.e. $N(S'') = (T'')^{-1}(\overline{T}(0)^{\perp\perp}) \subset \hat{E}$, T also satisfies (i).

Accordingly, assume that T is closable. In that case the subspace $D(T')$ is a separating subspace of Y' and hence the seminorm $\|\cdot\|_0$ defined on Y_0 in Definition VIII.5.5 is a norm, and the relation T_0 is a continuous linear operator between normed spaces.

Let T satisfy (ii). We shall verify that T_0 satisfies the weaker property (iii). Let $\{x_n\}$ be a bounded sequence in $D(T_0) = D(T)$ for which $\lim \|T_0 x_n\|_{D(T_0')} = 0$. Noting that $D(T_0') = Y_0'$ as T_0 is single valued and continuous, we have $\lim \|T_0 x_n\| = 0$ (norm convergence in Y_0). Thus

$$\sup_{y' \in D(T')} \frac{|y'Tx_n|}{\|y'\| + \|T'y'\|} \to 0.$$

Hence $y'Tx_n \to 0$ for each $y' \in D(T')$. In other words, $\sigma(Y, D(T')) - \lim Tx_n = 0$. Consequently, $\{x_n\}$ has a $\sigma(E, X')$ convergent subsequence.

Since T_0 satisfies (iii), we conclude that $N(T_0'') \subset \hat{E}$. The proof is completed by verifying that $N(T'') \subset N(T_0'')$. Since $D(T_0') \subset D(T')$, we have for $x \in D(T)$ and $y' \in D(T_0')$ that $T_0'y'x = y'T_0x = y'Tx = T'y'x$. Hence if $x'' \in N(T'')$, then $0 = T''x''y' = x''T'y' = x''T_0'y' = T_0''x''y'$ for $y' \in D(T_0')$, i.e. $T_0''x'' \subset D(T_0')^{\perp} = T_0''(0)$, whence $x'' \in N(T_0'')$, as required. □

VIII.5.8 Corollary.

(a) *If T is Tauberian, then $N(T)$ is prereflexive.*

(b) *If T is strongly Tauberian, then $N(\overline{T})$, $\overline{N(T)}$ and $\overline{N(T)} \cap D(T)$ are reflexive.*

PROOF. (a) Let T be Tauberian. Then $N(T'') \subset \tilde{D}(T)\hat{}$. Hence by Theorem VIII.5.7, every bounded sequence $\{x_n\}$ in $N(T)$ has a $\sigma(X, X')$ convergent subsequence, and the same is clearly true for the completion $N(T)^{\sim}$ of $N(T)$. Therefore $N(T)^{\sim}$ is reflexive.

(b) Let T be strongly Tauberian. By Theorem VIII.5.7, since $N(T'') = N(\overline{T}'') \subset D(T)\hat{} \subset D(\overline{T})\hat{}$, every bounded sequence $\{x_n\}$ in $N(\overline{T})$ has a $\sigma(X, X')$-convergent subsequence. Therefore $N(\overline{T})$ is reflexive, since $N(\overline{T})$ is closed. Hence $\overline{N(T)}$ is reflexive. Again, considering such a bounded sequence in $N(T)$, it is seen to have a $\sigma(D(T), X')$-convergent subsequence, with limit $x \in D(T)$, say. Then $x \in \overline{N(T)} \cap D(T)$. It follows that $\overline{N(T)} \cap D(T)$ is reflexive. □

VIII.5.9 Lemma. *Let E be a linear subspace of $\tilde{D}(T)$ containing $D(T)$. If $N(T'') \subset \hat{E}$ and if $(\sigma(Y, D(T')) - \mathrm{cl}(TB'))\hat{} \subset T''(\hat{E})$, then T is E-Tauberian.*

PROOF. Assume the given conditions hold and let $x'' \in (T'')^{-1}(\hat{y})$, i.e. $T''x'' = \hat{y} + T''(0)$, where $x'' \in B_{D(T'')}$. By Proposition VIII.1.9, $y \in \sigma(Y, D(T'))-\mathrm{cl}(TB')$, so $\hat{y} \in T''(\hat{E})$, whence $(T'')^{-1}(\hat{y}) \subset \hat{E} + N(T'') = \hat{E}$. □

VIII.5.10 Theorem. *Let E be a linear subspace of $\tilde{D}(T)$ containing $D(T)$. The following properties are equivalent:*

(i) *T is E-Tauberian*

(ii) *For every bounded subset B of $D(T)$, if $T(B)$ is relatively $\sigma(Y, D(T'))$ compact, then B is relatively $\sigma(E, X')$ compact.*

If T is continuous, then (i) and (ii) are equivalent to:

(iii) *For every bounded subset B of $D(T)$, if $T(B)$ is relatively $D(T')$-seminorm compact, then B is relatively $\sigma(E, X')$ compact.*

PROOF. As in the proof of Theorem VIII.5.7, we may suppose that T is single valued.

(i) $\implies$ (ii). Let $B \subset D(T)$ be bounded, and let $\overline{TB}$ be $\sigma(Y, D(T'))$-compact. Let $\{x_\alpha\}$ be a net in B. Then $\{\hat{x}_\alpha\}$ has a $\sigma(X'', X')$-convergent subnet, which we assume for simplicity to be $\{\hat{x}_\alpha\}$ itself. Let $x'' = \sigma(X'', X') - \lim \hat{x}_\alpha$. Let $y_\alpha = Tx_\alpha$. The net $\{y_\alpha\}$ has a subnet (again assumed to be itself) which is $\sigma(Y, D(T'))$-convergent to some point $y \in Y$. We have for $y' \in D(T)$,

$$y'y_\alpha = y'Tx_\alpha = Q_2(Tx_\alpha)\hat{}y' = (Q_2T''\hat{x}_\alpha)y' \to Q\hat{y}y' = y'y.$$

But by Proposition VIII.1.8,

$$(Q_2T''\hat{x}_\alpha)y' \to (Q_2T''x'')y'.$$

Thus $(\hat{x}_\alpha, \hat{y}_\alpha) \to (x'', \hat{y})$ in the topology $\sigma(X'', X') \times \sigma(Y'', D(T'))$ of $X'' \times Y''$. Hence by Proposition VIII.1.10, $(x'', \hat{y}) \in G(T'')$. Therefore $T''x'' = \hat{y} + T''(0)$, or equivalently, $x'' \in (T'')^{-1}(\hat{y})$. Condition (i) now gives $x'' = \hat{e}$ for some $e \in E$, proving that $\{x_\alpha\}$ has a $\sigma(E, X')$ convergent subnet.

(ii) $\implies$ (i). Assume (ii) holds. As in the second part of the proof of Theorem VIII.5.7, we first suppose that T is continuous. We shall verify that

$$\overline{TB'}\hat{} \subset T''(\hat{E}). \tag{1}$$

Let $y \in \overline{TB'}$ and choose a sequence $\{x_n\}$ in $B_{D(T)}$ such that $Tx_n \to y$. Then $\{Tx_n\}$ is relatively compact. By (ii), $\{x_n\}$ is relatively $\sigma(E, X')$-compact and hence relatively $\sigma(E, X')$-sequentially compact. Thus there exists $e \in B_E$ and a subsequence of $\{x_n\}$ which is $\sigma(E, X')$-convergent to e. Then by continuity, $T''\hat{e} = \hat{y}$ (as $D(T') = Y'$, so T'' is single valued) and so (1) is established.

Now the norm and $\sigma(Y, Y')$ closures of TB' coincide. Hence by Proposition VIII.1.13,

$$\hat{Y} \cap T''B_{D(T'')} \subset (\sigma(Y, Y') - \operatorname{cl} TB_{D(T)})\hat{} = \overline{TB'}\hat{}. \tag{2}$$

From (1) and (2), we have

$$\hat{Y} \cap R(T'') \subset T''(\hat{E}) \tag{3}$$

where by I.3.1(d), $T''(T'')^{-1}(\hat{Y}) = \hat{Y} \cap R(T'') + T''(0)$. Hence

$$(T'')^{-1}(\hat{Y}) \subset \hat{E} + N(T''). \tag{4}$$

Therefore to establish (i) it only remains to prove that $N(T'') \subset \hat{E}$. We use the characterization of Theorem VIII.5.7. Let $\{z_n\}$ be a bounded sequence in $D(T)$ such that $\{Tz_n\}$ is weakly convergent to zero. Then by (ii), $\{z_n\}$ is relatively $\sigma(E, X')$ compact, and hence arguing as above, $\{z_n\}$ has a $\sigma(E, X')$-convergent subsequence. Therefore $N(T'') \subset \hat{E}$ (Theorem VIII.5.7), showing that (ii) $\Longrightarrow$ (i) in the case when T is continuous.

Now let T be an arbitrary single valued relation. It will be assumed without loss of generality that T is closable (as was done in the proof of VIII.5.7), so that Y_0 becomes a normed space. Let B be a bounded subset of $D(T)$ such that T_0B is relatively compact (in the Y_0-norm topology). Let $\{x_\alpha\}$ be a net in B. Then $\{x_\alpha\}$ has a subnet, assumed for simplicity to be itself, such that for some $y \in Y$,

$$\sup_{y' \in D(T')} \frac{|y'(Tx_\alpha - y)|}{\|y'\| + \|T'y'\|} \to 0.$$

Hence

$$y'(Tx_\alpha - y) \to 0 \qquad (y' \in D(T'))$$

i.e. $\{Tx_\alpha\}$ is $\sigma(Y, D(T'))$-convergent to y. Therefore by (ii), B is relatively $\sigma(E, X')$ compact showing that the continuous linear operator T_0 satisfies the weaker condition (iii). (To see this note that since T_0 is single valued and continuous, we have $\|\cdot\| = \|\cdot\|_{D(T_0')}$.) Assuming for the moment that (iii) $\Longrightarrow$ (i) holds we conclude that

$$(T_0'')^{-1}(\hat{Y}) \subset \hat{E}.$$

It only remains to verify that $(T'')^{-1}(\hat{Y}) \subset (T_0'')^{-1}(\hat{Y})$. Let $x'' \in (T'')^{-1}(\hat{y})$ (so $T''x'' = \hat{y} + T''(0)$). Then for $y' \in Y_0' = D(T_0') \subset D(T')$, we have $y'y = \hat{y}y' = T''x''y' = x''T'y' = x''T_0'y'$ (as $T_0' = T'J_{Y_0'}^{Y'}$) $= T_0''x''y'$. Therefore $(\hat{y} - T_0''x'')y' =$

$0 \ (y' \in D(T_0'))$, i.e. $T_0''x'' = \hat{y}$, or equivalently, $x'' \in (T_0'')^{-1}(\hat{y})$. This completes the proof of (ii) $\Longrightarrow$ (i).

Finally, consider the case when T is continuous. Since (ii) obviously implies (iii), it only remains to verify the implication (iii) $\Longrightarrow$ (i). Assume that (iii) holds. Then, as in the above proof of (ii) $\Longrightarrow$ (i) (where it was assumed that T is continuous), we obtain the inclusion (4). Then by the characterisation (iii) of Theorem VIII.5.7, we have $N(T'') \subset \hat{E}$. Therefore T is E-Tauberian, as required. □

VIII.5.11 Corollary. *The following properties are equivalent:*

(i) *T is Tauberian (strongly Tauberian)*

(ii) *T maps bounded non relatively $\sigma(\tilde{X}, X')$ (resp. $\sigma(D(T), X')$) compact sets onto non relatively $\sigma(Y, D(T'))$ compact sets*

VIII.5.12 Corollary. *Let T be Tauberian and let $S \in LR(Z, X)$ be a bounded linear operator such that $J_X^{\tilde{X}} S$ is Tauberian. Then TS is Tauberian.*

PROOF. Let B be a bounded subset of $D(TS)$ such that $TS(B)$ is relatively $\sigma(Y, D((TS)'))$ compact. Now $D(T') \subset D((TS)')$; indeed, if $y' \in D(T')$, then $y'T$ is single valued and continuous on $D(T)$ and hence $y'TS$ is single valued and continuous on $D(TS)$, so $y' \in D((TS)')$. Hence $TS(B)$ is relatively $\sigma(Y, D(T'))$ compact. Therefore $S(B)$ is relatively $\sigma(\tilde{X}, X')$ compact by Corollary VIII.5.11. Since $J_X^{\tilde{X}} S$ is Tauberian, it follows that B is $\sigma(\tilde{Z}, Z')$ compact. Therefore TS is Tauberian (Corollary VIII.5.11). □

VIII.5.13 Corollary. *Let T be strongly Tauberian and $S \in LR(Z, X)$ a bounded strongly Tauberian linear operator. Then TS is strongly Tauberian.*

VIII.5.14 Corollary. *Let $S \in LR(Y, Z)$ be a bounded linear operator. If ST is E-Tauberian, then so is T.*

In particular if $Y \subset Z$ and $J_Y^Z T$ is E-Tauberian, then T is E-Tauberian.

PROOF. Let ST be E-Tauberian and let B be a bounded set such that $T(B)$ is relatively $\sigma(Y, D(T'))$ compact. We claim that $ST(B)$ is relatively $\sigma(S(Y), D((ST)'))$ compact. Let $\{x_\alpha\}$ be a net in B. By assumption, $\{QTx_\alpha\}$ has a subnet, assumed for simplicity to be itself, such that $y'Tx_\alpha (= y'QTx_\alpha) \to y'y$ for some $y \in Y$ and every $y' \in D(T')$. Now $(ST)' = T'S'$ since S is bounded and single valued (see III.1.6(a)) and therefore $D((ST)') = (S')^{-1}(D(T'))$. Hence for $z' \in D((ST)')$ we have $S'z' \in D(T')$ and so

$$S'z'Tx_\alpha = z'STx_\alpha \to S'z'y = z'Sy$$

showing that $ST(B)$ is $\sigma(SY, D((ST)'))$ relatively compact as claimed. Therefore B is $\sigma(E, X')$ relatively compact, since ST is E-Tauberian. The result now follows from Theorem VIII.5.10. □

VIII.5.15 Corollary. *Let $N(T)$ be relatively closed in $D(T)$. If T is a continuous E-Tauberian relation, then $Q_{N(T)}^{D(T)}$ is E-Tauberian.*

VIII.5.16 Corollary. *The operator T is E-Tauberian if and only if its regular contraction $Q_{D(T')^\top}T$ is E-Tauberian.*

PROOF. Sufficiency is contained in Corollary VIII.5.14. For the converse, assume that T is E-Tauberian, set $S := Q_{D(T')^\top}T$. Let B be a bounded subset of $D(S) = D(T)$ for which $S(B)$ is $\sigma(Q_{D(T')^\top}Y, D(S'))$ relatively compact and let $\{x_\alpha\}$ be a net in B. Then $\{x_\alpha\}$ has a subnet, for simplicity assumed to be itself, such that $\{Sx_\alpha\}$ is $\sigma(Q_{D(T')^\top}Y, D(S'))$ convergent to some point $Q_{D(T')^\top}y$, where $y \in Y$. Then for $y' \in D(T') = D(S')$ we have

$$y'Tx_\alpha = y'Sx_\alpha \text{ (as } y' \in D(T')^{\top\perp}) \to y'y.$$

If $y_\alpha \in Tx_\alpha$, then $y'Tx_\alpha = y'y_\alpha$, and thus $\sigma(Y, D(T') - \lim y_\alpha = y$. Therefore $T(B)$ is $\sigma(Y, D(T'))$ relatively compact, whence B is $\sigma(E, X')$ relatively compact. Consequently, S is E-Tauberian by Theorem VIII.5.10. □

VIII.5.17 Exercise Show that

$$\tilde{T} \text{ Tauberian} \implies T \text{ Tauberian}.$$

VIII.5.18 Lemma. *Let M be a linear subspace of X. Then*

(a) *$J_M^{,}$ is Tauberian*

(b) *if M is closed, then $J_M^{,}$ is strongly Tauberian.*

PROOF. Write $J := J_M^{,}$ and let $J''m'' = \hat{x}$ where $m'' \in M''$ and $x \in X$. Then $x \in \overline{R(J)} = \overline{M}$. Let $\{m_k\}$ be a sequence in M such that $m_k \to x$. Then $J''(\hat{m}_k) = \hat{m}_k \to \hat{x} = J''m''$, whence $J''(\hat{m}_k - m'') \to 0$. But J'' is injective and open (see e.g. the State Diagram III.6.3). Hence $\hat{m}_k - m'' \to 0$. Therefore $m'' = \hat{x}$, showing that J is Tauberian, and strongly Tauberian in case M is closed. □

VIII.5.19 Proposition.

(a) *The restriction of a Tauberian relation to any linear subspace is Tauberian.*

(b) *The restriction of a strongly Tauberian relation to any closed subspace is strongly Tauberian.*

PROOF. Let M be a linear subspace of X. Then $J_{M\cap D(T)}^{\sim}$ is Tauberian by Lemma VIII.5.18. Hence if T is Tauberian, then so is $TJ_{M\cap D(T)}$ by Corollary VIII.5.12. An easy check shows that $D((T|_M)') = D((TJ_{M\cap D(T)})')$. It now follows from Theorem VIII.5.10 that $T|_M$ is Tauberian. □

VIII.5.20 Proposition. *Let X be prereflexive (resp. $D(T)$ reflexive) and Y complete. Then T is Tauberian (resp. strongly Tauberian).*

PROOF. We have $(T'')^{-1}(\hat{Y}) \subset D(T'') \subset D(T)^{\perp\perp} = \tilde{D}(T)\hat{}$ (resp., $D(T)\hat{}$). □

VIII.5.21 Corollary. *Let $D(T)$ be prereflexive. Then*

$$R(T'') \cap \hat{Y} \subset T''(\tilde{D}(T)\hat{}).$$

PROOF. By Proposition VIII.5.20, T is Tauberian, i.e.

$$(T'')^{-1}(\hat{Y}) \subset \tilde{D}(T)\hat{}.$$

Hence by Proposition I.3.1(d)

$$R(T'') \cap \hat{Y} + T''(0) \subset T''(\tilde{D}(T)\hat{}). \quad \square$$

VIII.5.22 Exercise Let Y be reflexive. Show that T is strongly Tauberian if and only if $D(T)$ is a reflexive Banach space.

VIII.6 GEOMETRIC PROPERTIES OF TAUBERIAN RELATIONS

VIII.6.1 Theorem. *Let $D(T) \subset E \subset \tilde{D}(T)$. If T is an E-Tauberian relation, and if C is a bounded convex subset of $D(T)$ closed in E, then $T(C)$ is $\sigma(Y, D(T'))$- sequentially closed.*

PROOF. Since T is E-Tauberian if and only if QT is E-Tauberian, and since TC is $\sigma(Y, D(T'))$ sequentially closed if and only if $QT(C)$ is $\sigma(QY, D(T'))$ sequentially closed (as $y'QTx = y'Tx$ for $y' \in D(T')$ and $x \in D(T)$), we shall suppose without loss of generality that T is single valued.

First assume that T is closable. Let $\{x_n\}$ be a sequence in C such that $\{Tx_n\}$ is $\sigma(Y, D(T'))$ convergent to some point $y \in Y$. Since $\{Tx_n\}$ is a relatively $\sigma(Y, D(T'))$ compact subset of Y, it follows from Theorem VIII.5.10 that $\{x_n\}$ is relatively $\sigma(E, X')$ compact and hence relatively sequentially $\sigma(E, X')$ compact. Accordingly $\{x_n\}$ has a subsequence, for simplicity assumed to be itself, which is $\sigma(E, X')$ convergent to some point $x \in E$. Since C is convex and closed in E, we have $x \in C$. Choose $y_n'' \in T''\hat{x}_n$ arbitrarily. Then $(\hat{x}_n, y_n'')$ is a sequence in $G(T'')$ which is $\sigma(X'', X') \times \sigma(Y'', D(T'))$ convergent to $(\hat{x}, \hat{y})$. Hence by Proposition VIII.1.10, $(\hat{x}, \hat{y}) \in G(T'')$. Thus $\hat{x} \in D(T'')$ and $\hat{y} \in T''\hat{x}$, i.e. $T''\hat{x} = \hat{y} + T''(0)$. Since $x \in D(T)$, it easily follows that $y'Tx = y'y$ for $y' \in D(T')$. Hence $Tx - y \in D(T')^{\top} = T(0) = 0$, since T is closable. Thus $y = Tx \in T(C)$.

Now let T be an arbitrary single valued relation and let $S := Q_{D(T')^{\top}}T$. By Corollary VIII.5.16, S is E-Tauberian. Since S is closable (by III.4.11), the set $S(C)$ is $\sigma(Q_{D(T')^{\top}}Y, D(T'))$-sequentially closed. The theorem now follows from the observation that $y'Sx = y'Tx$ for $y' \in D(T')$ and $x \in D(T)$. □

VIII.6.2 Proposition. *Let $D(T) \subset E \subset \tilde{D}(T)$, let $T(B$') be $\sigma(Y, D(T'))$ sequentially closed, and let $N(T'') \subset \hat{E}$. Then T is E-Tauberian.*

PROOF. Let B be a subset of $B_{D(T)}$ such that $T(B)$ is relatively $\sigma(Y, D(T'))$ compact. Let $\{x_n\}$ be a sequence in B, and let $y_n \in Tx_n$ $(n \in \mathbb{N})$. Passing to a subsequence if necessary, we may assume that $\{y_n\}$ is $\sigma(Y, D(T'))$ convergent. Then there exists $x \in B_{D(T)}$ such that $y'(Tx_n - Tx) \to 0$ for all $y' \in D(T')$.

Hence by Theorem VIII.5.7, $\{x_n - x\}$ has a $\sigma(E, X')$ convergent subsequence. Thus B is relatively sequentially $\sigma(E, X')$ compact, and hence relatively $\sigma(E, X')$ compact (see Floret [1]). Therefore T is E-Tauberian, by Theorem VIII.5.10. □

VIII.6.3 Theorem. *The following properties are equivalent:*

(i) *T is strongly Tauberian*
(ii) *$N(T'') \subset D(T)\hat{}$, and for every bounded convex relatively closed subset of $D(T)$, $T(C)$ is $\sigma(Y, D(T'))$ sequentially closed*
(iii) *$N(T'') \subset D(T)\hat{}$, and $T(B')$ is $\sigma(Y, D(T'))$ sequentially closed.*

PROOF. We have (i) $\Longrightarrow$ (ii) by TheoremVIII.6.1, and (iii) $\Longrightarrow$ (i) by Proposition VIII.6.2, while the implication (ii) $\Longrightarrow$ (iii) is trivial. □

VIII.6.4 Corollary. *Let T be strongly Tauberian. Then $T(B')$ is closed.*

VIII.6.5 Corollary. *T' is strongly Tauberian if and only if $N(T''') \subset D(T')\hat{}$.*

PROOF. By Corollary VIII.1.11, $T'B_{D(T')}$ is $\sigma(X', D(T))$ closed, therefore $\sigma(X', D(T''))$ closed. Hence the result follows by Theorem VIII.6.3. □

VIII.7 OPEN TAUBERIAN RELATIONS

In this section we shall be studying Tauberian properties under the additional assumption that $\gamma(T) > 0$ (or $\gamma(T') > 0$). Great simplification results. For closed relations in Banach spaces, either assumption is equivalent to assuming that the range of T is complete.

VIII.7.1 Theorem. *Let $\gamma(T') > 0$. Then T is Tauberian if and only if $N(T'') \subset \tilde{D}(T)\hat{}$.*

PROOF. Only sufficiency remains to be established. Assume that $N(T'') \subset \tilde{D}(T)\hat{}$ and let $x'' \in (T'')^{-1}(\hat{y})$ where $y \in Y$. Since $\hat{y} \in T''x''$, there exists (by VIII.1.14) a sequence $\{x_n\}$ in $D(T)$, and $y_n \in Tx_n$ $(n \in \mathbb{N})$, such that $\|y_n - y\| \to 0$. But $\gamma(T'') = \gamma(T') > 0$, so $(T'')^{-1}$ is continuous. Consequently $\lim \|(T'')^{-1}(\hat{y}_n - \hat{y})\| = 0$. Therefore

$$(1) \qquad \lim d(\hat{x}_n - x'', N(T'')) = 0.$$

But $N(T'') \subset \tilde{D}(T)\hat{}$. Hence $x'' \in \tilde{D}(T)\hat{}$, showing that T is E-Tauberian. □

If $R(T)$ is complete, then $\gamma(T') > 0$ and so Theorem VIII.7.1 is applicable. However, this case is covered under a weaker assumption.

VIII.7.2 Proposition. *Let T have closed range. Then T is E-Tauberian if and only if $N(T'') \subset \hat{E}$ (where $D(T) \subset E \subset \tilde{D}(T)$).*

PROOF. Necessity is immediate. Conversely assume that $N(T'') \subset \hat{E}$, and let $x'' \in (T'')^{-1}(\hat{y})$ where $y \in Y$. Then $\hat{y} \in T''x''$, and so by Lemma VIII.1.14, $\hat{y} \in R(T)\hat{}$ (since $R(T)$ is closed). Thus $y \in Tx$ for some $x \in D(T)$. Then $Q_2T''x'' = Q_2\hat{y} = Q_2(Tx)\hat{} = Q_2T''\hat{x}$, whence $Q_2T''(x'' - \hat{x}) = 0$, i.e. $x'' - \hat{x} \in N(T'') \subset \hat{E}$. Therefore $x'' \in \hat{E}$, showing that T is E-Tauberian. □

VIII.7.3 Exercise Verify VIII.7.2 indirectly by taking $R(T) = Y$, and then again by considering $T := JT_1$ where $J := J^Y_{R(T)}$ and $T_1 = J^{-1}T$.

VIII.7.4 Corollary. *Let N be a closed subspace of X. Then*
(a) Q'_N *is Tauberian if and only if N is prereflexive*
(b) Q'_N *is strongly Tauberian if and only if N is reflexive.*

PROOF. (a) We have $(Q'_N)'' = Q'''_{N^{\perp\perp}}$ with kernel $N^{\perp\perp}$. Now N is prereflexive if and only if $\tilde{N}\hat{} = N^{\perp\perp}$. Hence if N is prereflexive, then Q'_NT is Tauberian, by Proposition VIII.7.2. The converse follows from Corollary VIII.5.8.

The proof of (b) is similar. □

VIII.7.5 Proposition. *We have the implications*
$N(T'') \subset D(T)\hat{} \implies T^{-1}(\overline{T}(0))$ *is reflexive* $\implies \overline{N(T)} \cap D(T)$ *is reflexive.*

PROOF. Write $N := T^{-1}(\overline{T}(0))$. Suppose that $N(T'') \subset D(T)\hat{}$ and let $\{x_\alpha\}$ be a net in B_N. Then for $y' \in D(T')$ we have

$$0 = y'QTx_\alpha = y'Tx_\alpha = T''\hat{x}_\alpha y' \tag{1}$$

whence $T''\hat{x}_\alpha \subset D(T')^\perp = T''(0)$, i.e. $\hat{x}_\alpha \in N(T'')$. Since $\{\hat{x}_\alpha\}$ is bounded, it has a $\sigma(X'', X')$ convergent subnet (by the Banach-Alaoglu Theorem) which for simplicity we assume to be itself. Let $\sigma(X'', X') - \lim \hat{x}_\alpha = x'' \in B_{X''}$. Since $N(T'')$ is $\sigma(X'', X')$ closed, we have $x'' \in N(T'') \subset D(T)\hat{}$. Thus $x'' = \hat{x}$ where $x \in D(T)$. Passing to the limit in (1) now gives

$$\lim y'Tx_\alpha = 0 = (T''x'')y' = y'Tx.$$

Hence $Tx \subset D(T')^\top = \overline{T}(0)$, whence $x \in T^{-1}(\overline{T}(0))$. Therefore $x \in B_N$, where $x = \sigma(X, X') - \lim x_\alpha$. Thus B_N is weakly compact. Hence N is reflexive.

Since $N(T)$ is a subspace of $N(QT)$, it follows that if $T^{-1}(\overline{T}(0))$ is reflexive, then the relative closure of $N(T)$ in $D(T)$ is reflexive. □

VIII.7.6 Proposition. *Let $\gamma(T) > 0$. Then the following properties are equivalent:*

(i) *$N(T)$ is prereflexive (resp. $N(\tilde{T})$ is reflexive)*
(ii) *$N(T'') = \tilde{N}(T)^\wedge$ (resp. $N(\tilde{T})^\wedge$)*
(iii) *$N(T'') \subset \tilde{D}(T)^\wedge$ (resp. $D(\tilde{T})^\wedge$).*

PROOF. We have $R(T') = N(T)^\perp$ (Proposition III.4.6(b)). Hence the equivalence

$$N(T) \text{ prereflexive} \iff \tilde{N}(T)^\wedge = N(T)^{\perp\perp} = R(T')^\perp = N(T'').$$

Therefore (i) $\iff$ (ii).

Next, (iii) $\implies$ (i) by Theorem VIII.7.1 (resp. Proposition VIII.7.2) and Corollary VIII.5.8.

Since (ii) $\implies$ (iii) trivially, the proof is complete. □

VIII.7.7 Theorem. *Let $\gamma(T) > 0$ (resp. $\gamma(\tilde{T}) > 0$). Then T is Tauberian (resp. $\tilde{T}$ is strongly Tauberian) if and only if $N(T)$ is prereflexive (resp. $N(\tilde{T})$ is reflexive). In that case all the even adjoints $T'', T^{(4)}, T^{(6)}, \ldots$ are strongly Tauberian.*

PROOF. Only sufficiency remains to be established. Let $N(T)$ be prereflexive. Then T is Tauberian, by Proposition VIII.7.6 and Theorem VIII.7.1. The second case is similar. Since $\gamma(T) > 0$ implies that $\gamma(T'') > 0$, the final assertion is clear. □

An example of a non Tauberian bounded linear operator whose second adjoint is Tauberian has been constructed by T. Alvarez and M. Gonzalez [AG].

VIII.7.8 Corollary. *Let $\gamma(T') > 0$. Then T' is Tauberian if and only if T' is strongly Tauberian.*

VIII.7.9 Proposition. *Let $\overline{R(T')}$ be weak*-closed. Then the following properties are equivalent:*

(i) *$N(T'') \subset D(T)^\wedge$*
(ii) *$T^{-1}(\overline{T}(0))$ is reflexive*
(iii) *$N(T'') = (T^{-1}\overline{T}(0))^\wedge$.*

PROOF. (i) $\Longrightarrow$ (ii). This is contained in Proposition VIII.7.5.

(ii) $\Longrightarrow$ (iii). Let $T^{-1}\overline{T}(0)$ be reflexive. Since $(TJ_{D(T)})'' = T''J_{D(T'')^{\top\perp}}$, we may suppose that $D(T) = X$. Then we have (see III.2.3)

$$T^{-1}(\overline{T}(0)) = T^{-1}(D(T')^{\top}) = (T'D(T'))^{\top} \cap D(T) = R(T')^{\top} \text{ (as } D(T) = X \text{).}$$

Hence

$$(T^{-1}\overline{T}(0))\hat{} = (T^{-1}\overline{T}(0))^{\perp\perp} \text{ (by reflexivity) } = R(T')^{\top\perp\perp} = R(T')^{\perp}$$

(since $\overline{R(T')}$ is weak*-closed) $= N(T'')$.

Since (iii) $\Longrightarrow$ (i) trivially, the proof is complete. □

VIII.7.10 Exercise Let P be a linear projection defined on a subspace of X. Then P is Tauberian if and only if the kernel of P is prereflexive.

VIII.7.11 Exercise Let $X = \ell_1 \times \ell_2$ and let P be the bounded projection defined on X with kernel $\{0\} \times \ell_2$ and range $\ell_1 \times \{0\}$. Show that P is Tauberian and $I - P$ is weakly compact.

VIII.7.12 Exercise Let S and T be linear operators in $LR(X)$ such that T is open, $S = T - I$ and S has prereflexive range. Show that T is Tauberian and S is weakly compact.

VIII.7.13 Exercise Let $\gamma(T) > 0$, and let $N(T)$ be prereflexive. Show that $\tilde{N}(T)\hat{} = N(\tilde{T})\hat{} = N(T'')$.

VIII.8 FURTHER TAUBERIAN PROPERTIES

VIII.8.1 Theorem. *Let T be a bounded strongly Tauberian relation, and let R be a reflexive subspace of Y. Then $T^{-1}(R)$ is reflexive.*

PROOF. (a) First suppose that T is single valued. Since Q_R^Y is strongly Tauberian by Theorem VIII.7.4, therefore $Q_R^Y T$ is strongly Tauberian by Corollary VIII.5.13. Hence $\overline{N(Q_R^Y T)}$ is reflexive (Theorem VIII.5.8), i.e. $\overline{T^{-1}(R)}$ is reflexive.

For the general case, since QT is bounded and strongly Tauberian, we have $(QT)^{-1}(QR)$ reflexive. It follows that $T^{-1}(R)$ is prereflexive. Finally since T is bounded, $T^{-1}(R)$ is closed. Therefore $T^{-1}(R)$ is reflexive. □

VIII.8.2 Exercise Show that in Theorem VIII.8.1, if R is assumed only prereflexive, and Y is complete, then $T^{-1}(R)$ will be reflexive.

VIII.8.3 Remark The converse of Theorem VIII.8.1 is false even for bounded linear operators in Banach spaces. For example, let T be a compact injective operator from c_0 into itself. Then T has the stated property but is not Tauberian since $T''(\ell_\infty) \subset c_0$.

VIII.8.4 Theorem. *Let T be an F_+-relation. Then T is Tauberian.*

PROOF. If $\gamma(T) > 0$, then T is Tauberian, by Theorem VIII.7.7 (see Exercise VIII.7.13). Passing now to the general case, we may suppose without loss of generality that $\operatorname{codim} R(T) = \infty$ (see Corollary VIII.5.14). By Lemma V.10.1(b) and Corollary II.6.4, there exists a bounded linear operator F with finite dimensional range such that $\gamma(T + F) > 0$. Since $T + F \in F_+$, therefore $T + F$ is Tauberian by what has been proved. Hence T is Tauberian, by the perturbation result VIII.5.4. □

VIII.8.5 Definition A normed space X is said to be **very irreflexive** (VIR) if no infinite dimensional subspace of X contains an infinite dimensional prereflexive subspace.

VIII.8.6 Proposition. *Let $D(T)$ be VIR and let Y be complete. Then the following properties are equivalent:*

(i) *T is Tauberian*
(ii) *$T \in F_+$*

PROOF. Assume that $T \notin F_+$. Then there exists an infinite dimensional subspace M of $D(T)$ for which $T|_M$ is precompact (Theorem V.1.6). Hence TB_M is a relatively compact subset of Y, hence relatively $\sigma(Y, D(T'))$ compact. But B_M is not relatively weakly compact, since $\tilde{M}$ is not reflexive. Consequently $T|_M$, and hence T, is not Tauberian by Corollary VIII.5.11. Hence if T is Tauberian then $T \in F_+$. The converse implication follows from Theorem VIII.8.4. □

VIII.8.7 Theorem. *Consider the following properties:*

(i) *T is nowhere Tauberian*
(ii) *T is strictly singular and $D(T)$ is VIR.*

Then (i) $\Longrightarrow$ (ii), and if Y is complete, then (i) and (ii) are equivalent.

PROOF. (i) $\Longrightarrow$ (ii). Let T be nowhere Tauberian. Then clearly $D(T)$ is VIR. Let M be an infinite dimensional subspace of $D(T)$. Then $T|_M$ is not an F_+-relation, by Theorem VIII.8.4. Hence M contains an infinite dimensional subspace N such that $T|_N$ is precompact (Theorem V.1.6). Thus T is ubiquitously precompact, hence strictly singular by Theorem V.4.5.

Now let Y be complete and assume that (ii) holds. Let M be an infinite dimensional subspace of $D(T)$ for which $T|_M$ is Tauberian. By Proposition VIII.8.6, $T|_M \in F_+$. The strict singularity of T now implies that M is finite dimensional. Hence (ii) $\Longrightarrow$ (i). □

VIII.8.8 Proposition. *Let $D(\overline{T}) \subset E \subset \tilde{D}(T)$. Then T is E-Tauberian if and only if $\overline{T}$ is E-Tauberian.*

PROOF. Follows immediately from the equality $T'' = \overline{T}''$. □

VIII.8.9 Corollary. *T is Tauberian if and only if $\overline{T}$ is Tauberian.*

VIII.8.10 Corollary. *Let Y be complete. Then T is Tauberian if and only if $\tilde{T}$ is Tauberian.*

VIII.8.11 Exercise Verify that if T is strongly Tauberian and Y is complete then $\tilde{T}$ is strongly Tauberian.

Our next objective is to obtain a perturbative characterisation of continuous Tauberian relations in Banach spaces (Theorem VIII.8.15). For this we shall need to recall the notion of "basic sequence" which we now define.

VIII.8.12 Definitions A sequence $\{b_n\}$ in X is called a **Schauder basis** (or **basis**) for X if each $x \in X$ has a unique representation $x = \sum_{n=1}^{\infty} \alpha_n b_n$.

If $\{b_n\}$ is a basis and $x = \sum_{n=1}^{\infty} \alpha_n b_n$, then the linear functionals $b'_n : X \to \mathbb{K}$ defined by $b'_n(x) = \alpha_n \quad (n \in \mathbb{N})$ are called the **coefficient functionals** (or **biorthogonal functionals**) associated with $\{b_n\}$. We clearly have

$$b'_n(b_k) = \delta_{nk} \tag{1}$$

$$x = \sum_{n=1}^{\infty} b'_n(x) b_n. \tag{2}$$

In case the space X is complete, the coefficient functionals $\{b'_n\}$ are continuous (in fact, there is a constant K, called the "basis constant", such that $\|b'_n\| \leq 2K / \|b_n\| \quad (n \in \mathbb{N})$) (see Lindenstrauss and Tzafriri [1; p.7]).

A sequence $\{x_n\}$ in X which is a basis for its closed linear span is called a **basic sequence**. The following result is due to Kadec and Pelczynski [1] (see e.g. Singer [3; p.53]).

VIII.8.13 Lemma. *Let X be a Banach space. If $\{x_n\}$ is a bounded sequence in X having no weakly convergent subsequence, then $\{x_n\}$ has a basic subsequence.*

VIII.8.14 Definitions Let X and Y be seminormed spaces. A bounded linear operator $S : X \to Y$ is called **nuclear** if there exists a sequence $\{x'_n\}$ in X' and a sequence $\{y_n\}$ in Y such that

$$\sum_{n=1}^{\infty} \|x'_n\| \, \|y_n\| < \infty \tag{1}$$

$$Sx = \sum_{n=1}^{\infty} x'_n(x) y_n \qquad (x \in X) \tag{2}$$

We say that S is **precompact** if $S(B')$ is a totally bounded subset of Y (as for the normed space case). It is easy to see that if S is nuclear then S is uniformly approximable by the bounded finite rank operators S_n defined by

$$S_n x := \sum_{k=1}^{n} x_k^*(x) y_k.$$

Consequently, if S is nuclear then S is precompact (see for example the proof of Goldberg [3; III.1.5]). Again, as for the normed space case, the classes of compact and precompact linear operators coincide in the case when Y is complete.

VIII.8.15 Theorem. *Let Y be complete and let T be continuous. Then the following properties are equivalent:*

(i) *T is Tauberian*

(ii) *$N(T+S)$ is prereflexive for every bounded weakly compact relation S with $S(0) \subset T(0)$*

(iii) *$N(T+S)$ is prereflexive for every bounded linear relation S with $S(0) \subset T(0)$ which is nuclear when considered as a relation from X into $Y_{D(T')}$ (i.e., Y equipped with the $D(T')$-seminorm).*

PROOF. We have (i) $\Longrightarrow$ (ii) by Theorem VIII.5.4 and Corollary VIII.5.8. Since nuclear operators are compact, and hence weakly compact and since $Q_T S$ is weakly compact whenever S is weakly compact with $S(0) \subset T(0)$, we have (ii) $\Longrightarrow$ (iii).

(iii) $\Longrightarrow$ (i). Suppose that T is not Tauberian. Then by Theorem VIII.5.10, there exists a bounded non relatively $\sigma(\tilde{X}, X')$ compact subset B of $D(T)$ for which $T(B)$ is relatively $D(T')$-seminorm compact. Let $\{x_n\}$ be a bounded sequence in B without $\sigma(\tilde{X}, X')$ convergent subsequences such that

$$\sup_{y' \in B_{D(T')}} |y'Tx_n - y'y| \to 0.$$

Let P be a linear projection defined on $R(T)$ with kernel $T(0)$. Then PT is a selection of T, and so $PTx \in Tx$ for each $x \in D(T)$ (see Proposition I.5.2). We now have

$$\lim \|PTx_n - y\|_{D(T')} = 0. \tag{1}$$

By Lemma VIII.8.13, $\{x_n\}$ has a basic subsequence, which for simplicty we again denote by $\{x_n\}$. Let $\{x'_n\}$ be the sequence of Hahn-Banach extensions in X' of the associated sequence of coefficient functionals. Since evidently $\inf_{n \in \mathbb{N}} \|x_n\| > 0$, the sequence $\{x'_n\}$ is bounded. Hence, by passing to a subsequence if necessary, we may suppose that

$$\|PTx_n - y\|_{D(T')} \|x'_n\| < 2^{-n}. \tag{2}$$

Define the linear relation A by

$$Ax := \sum_{n=1}^{\infty} x'_n(x)(PTx_n - y) \qquad (x \in X) \tag{3}$$

Then A is everywhere defined since $Y_{D(T')}$ is complete. (To see this observe that $QY = Y_{D(T')}$ by III.1.4(d) and II.5.5). Furthermore it is clear from (2) that A is nuclear when considered as a relation from X into $Y_{D(T')}$. We have $A(0) = \lim_{D(T')} 0 = T(0)$ with in addition

$$Q(PT - A)x_n = Qy \qquad (n \in \mathbb{N}) \tag{4}$$

Let QL be a Hahn-Banach extension of the one (or zero) dimensional operator $Q(PT - A)|_{\mathrm{sp}\{x_n\}}$ with range $\mathrm{sp}\{Qy\}$. Put

$$K := -L - A. \tag{5}$$

Then $Q(PT + K)x_n = 0$. Hence $\mathrm{sp}\{x_n\} \subset N(PT + K)$, and $N(PT + K)$ is thus non prereflexive since $\{x_n\}$ has no $\sigma(\tilde{X}, X')$ convergent subsequence. Moreover since by construction $K(0) = T(0)$ and since $Y_{D(T')} \equiv QY$ (III.1.4(d)), K is in fact continuous as a map into Y.

Now for $x \in D(T)$ we have

$$(PT + K)x = 0 \iff (T + K)x = T(0) = (T + K)(0).$$

Hence $N(PT+K) = N(T+K)$. Thus $N(T+K)$ is non prereflexive, therefore (iii) $\Longrightarrow$ (i). $\square$

VIII.8.16 Exercise Let Y be complete, and let T be continuous and single valued. Show that T is Tauberian if and only if $N(T+S)$ is prereflexive for every bounded compact linear operators S.

VIII.9 NOTES AND REMARKS

Unbounded weakly compact operators were introduced by A.I. Gouveia [1] and Cross and Gouveia [1]. The results of Sections VIII.2, VIII.3 and VIII.4 are generalisations of Alvarez, Cross and Gouveia [1] and Cross [10].

Theorem VIII.2.6 generalising Gantmacher s theorem (see Dunford and Schwartz [1; VI.4.8], or C. Swartz [1; p.426]) characterizes in particular, arbitrary densely defined linear operators having a weakly compact adjoint.

Bounded Tauberian operators in Banach spaces were studied by Kalton and Wilansky [1] (see Wilansky [2] for an elementary introduction). As remarked by Gonzalez and Onieva [3], these operators appear in summability (see Garling and Wilansky [1]), factorization of operators (see Davis, Figiel, Johnson and Pelczynski [1], and Neidinger [1]), preservation of isomorphic properties of Banach spaces (Neidinger [1]), the preservation of the closedness of images of closed sets (Neidinger and Rosenthal [1]), the equivalence between the Radon-Nikodym property and the Krein-Milman property (Schachermayer [1]), and generalised Fredholm operators (Tacon [1], Yang [1]). Other recent works are Alvarez and Gonzalez [1] (which answers a question raised in Kalton and Wilansky [1]), Gonzalez [3], Gonzalez and Martinez-Abejon [1, 2], Gonzalez and Onieva [1, 2, 3], Gonzalez, Saksman and Tylli [1], J.R. Holub [2], and Martinez-Maurica and Pellon [1]. The article by M. Gonzalez [3] surveys some applications of Tauberian operators. Tauberian operators between Fréchet spaces are discussed in Bonet and Ramanujan [1]. Unbounded Tauberian operators were introduced in Cross [9, 12] and are used in Alvarez, Cross and Gonzalez [1]. Sections VIII.5-8 generalises the results of Cross [12] to linear relations.

The perturbative characterisation of continuous Tauberian relations VIII.8.15 is the multivalued version of the corresponding result of Gonzalez and Onieva [3], namely that if X and Y are Banach spaces and $T : X \to Y$ is a bounded linear operator, then T is Tauberian if and only if $N(T+S)$ is reflexive for every compact (everywhere defined) linear operator S. (This is analogous to the corresponding characterisation of upper semi-Fredholm operators of Corollary V.7.11.)

Bibliography

AIENA, P., and GONZALEZ, M.

[1] Perturbation classes for Fredholm and semi-Fredholm operators, *Rend. Circ. Math. Palermo. Suppl.* 40(1996), 37-46.

AKIN, E.,

[1] The General Topology of Dynamical Systems, *Amer. Math. Soc.,* 1993.

ALPAY, D., BRUINSMA, P., DIJKSMA, A., and DE SNOO, H.S.V.

[1] A Hilbert space associated with Nevanlinna function, in: *Signal Processing, Scattering and Operator Theory. Proceedings International Symposium MTNS-89, Vol.III*; Editors: M.A. Kaashoek, J.H. van Schuppen, A.C.M. Ran. Progress in Systems and Control Theory, Birkhauser, Boston-Basel-Berlin, 1990, pp.115-122.

[2] Interpolation problems, extensions of symmetric operators and reproducing kernels II. *Integral Equations and Operator Theory*, 14(1991), 465-500.

[3] Interpolation problems, extensions of symmetric operators and reproducing kernel spaces II (missing section 3), *ibid* 15(1991), 378-388.

ALPAY, D., DIJKSMA, A., DE SNOO, H.S.V., and ROVNYAK, J.

[2] Schur functions, operator colligations, and reproducing kernel Pontryagin spaces, in preparation.

ALVAREZ, J., ALVAREZ, T., and GONZALEZ, M.

[1] The gap between subspaces and perturbation of non semi-Fredholm operators, *Bull. Austral. Math. Soc.*, 45(1992), 369-376.

ALVAREZ, T.

[1] State diagram for operators with null space or conull space in an ideal of Banach spaces, *Pub. Mat. UAB*, 30, no.1 (1986),85-90.

[2] A note on generalized Tauberian operators, *Quaestiones Math.*, 10(1987), 377-390.

ALVAREZ, T., and CROSS, R.W.

[1] Operational quantities characterizing upper and lower semi-Fredholm type operators, preprint.

ALVAREZ, T., CROSS, R.W., and GONZALEZ, M.

[1] Factorization of unbounded thin and cothin operators, preprint.

ALVAREZ, T., CROSS, R.W., and GOUVEIA, A.I.

[1] Adjoint characterizations of unbounded weakly compact, weakly completely continuous and unconditionally converging operators, *Studia Math.*, 113(3)(1995), 283-298.

ALVAREZ, T., and GONZALEZ, M.

[1] Some examples of Tauberian operators, *Proc. Amer. Math. Soc.*, 111(1991), 1023-1027.

ALVAREZ, T., and ONIEVA, V.M.,

[1] A note on three-space ideals of Banach spaces, in: *Proceedings of the Tenth Spanish-Portuguese Conference on Mathematics*, III (Murcia, 1985), Univ. Murcia, 1985, pp.251-254.

[2] Generalized Fredholm operators, *Arch. Math.* 44(1985), 270-277.

ANDREW, A.D., and PATTERSON, W.M.

[1] Range Inclusion and Factorization of Operators on Classical Banach Spaces, *J. Math. Anal. Appl.*, 156(1991), 40-43.

ARENS, R.

[1] Operational calculus of linear relations, *Pacific J. Math.*, 11(1961), 9-23.

ASTALA, K.

[1] On measures of noncompactness and ideal variations in Banach spaces, *Ann. Acad. Sci. Fenn. Ser. A1 Math. Dissertations*, 29(1980), 1-46.

ASTALA, K., and TYLLI, H.-O.

[1] On the bounded compact approximation property and measures of noncompactness, *J. Funct. Anal.*, 70(1987), 388-401.

[2] Seminorms related to weak compactness and to Tauberian operators, *Math. Proc. Cambridge Phil. Soc.*, 107(1990), 367-375.

ATKINSON, F.V.

[1] The normal solvability of linear equations in normed spaces, *Mat. Sbornik*, 28,(70) (1951), 3-14 (Russian).

[2] A spectral problem for completely continuous operators, *Acta Math. Acad. Sci. Hungar.*, 3(1952), 53-60.

AUBIN, J.P., and CELLINA, A.

[1] *Differential Inclusions*, Springer-Verlag, New York, 1984.

AUBIN, J.P., and FRANKOWSKA, H.

[1] *Set Valued Analysis*, Birkhauser, Boston, 1990.

AULT, D.A., DEUTSCH, F.R., and MORRIS, P.D.

[1] Interpolating subspaces in approximation theory, *J. Approx. Theory*, 3(1970), 164-182.

AUMANN, R.,

[1] Integrals of set-valued functions, *J. Math. Anal. Appl.*, 12(1965), 1-12.

AZIZOV, T.Ya., and IOKHVIDOV, I.S.

[1] *Linear Operators in Spaces with Indefinite Metric*, Wiley, 1989.

AZIZOV, T.Ya., and DIJKSMA, A.

[1] Contractive relations in Pontryagin spaces, preprint.

BANACH, S.

[1] *Theorie des Operations Lineares*, (Monografje Matematyczne, t.1), Warszawa, 1932.

BANAS, J.

[1] Applications of measures of noncompactness to various problems, *Zesz. Nauk. Polit. Rzeszow., Mat. Fiz.z.5*, No.34 (1987).

BANAS, J., and GOEBEL, K.

[1] *Measures of Noncompactness in Banach Spaces*, Dekker, New York, 1981.

BANAS, J., and KNAP, Z.

[1] Measures of weak noncompactness and nonlinear integral equations of convolution type, *J. Math. Anal. Appl.*, 146(1990), 353-362.

BANAS, J., and RIVERO, J.

[1] On measures of weak noncompactness, *Ann. Mat. Pura Appl.*, (IV) 151(1988), 213-224.

BENNEWITZ, C.

[1] Symmetric relations on a Hilbert space, *Conference on the Theory of Ordinary and Partial Differential Equations, Dundee, Scotland, 1972*. Lecture Notes in Mathematics 280, Springer-Verlag, Berlin, 1972, 212-218.

BERG, I.D.

[1] A Banach algebra criterion for Tauberian theorems, *Proc. Amer. Math. Soc.*, 15(1964), 648-652.

BERGE, C.

[1] *Topological Spaces including a treatment of Multi-Valued Functions*, Vector Spaces and Convexity, Oliver and Boyd, London, 1963.

BERKSON, E.

[1] Some Metrics on the Subspaces of a Banach Space, *Pacific J. Math.*, 13(1963), 7-22.

BERMUDEZ, T.

[1] *A local functional calculus for operators* (Spanish), Doctoral dissertation, Univ. La Laguna, 1996.

BERMUDEZ, T., GONZALEZ, M., and MARTINON, A.

[1] Stability of the local spectrum, *Proc. Amer. Math. Soc.*, 125(1997), 417-425.

[2] On the poles of the local resolvent, *Math. Nachr.* (to appear).

[3] Properties and applications of the local functional calculus, preprint (1996).

BESPALOV, Yu.N., SAMOILENKO, Yu.S., and SHULMAN, V.S.

[1] Sets of operators connected by semilinear relations (Russian), *Applications of functional analysis in Mathematical physics* (Russian), 28-51, Akad. Nauk Ukrainy, Inst. Mat., Kiev, 1991.

BESSAGA, C., and PELCZYNSKI, A.

[1] On bases and unconditional convergence of series in Banach spaces, *Studia Math.*, 17(1958), 151-164.

BEUCHER, O.J.

[1] On certain quantities in Fredholm-operator theory and Milman's isometry spectrum, *Suppl. Rendiconti Circolo Mat. Palermo*, Serie II, 10(1985), 17-24.

BOGACHEV., V.I., KIRCHEIM, B., and SCHACHERMEYER, S.

[1] Continuous Restrictions of Linear Maps between Banach Spaces, *Acta Univ. Carolin. Math. Phys.*, 30(1989), 31-35.

BONET, J., and RAMANUJAN, M.S.

[1] Two classes of operators between Fréchet spaces, *Functional Analysis (Trier,* 1994), 53-68, de Gruyter, Berlin, 1996.

BORISOVICH, Yu.G., GELMAN, B.D., MYSHKIS, A.D., and OBUKHOVSKII, V.V.

[1] Multivalued analysis and operator inclusions (Russian). *Current problems in mathematics. Newest results*, Vol.29 (Russian), 151-211, 215, Itogi Nauki i Tekhniki, Akad. Nauk SSSR, Vsesoyuz. Inst. Nauchn. Inform., Moscow, 1986.

BORWEIN, J.M., and ZHUANG, D.M.

[1] Verifiable Necessary and Sufficient Conditions for Openness and Regularity of Set-Valued and Single-Valued Maps, *J. Math. Anal. Appl.*, 134(1988), no.2, 441-459.

BOULMAAROUF, Z., and LABROUSSE, J.-Ph.

[1] The Cayley transform of linear relations, *J. Egyptian Math. Soc.*, 2(1994), 53-65.

BOURGAIN, J., and DIESTEL, J.

[1] Limited operators and strict cosingularity, *Math. Nachr.*, 119(1984), 55-58.

BRACE, J.W., and ROYCE-KNEECE, R.

[1] Approximation of strictly singular and strictly cosingular operators using nonstandard analysis, *Trans. Amer. Math. Soc.*, 168(1972), 483-486.

BREZIS, H., and BROWDER, F.E.

[1] Linear maximal monotone operators and singular nonlinear integral equations of Hammerstein type, in: *Nonlinear Analysis, A Collection of Papers in Honor of Erich H. Rothe* (L. Cesari, R. Kannan, and H.F. Weinberger, Eds.), Academic, New York, 1978, pp.31-42.

BROWDER, F.E.

[1] Functional analysis and partial differential equations. I., *Math. Ann.*, 138(1959), 55-59.

[2] On the spectral theory of elliptic differential operators. I., *Math. Ann.*, 142(1961), 22-130.

[3] Functional analysis and partial differential operators. II., *Math. Ann.*, 145(1962), 81-226.

BROWN, A.

[1] On the adjoint of a closed transformation, *Proc. Amer. Math. Soc.*, 15(1964), 239-240.

BROWN, R.C.

[1] Notes on generalized boundary value problems in Banach spaces II. Infinite dimensional extension theory, *Pacific J. Math.*, 99(1982), 271-302.

BRUINSMA, P., DIJKSMA, A., and DE SNOO, H.S.V.

[1] Modes for generalized Caratheodory and Nevanlinna functions, in: *Challenges of a generalized system theory* (P. De Wilde, M.A. Kaashoek, and M. Verhaegen, Eds.), Koninklijke Nederlandse Akademie van Wetenschappen, Verhandelingen, Afd. Natuurkunde, Eerste Reeks, deel 40, Amsterdam, 1993, 161- 178.

CALKIN, J.W.

[1] Abstract symmetric boundary conditions, *Trans. Amer. Math. Soc.*, 45(1939), 369-442.

[2] Two-sided ideals and congruences in the ring of bounded operators in Hilbert space, *Annals Math.*, 42(1941), 839-873.

CARADUS, S.R., PFAFFENBERGER, W.E., and YOOD, B.

[1] *Calkin Algebras and algebras of operators on Banach spaces*, Lecture Notes in Pure and Applied Mathematics, Vol.9, Marcel Dekker, New York, 1974.

CASTAING, C.H., and VALADIER, M.

[1] *Convex Analysis and Measurable Multifunctions*, Lecture Notes in Mathematics 580, Springer, Berlin, 1977.

CECH, E.

[1] *Point sets*. Translation from the Czech Bodovu Mnoziny into English by Ales Pultr; Preface: M. Katetov, Academia, Prague, 1969.

CHAPMAN, G.R., and LEE, S.J.

[1] Boundary conditions for linear manifolds II: Algebraic properties and applications, *J. Math. Anal. Appl.* 77(1980), 351-363.

CHENEY, E.W.

[1] *Introduction to Approximation Theory*, McGraw-Hill, New York, 1966.

CODDINGTON, E.A.

[1] Multi-valued operators and boundary value problems, in: *Analytic theory of differential equations (Pro. Conf., Western Michigan Univ., Kalamazoo, Mich., 1970)*, pp.2-8. Lecture Notes in Mathematics 183, Springer, Berlin, 1971.

[2] Selfadjoint subspace extensions of non densely defined symmetric operators, *Bull. Amer. Math. Soc.*, 79(1973), 712-715.

[3] Eigenfunction expansions for nondensely defined operators generated by symmetric ordinary differential expressions, *Bull. Amer. Math. Soc.*, 79(1973), 964-968.

[4] Eigenfunction expansions for self adjoint subspaces generated by symmetric ordinary differential expressions, *International Conference on Differential Equations (Proc. Univ. California, Los Angeles, Calif., 1974)*, pp.185-192, Academic Press, New York, 1975.

[5] Self-adjoint subspace extensions of nondensely defined symmetric operators, *Advances in Math.*, 14(1974), 309-332.

[6] Self-adjoint problems for non-densely defined ordinary differential operators and their eigenfunction expansions, *Advances in Math.*, 15(1975), 1-40.

[7] Spectral theory of ordinary differential operators, in: *Spectral theory and differential equations (Proc. Symp., Dundee, 1974; dedicated to Konrad Jörgens)*, pp.1-24. Lecture Notes in Mathematics 448, Springer, Berlin, 1975.

[8] *Extension theory of formally normal and symmetric subspaces*, Memoirs of the American Mathematical Society, 134, American Mathematical Society, Providence, Rhode Island, 1973.

CODDINGTON, E.A., and DE SNOO, H.S.V.

[1] Positive selfadjoint extensions of positive symmetric subspaces, *Math. Z.*, 159(1978), 203-214.

CODDINGTON, E.A., and DIJKSMA, A.

[1] Selfadjoint subspaces and eigenfunction expansions for ordinary differential subspaces, *J. Differential Equations*, 20(2)(1976), 473-526.

[2] Adjoint subspaces in Banach spaces with applications to ordinary differential subspaces, *Ann. Mat. Pura Appl. (IV)*, CXVIII(1978), 1-118.

CODDINGTON, E.A., DIJKSMA, A., and DE SNOO, H.S.V.

[1] Pointwise convergence of eigenfunction expansions associated with a pair of ordinary differential expressions, *Proc. Roy. Soc. Edinburgh*, 97(A)(1984), 59-71.

COHN, D.L.

[1] *Measure Theory*, Birkhauser, Boston, 1980.

COPPING, J.

[1] Mercerian theorems and inverse transformations, *Studia Math.*, 21(1961/62), 177-194.

CORDES, H.O., and LABROUSSE, J.P.

[1] The invariance of the index in the metric space of closed operators, *J. Math. Mech.*, 12(1963), 693-720.

CRAWFORD, J.P.

[1] *Transformations in Banach spaces*, Dissertation, Lehigh University, 1966.

CROSS, R.W.

[1] On the continuous linear image of a Banach space, *J. Austral. Math. Soc. (Ser A)*, 29(1980), 219-234.

[2] Unilateral quasi-regularizers of closed operators, *Math. Proc. Camb. Phil. Soc.*, 87 (1980), 471-480.

[3] Some continuity properties of linear tranformations in normed spaces, *Glasgow Math. J.*, 30 (1988), 243-247.

[4] Unbounded strictly singular operators, *Nederl. Akad. Wetensch. Indag. Math. Ser. A*, 91 (1988), 245-248.

[5] Properties of some norm related functions of unbounded linear operators, *Math. Z.*, 199 (1988), 285-302.

[6] The essential state diagram of a linear operator, *Ann. Acad. Sc. Fenn.*, Ser A I Math. 15 (1990), 283-291.

[7] Unbounded linear transformations of upper semi- Fredholm type in normed spaces, *Portugal. Math.*, 47, No.1 (1990), 61-79.

[8] On the perturbation of unbounded linear operators with topologically complemented ranges, *J. Funct. Anal.*, 92 (1990), 468-473.

[9] Linear transformations of Tauberian type in normed spaces, *Note Mat.*, volume dedicated to the memory of Prof. Gottfried M. Köthe, Suppl. No 1 (1990), 193-203.

[10] Adjoints of non densely defined linear operators, *Aportaciones Matematicas*, volume dedicated to the memory of Prof. Victor Onieva, Univ. Cantabria, Santander (1991), 117-136.

[11] A characterisation of almost reflexive normed spaces, *Proc. R. Ir. Acad.*, 92A (1992), 225-228.

[12] On a theorem of Kalton and Wilansky concerning Tauberian operators, *J. Math. Anal. Appl.*, 171, No.1 (1992), 156-170.

[13] F_+-operators are Tauberian, *Quaestiones Math.*, 16 (1993), 129-132.

[14] Relatively open operators and the ubiquitous concept, *Publ. Mat. UAB*, 38 (1994), 69-79.

[15] Note on some characterisations of unbounded weakly compact operators, *Quaestiones Math.*, 18 (1995), 257-260.

[16] On certain densely invariant quantities of linear operators, *Math. Nachr.*, 178(1996), 103-116.

[17] Closed operators and the Calkin property, *Proc. R. Ir. Acad.*, 95A, No.1 (1995), 109-111.

[18] The essential state diagram for linear relations in normed spaces, *Quaestiones Math.*, 20 (1)(1997), 95-107.

[19] An Index theorem for the product of linear relations, *Linear Algebra Appl.*

[20] Essential State Diagrams for closed linear relations, *Sixth International Colloquium on Differential Equations, Plovdiv, Bulgaria, August 18-23*, 1995, Vol.I, Editor: D. Bainov; Impuls-M, Sofia; 123-130.

[21] Note on the canonical factorization of a multivalued linear operator, *Communications in Applied Analysis.*

CROSS, R.W., and GOUVEIA, A.I.

[1] Unbounded weakly compact operators, Research Report 115(1991), Dept. Math. Univ. Cape Town, 1991.

CROSS, R.W., and LABUSCHAGNE, L.E.

[1] Partially continuous and semicontinuous linear operators in normed spaces, *Expo. Math.*, 7 (1989), 189-191.

[2] Characterisations of operators of lower semi-Fredholm type in normed linear spaces, *Quaestiones Math.* 15 (1992), 151-173.

CROSS, R.W., OSTROVSKII, M.I., and SHEVCHIK, V.V.

[1] Operator ranges in Banach spaces I, *Math. Nachr.*, 173 (1995), 91-114.

CROSS, R.W., and PILLAY, P.

[1] Ubiquitous properties of certain operational quantities of linear relations, *Quaestiones Math.*, 17 (1994), 487-498.

CROSS, R.W., and SHEVCHIK, V.V.

[1] Disjointness of operator ranges in Banach spaces, *Quaestiones Math.*

CURGUS, B., DIJKSMA, A., LANGER, H., and DE SNOO, H.S.V.

[1] Characteristic functions of unitary colligations and of bounded operators in Krein spaces, *Operator Theory: Adv. Appl.*, 41(1989), 125-152.

DAMM, W.

[1] Beiträge zur Theorie der Lineare operatoren und Nichtkompaktheitsmasse, Dissertation, Univ. Karlsruhe, 1981.

DANES, J.,

[1] On the Istratescu's measure of noncompactness, *Bull. Math. Soc. Sci. Math. R.S. Roumanie*, 16(64)(1972), 403-406.

DAVIS, C.

[1] The shell of a Hilbert space operator, *Acta Scientarum Mathematicarum*, 29(1968), 69-86.

[2] The shell of Hilbert space operator II, *ibid*, 31(1970), 301-318.

[3] Matrix-valued shell of an operator or relation, *Integral Equations and Operator Theory*, 1(1978), 334-363.

[4] Extending the Kantorovic inequality to normal matrices, *Linear Algebra Appl.*, 31(1980), 173-177.

DAVIS, P.J.

[1] *Interpolation and Approximation*, Blaisdell, New York, 1963.

DAVIS, W.J., FIGIEL, T., JOHNSON, W.B., and PELCZYNSKI, A.

[1] Factoring weakly compact operators, *J. Funct. Anal.*, 17(1974), 311-327.

DAY, M.M.

[1] *Normed linear spaces*, Springer-Verlag, Berlin, 1962.

DE BLASI, F.S.

[1] On a property of the unit sphere in a Banach space, *Bull. Mat. Sci. Math. Roumanie*, 21(69)(1977), 259-262.

DE WILDE, M.

[1] *Closed graph theorems and webbed spaces*, Pitman, London, 1978.

DERKACH, V.A., and MALAMUD, M.M.

[1] Characteristic functions of linear operators, *Russian Acad. Dokl. Math.*, 45(1992), no.2, 417-424.

DEUTSCH, F.

[1] On uniform approximation with interpolatory constraints, *J. Math. Anal. Appl.*, 24(1968), 62-79.

[2] Linear selections for metric projection, *J. Funct. Anal.*, 49(1982), 269-292.

[3] A survey of metric selections, *Contemporary Math.*, 18(1983), 49-71.

[4] An exposition of recent results on continuous metric selections, *International Series Numer. Math.*, 81(1987), 67-79.

[5] Selections for metric projections, in: *Approximation Theory, Spline Functions and Applications*, (S.P. Singh, Ed.), Kluwer Academic Publ., 1992, pp.123-137.

DEUTSCH, F., LI, W., and MABIZELA, S.

[1] Helly Extensions and Best Approximation, in: *Parametric Optimization and Related Topics III*, (J. Guddat, H. Th. Jongen, B. Kummer, and F. Nizicka, Eds.), Approximation and Optimization, Vol.3, Verlag Peter Lang, Frankfurt, 1993, pp.107-120.

DEUTSCH, F., LI, W., and PARK, S.-H.

[1] Characterizations of continuous and Lipschitz continuous metric selections in normed linear spaces, *J. Approx. Theory*, 58(1989), 297-314.

DEUTSCH, F., and MABIZELA, S.

[1] Best interpolatory approximation in normed spaces, preprint.

DEUTSCH, F., POLLUL, W., and SINGER, I.

[1] On set-valued metric projections, Hahn-Banach extension maps, and spherical image maps, *Duke Math. J.*, 40(1973), 355-370.

DEUTSCH, F., and SINGER, I.

[1] On Single-Valuedness of Convex Set-Valued Maps, *Set-Valued Anal.*, 1(1993), 97-103.

DIESTEL, J.

[1] *Sequences and series in Banach spaces*, Graduate Texts in Mathematics 92, Springer-Verlag, Berlin, 1984.

DIESTEL, J. and UHL, J.

[1] *Vector Measures*, Amer. Math. Soc. Surveys # 15, Providence, 1977.

DIEUDONNE, J.

[1] Sur les homomorphismes d'espaces normes, *Bull. Sci. Math.*, 67(1943), 72-84.

DIJKSMA, A.

[1] Integral-ordinary differential-boundary subspaces and spectral theory, *New Developments in Differential Equations, Proceedings of the Second Scheveningen Conference, 1975*, Math. Studies 21, North-Holland, Amsterdam, 1976, 199-211.

[2] The generalized Green's function for regular ordinary differential subspaces in $L_2[a, b]$, *Differential Equations and Applications (Proceedings of the Third Scheveningen Conference, 1977)*, Math. Studies 31, North-Holland, 1978, 199-221.

[3] Eigenfunction expansions for a class of J-selfadjoint ordinary differential operators with boundary conditions containing the eigenvalue parameter, *Proc. Roy. Soc. Edinburgh*, 86A(1980), 1-27.

[4] Hamiltonian systems, Titchmarsh-Weyl coefficients, and models, preprint.

DIJKSMA, A., EL SABBAH, A., and DE SNOO, H.S.V.

[1] Selfadjoint extensions of regular canonical systems with Stieltjes boundary conditions, *J. Math. Anal. Appl.*, 152(1990), 546- 583.

DIJKSMA, A., and GHEONDEA, A.

[1] Index formulae for subspaces of Krein spaces, *Integral Equations Operator Theory*, (to appear).

DIJKSMA, A., LANGER, H., and DE SNOO, H.S.V.

[1] Self adjoint π_k-extensions of symmetric subspaces: an abstract approach to boundary problems with spectral parameter in the boundary conditions, *Integral Equations Operator Theory*, 7(1984), 459-515.

[2] Addendum, *ibid*, 905.

[3] Unitary colligations in π_k-spaces, characteristic functions and Shtraus extensions, *Pacific J. Math.*, 125(2)(1986), 347-362.

[4] Characteristic functions of unitary operator colligations in π_k-spaces, *Operator Theory: Adv. Appl.*, 19(1986), 125-194.

[5] Unitary colligations in Krein spaces and their role in extension theory of isometries and symmetric relations in Hilbert spaces, *Functional Analysis II, Proceedings Dubrovnik 1985*, Lecture Notes in Mathematics 1242, Springer-Verlag, Berlin, 1987, 1-42.

[6] Some representations of holomorphic operator functions by means of resolvents of unitary or selfadjoint operators in Krein spaces, *Operator Theory: Adv. Appl.*, 24(1987), 123-143.

[7] Hamiltonian systems with eigenvalue depending boundary conditions, *Operator Theory: Adv. Appl.*, 35(1988), 37-83.

[8] Generalised coresolvents of standard isometric operators and generalized resolvents of standard symmetric relations in Krein spaces, *Operator Theory: Adv. Appl.*, 48(1990), 261-274.

[9] Eigenvalues and pole functions of Hamiltonian systems with eigenvalue depending boundary conditions, *Math. Nachr.*, 161(1993), 107-154.

[10] Selfadjoint Krein space extensions and their parametrizations, unpublished manuscript.

[11] Selfadjoint Krein and Pontryagin space extensions and their parametrizations and the Titchmarch-Weyl coefficients of selfadjoint realizations of a Sturm-Liouville operator with interface coefficients, in preparation.

DIJKSMA, A., and DE SNOO, H.S.V.

[1] Selfadjoint extensions of symmetric subspaces, *Pacif. J. Math.*, 54(1)(1974), 71-100.

[2] Eigenfunction expansions for non densely defined differential operators, *J. Differential Equations*, 17(1)(1975), 198-219.

[3] Linear relations in indefinite inner product spaces, *Proceedings 1984 Workshop Spectral Theory of Sturm-Lionville Differential Operators, Argonne National Laboratory Report*, ANL-84-73(1984), 81-90.

[4] Eigenfunction expansions associated with pairs of ordinary differential expressions, *J. Differential Equations*, 60(1)(1985), 21-56.

[5] Symmetric and selfadjoint relations in Krein spaces I, *Operator Theory: Adv. Appl.*, Vol.24(1987), 145-166.

[6] Symmetric and selfadjoint relations in Krein spaces II, *Ann. Acad. Sci. Fenn.*, Ser.A1 Math.12(1987), 199-216.

DIJKSMA, A., DE SNOO, H.S.V., and SHTRAUS, A.V.

[1] Symmetric relations and module spaces, *Integral Equations and Operator Theory*, 18(1994), 154-165.

DIXMIER, J.,

[1] Sur un theoreme de Banach, *Duke Math. J.*, 15(1948), 1057-1071.

[2] Etude sur les Varietes et les Operateurs de Julia, *Bull. Soc. Math. France*, 77(1949), 11-101.

DOUADY, R.

[1] Petite perturbations d'une suite exact et d'une espace quasiexacte, *Seminaire d'Analyse*, Nice, 1965, pp.21-34.

DOUGLAS, R.G.

[1] *Banach algebra techniques in operator theory*, Academic Press, New York and London, 1972.

DREWNOWSKI, L.

[1] Some characterisations of semi Fredholm operators, *Comment. Math. Prace. Math.*, 24(1984), 215-218.

[2] Any two norms are somewhere comparable, *Funct. Approx. Comment. Math.*, 7(1979), 13-14.

DULST, D. VAN

[1] Perturbation theory and strictly singular operators in locally convex spaces, *Studia Math.*, 38(1970), 341-372.

[2] On strictly singular operators, *Compositio Math.*, 23(1971), 169-183.

DUNFORD, N., and SCHWARTZ, J.T.

[1] *Linear Operators, Part I: General Theory*, Interscience, New York, 1967.

EDGAR, G., ERNEST, J., and LEE, S.G.

[1] Weighing operator spectra, *Indiana Univ. Math. J.*, 21(1971), 61-79.

EDMUNDS, D.E., and EVANS, W.D.

[1] *Spectral Theory and Differential Operators*, Clarendon, Oxford, 1987.

EDMUNDS, D.E., and TYLLI, H.-O.

[1] On the entropy numbers of an operator and its adjoint, *Math. Nachr.*, 126(1986), 231-239.

EMBRY, M.R.

[1] Factorization of Operators on Banach Spaces, *Proc. Amer. Math. Soc.*, 38(1973), 587-590.

EMMANUELE, G.

[1] Measure of weak noncompactness and fixed point theorems, *Bull. Math. Soc. Sci. Math. R.S. Roumanie*, 25(73)(1981), 353-358.

EWERT, J

[1] Multivalued contractions with respect to the weak measure of noncompactness, *Bull. Polish Acad. Sci. Math.*, 34(1986), 457-463.

FAJNSTEJN, A.S.

[1] On measures of non compactness of linear operators and analogs of the minimal modulus of semi-Fredholm operators (Russian), *Spektr. Teor. Oper.*, 6(1985), 182-195. Zbl.634#47010.

FAVINI, A., and YAGI, A.

[1] Multivalued linear operators and degenerate evolution equations, *Ann. Mat. Pura. Appl.*, (4)163(1993), 353-384.

FILLMORE, P.A., and WILLIAMS, J.P.

[1] On Operator Ranges, *Adv. Math.*, 7(1971), 254-281.

FLEISCHMANN, W. (Editor)

[1] *Set-Valued Mappings, Selections and Topological Properties of* 2^X, Lecture Notes in Mathematics 171, Springer-Verlag, Berlin, 1970.

FLORET, K.

[1] *Weakly compact sets*, Lecture Notes in Mathematics 801, Springer, Berlin, 1980.

FONF, V.P.

[1] One property of families of embedded Banach spaces, *J. Soviet Math.*, 59(1992), 687-690.

FONF, V.P., and SHEVCHIK, V.V.

[1] Operators of dense embedding and quasicomplements of subspaces of Banach spaces, *Archiv. Math.*, 62(1994), 539-549.

[2] On the sums of operator ranges in Banach spaces (Russian), *Funk. Anal. and Appl.*, 29, no.3(1995), 91-93.

FÖRSTER, K.-H., and LIEBETRAU, E.O.

[1] Semi-Fredholm operators and sequence conditions, *Manuscripta Math.*, 44(1983), 35-44.

FRIEDMAN, B.

[1] Operators with closed range, *Comm. Pure Appl. Math.*, 8(1955), 539-550.

FUHRMAN, M.

[1] Sums of Linear Operators of Parabolic Type: A priori Estimates and Strong Solutions, *Ann. Mat. Pura. Appl.*, (4)164(1963), 229-257.

[2] Sums of generators of analytic semigroups and multivalued linear operators, preprint.

GALAZ-FONTES, F.,

[1] Measures of noncompactness and upper semi-Fredholm perturbation theorems, *Proc. Amer. Math. Soc.*, 118(1993), 891-897.

GARLING, D.J.H., and WILANSKY, A.

[1] On a summability theorem of Berg, Crawford and Whitley, *Math. Proc. Cambridge Phil. Soc.*, 71(1972), 495-497.

GOHBERG, I., GOLDBERG, S., and KAASHOEK, M.

[1] *Classes of Linear Operators*, Vol.1, Birkhauser, Basel, 1990.

GOHBERG, I., and KREIN, M.G.

[1] The basic propositions on defect numbers, root numbers, and indices of linear operators, *Amer. Math. Soc. Transl.*, (2)13(1964), 185-264; Russian original: *Uspechi Mat. Nauk*, 12(1957), 2(74), 43- 118.

GOHBERG, I., and MARKUS, A.S.

[1] Two theorems on the opening of subspaces of a Banach space, *Uspechi Mat. Nauk.*, 14(1959), 5(89), 135-140 (Russian).

GOHBERG, I.C., MARKUS, A.S., and FELDMAN, I.A.

[1] On normally solvable operators and related ideals, *Izv. Moldav. Filial Akad. Nauk SSSR 10*, (76)(1069), 51-69 (Russian).

GOLDBERG, S.

[1] Linear operators and their conjugates, *Pacific J. Math.*, 9(1959), 69-79.

[2] Closed Linear Operators and Associated Continuous Linear Operators, *Pacific J. Math.*, 12(1962), 183-186.

[3] *Unbounded linear operators*, McGraw-Hill, New York, 1966.

GOLDBERG, S., and THORPE, E.O.

[1] On some open questions concerning strictly singular operators, *Proc. Amer. Math. Soc.*, 13(1963), 334-336.

GOLDENSTEIN, L.S., and MARKUS, A.S.

[1] On the measure of non-compactness of bounded sets and of linear operators (Russian). Studies in Algebra and Math. Anal., *Izdat.'Karta Moldovenjaske'*, Kishinev (1965), 45-54. MR35#789

GOLDMAN, M.A.

[1] On the Stability of the Property of Normal Solvability of Linear Equations, *Dokl. Akad. Nauk SSSR*, (N.S)100(1955), 201-204.

GOLIGHTLY, G.O.

[1] Graph-dense linear transformations, *Pacific J. Math.*, 82(1979), 371-377.

GONZALEZ, M.

[1] The fine spectrum of the Cesaro operator in ℓ_p $(1 < p < \infty)$, *Arch. Math.*, 44(1985), 355-358.

[2] A perturbation result for generalized Fredholm operators in the boundary of the group of invertible operators, *Proc. Royal. Irish Acad.*, 86A(1986), 123-126.

[3] Properties and applications of Tauberian operators, *Extracta Math.*, 53(1990), (1990), 91-107.

[4] On the index of weakly Fredholm operators, *Panamerican Math. J.*, 2(1992), 57-66.

[5] On essentially incomparable Banach spaces, *Math. Z.*, 215(1994), 621-629.

GONZALEZ, M., and HARTE, R.

[1] The death of an index theorem, *Proc. Amer. Math. Soc.*, 108(1990), 151-156.

GONZALEZ, M., and MARTINEZ-ABEJON, A.

[1] Super Tauberian operators and perturbations, *Arch. Math.*, 64(1995), 423-433.

[2] Ultrapowers and semi-Fredholm operators, *Bolletino U.M.I.*, 11-13(1997), 415-433.

[3] Tauberian operators in $L_1(\mu)$ spaces, *Studia Math.*, 125(1997), 289-303.

GONZALEZ, M., and MARTINON, A.

[1] Note on the Hausdorff measure of noncompactness, *Rev. Acad. Canaria Ciencias*, I(1990), 119-123.

[2] Operational quantities derived from the norm and measures of noncompactness, *Proc. Royal Irish Acad.*, Ser A, 91A(1991), 63-70.

[3] Operational quantities derived from the norm and generalized Fredholm operators, *Comment. Math. Univ. Carolinae*, 32(1991), 645- 657.

[4] Note on operational quantities and Milman isometry spectrum, *Rev. Acad. Canaria Ciencias*, III.1(1991), 103-111.

[5] Fredholm theory and space ideals, *Bollettino U.M.I.*, 7- B(1993), 473-488.

[6] On incomparibility of Banach spaces, *Banach Center Publ.*, 30(1994), 161-174.

[7] Operational quantities characterizing semi-Fredholm operators, *Studia Math.*, 114(1995), 13-27.

GONZALEZ, M., and ONIEVA, V.M.

[1] On the spectral mapping theory for essential spectra, *Publ. Sem. Mat. U.A.B.*, 29(1985), 105-110.

[2] On the instability of non-semi-Fredholm operators under compact perturbations, *J. Math. Anal. Appl.*, 114(1986), 450-457.

[3] On the meromorphic and Schechter-Shapiro operational calculi, *J. Math. Anal. Appl.*, 116(1986), 363-377.

[4] On functions of relatively regular and normally solvable operators, *J. Math. Anal. Appl.*, 133(1988), 209-213.

[5] Semi-Fredholm operators and semigroups associated with some classical operator ideals, *Proc. Royal Irish Acad.*, Ser A, 88A(1988), 35-38.

[6] Semi-Fredholm operators and semigroups associated with some classical operator ideals II, *Proc. Royal Irish Acad.*, Ser.A, 88A(1988), 119-124.

[7] Characterizations of Tauberian operators and other semigroups of operators, *Proc. Amer. Math. Soc.*, 108(1990), 399-405.

GONZALEZ, M., SAKSMAN, E., and TYLLI, H.-O.

[1] Representing non-weakly compact operators, *Studia Math.*, 113(1995), 265-282.

GOUVEIA, A.I.

[1] *Unbounded linear operators in seminormed spaces*, Univ. Cape Town Thesis Reprints 9/1990.

GOWERS, W.T., and MAUREY, B.

[1] The unconditional basic sequence problem, *J. Amer. Math. Soc.*, 6(1993), 851-874.

GRAMSCH, B.

[1] Über analytische Störungen und den Index von Fredholmoperatoren auf Banachräumen, Dept. of Mathematics, *University of Maryland*, 1969.

GURARII, V.I.

[1] On openings and inclinations of subspaces of a Banach space, *Teor. Funksii, Funksional. Anal. i Prilozhen*, 1(1965), 194-204. (Russian).

HADWIN, D.

[1] An algebraic Version of the Multiplication Property of the Fredholm Index, *Linear Algebra Appl.*, 208/209 (1994), 229-230.

HARTE, R.

[1] *Invertibility and Singularity for Bounded Linear Operators*, Dekker, New York, 1988.

HARTE, R., and SHANNON, G.

[1] Closed subspaces of operator ranges, *Proc. Amer. Math. Soc.*, 118(1993), 171-173.

HASSI, S., KALTENBÄCK, M., and DE SNOO, H.S.V.

[1] Triplets of Hilbert spaces and Friedrichs extensions associated with the subclass $\mathbf{N}_1$ of Nevanlinna functions, submitted.

[2] A characterization of semibounded selfadjoint operators, *Proc. Amer. Math. Soc.*, to appear.

[3] Selfadjoint extensions and the Krein formula, submitted.

[4] Matrix Nevanlinna functions, tentative title, in preparation.

[5] Friedrichs and Von Neumann extensions, tentative title, in preparation.

HASSI, S., LANGER, H., and DE SNOO, H.S.V.

[1] Selfadjoint extensions for a class of symmetric operators with defect numbers $(1,1)$, *Proc. Oper. Theory Conf.*, 1(1995), to appear.

HASSI, S., and DE SNOO, H.S.V.

[1] On some subclasses of Nevanlinna functions, *Zeitschrift für Analysis und ihre Anwendungen*, 15(1996), 45-55.

[2] One-dimensional graph perturbations of selfadjoint relations, *Ann. Acad. Sci. Fenn. A.I. Meth.*, to appear

[3] Nevanlinna functions, perturbation formulas and triplets of Hilbert spaces, submitted.

[4] On rank one perturbations of selfadjoint operators, to be submitted.

HERRERA, I.

[1] *Boundary methods*, Pitman, Boston, 1984.

HIAI, F., and UMEGAKI, H.

[1] Integrals, conditional expectations, and martingales of multivalued functions, *Journal of multivariate analysis*, 7(1977), 149- 182.

HOLA, L.

[1] Some properties of almost continuous linear relations (Russian), *Acta Math. Comenian.*, 50/51(1987), 61-69.

HOLA, L., and KUPKA, I.

[1] Closed graph and open mapping theorems for linear relations, *Acta Math. Comenian.*, 46/47(1985), 157-162.

HOLA, L., and MALICKY, P.

[1] Continuous linear selectors of linear relations, *Acta Math. Comenian.*, 48/49(1986), 153-157.

HOLUB, J.R.

[1] On perturbation of operators with complemented range, *Acta Math. Hungar.*, 44(1984), 269-273.

[2] Characterizations of tauberian and related operators on Banach spaces, *J. Math. Anal. Appl.*, 178(1993), 280-288.

HOWARD, J., and MELENDEZ, K.

[1] Characterizing operators by their second adjoints, *Bull. Aca. Sinica*, 5(1977), 129-134.

IOFFE, A.D.

[1] Single-valued representation of set-valued mappings II; Application to differential inclusions, *SIAM J. Control and Optimization*, 21(1983), 641-651.

ISTRATESCU, V.I.

[1] On a measure of noncompactness, *Bull. Math. Soc. Sci. Roumanie*, 16(64)(1972), 195-197.

JACOBS, M.Q.

[1] Measurable multivalued mappings and Lusin's theorem, *Trans. Amer. Math. Soc.*, 134(1968), 471-481.

JEGANATHAN, P.,

[1] Tauberian property for linear relations, *Indian J. Math.* 38, No.1 (1966).

JUBERG, R.K.

[1] The measure of noncompactness in *Lp* for a class of integral operators, *Indiana Math. J.*, 23(1974), 925-936.

KADEC, M-I., and PELCZYNSKI, A.

[1] Basic sequences, biorthogonal systems and norming sets in Banach and Frechet spaces, *Studia Math.* , 25(1965), 297-323.

KALMAN, R.E.

[1] A theory for the identification of linear relations, in: *Frontiers in pure and applied mathematics*, Robert Dautry (Ed.), North Holland, Amsterdam, 1991, 117-132.

KALTON, N.J., and WILANSKY, A.

[1] Tauberian operators in Banach spaces, *Proc. Amer. Math. Soc.*, 57(1976), 251-255.

KANDILAKIS, D.A., and PAPAGEORGIOU, N.S.

[1] On the properties of the Aumann integral with applications to differential inclusions and control systems, *Czechoslovaki Math. J.*, 39(1989), No.1, 1-15.

KANIEL, S., and SCHECHTER, M.

[1] Spectral theory for Fredholm operators, *Comm. Pure Appl. Math.*, 16(1963), 423-448.

KASCIC, M.J.

[1] Polynomials in linear relations, *Pacific J. Math.* , 24(1968), 291-295.

[2] Polynomials in linear relations II, *Pacific J. Math.*, 29(1969), 593-607.

KATO, T.

[1] Perturbation theory for nullity, deficiency, and other quantities of linear operators, *J. Anal. Math.*, 6(1958), 273-322.

[2] *Perturbation theory for linear operators*, Springer- Verlag, Berlin, 1966.

KATS, I.S.

[1] Linear relations generated by canonical differential equations, *Funct. Anal. Appl.*, 17(1983), No.4, 315-317.

KINYON, M.K.

[1] The Adjoint Problem for Coupled Linear Operators, *Results Math.*, 18 (1990), 60-73.

[2] Density invariance of certain operational quantities of bounded linear operators in normed spaces, Indiana University South Bend Technical Report (1995).

KÖTHE, G.

[1] General Linear Transformations in Locally Convex Spaces, *Math. Annalen*, 159(1965), 309-328.

[2] *Topological Vector Spaces I*, Springer-Verlag, Berlin, 1969

KRALL, A.M.

[1] The development of general differential and general differential- boundary systems, *Rocky Mountain J. Math.*, 5(1975), 493-542.

[2] Stieltjes Differential-boundary operators III, *Pacific J. Math.*, 59(1975), 125-134.

KREIN, M.G., and KRASNOSELSKII, M.A.

[1] Stability of the index of an unbounded operator, *Mat. Sbornik*, 30(1952), 219-224 (Russian).

KREIN, S.G., PETUNIN, Ju.I., and SEMENOV, E.M.

[1] Interpolation of linear operators, *Transl. Monographs* 54, *Amer. Math. Soc.*, Providence, R.I., 1982.

KRISHNAMURTHY, V.

[1] On the State of a Linear Operator and Its Adjoint, *Math. Ann.*, 141(1960), 153-160.

KUPKA, I.A.K.

[1] Quasicontinuous selections for compact valued multifunctions, *Math. Slovaca*, 43(1993), 69-75.

KURATOWSKI, C.

[1] Sur les Espaces Complets, *Fundam. Math.*, 15(1930), 301- 309.

LABUSCHAGNE, L.E.

[1] *Functions of operators and the classes associated with them*, Ph.D. Thesis, Univ. Cape Town, 1988.

[2] On the instability of non semi-Fredholm closed operators under compact perturbations with applications to ordinary differential equations, *Proc. Roy. Soc. Edinburgh*, Sect.A 109(1988), 98-108.

[3] The instability of a class of unbounded non-semi-Fredholm type operators under compact perturbations, *Quaestiones Math.*, 13(1990), 39-65.

[4] Characterisations of partially continuous, strictly cosingular and ϕ_--type operators, *Glasgow Math. J.*, 33(1991), 203-212.

[5] On the minimum modulus of an arbitrary linear operator, *Quaestiones Math.*, 14(1991), 77-91.

[6] Certain norm related quantities of unbounded linear operators, *Math. Nachr.*, 157(1992), 137-162.

[7] The perturbation of relatively open operators with reduced index, *Math. Proc. Cambridge Phil. Soc.*, 112(1992), 385-402.

[8] Characteristic quantities of semi-Fredholm operators, *Proc. Royal Irish Acad.*, A 94(1994), 47-58.

[9] Unpublished note.

LABUSCHAGNE, L.E., STRÖH, A., and SWART, J.

[1] The uniqueness of operational quantities in von Neumann algebras, *Quaestiones Math.*, 18(1995), 167-183.

LACEY, E.

[1] *Generalizations of Compact Operators in Locally Convex Topological Linear Spaces*, Thesis, New Mexico State University, 1963.

LANGER, H.

[1] Zur Spektraltheorie verallgemeinerter gewöhnlicher Differentialoperatoren zweiter Ordnung mit einer nichtmonotonen Gewichtsfunktion, *Ber. Univ. Jyäskylä Math. Inst. Ber.*, 14(1972), 58 pp.

LANGER, H., and TEXTORIUS, B.

[1] On generalized resolvents and Q-functions of symmetric linear relations (subspaces) in Hilbert space, *Pacific J. Math.*, 72(1977), no.1, 135-165.

[2] A generalization of M.G. Krein's method of directing functionals to relations, *Proc. Roy. Soc. Edinburgh Sect. A*, 81(1978), no.3-4, 237-246.

[3] L-resolvent matrices of symmetric linear relations with equal defect numbers; applications to canonical differential relations, *Integral Equations Operator Theory*, 5(1982), 208-243.

[4] Spectral functions of a symmetric linear relation with a directed mapping I and II, *Proc. Roy. Soc. Edinburgh*, 97A(1984), 165- 176 and 101A(1985), 111-124.

LEBOW, A., and SCHECHTER, M.

[1] Semigroups of operators and measures of noncompactness, *J. Funct. Anal.*, 7(1971), 1-26.

LEE, S.J.

[1] Coordinated adjoint subspaces in Hilbert spaces, with applications to ordinary differential equations, *Proc. London Math. Soc.*, 40(1980), 138-160.

[2] Infinitely many boundary conditions, *J. Math. Anal. Appl.*, 105(1985), 301-321.

[3] Operator parts of ordinary differential subspaces with applications to nonlinear problems, in: *Trends in Theory and Practice of Nonlinear Differential Equations* (V. Lakshmikantham, Ed.), Marcel Dekker, New York, 1984, pp.315-321.

LEE, S.J. and NASHED, M.Z.

[1] Least-squares solutions of multi-valued linear operator equations in Hilbert spaces, *J. Approx. Theory*, 8(1983), 380-391.

[2] Gradient method for nondensely defined closed unbounded linear operators, *Proc. Amer. Math. Soc.*, 33(1983), 429-435.

[3] Generalized inverses for linear manifolds and applications to boundary valued problems, *C.R. Math. Rep. Acad. Sci. Canada*, 4(1982), 347-352.

[4] Constrained least-squares solutions of linear inclusions and singular control problems in Hilbert spaces, *Appl. Math. Optim.*, 19(1989), 225-242.

[5] Algebraic and topological selections of multi-valued linear relations, *Ann. Scuola Norm. Sup. Pisa*, 17(1990), 111-126.

[6] Normed linear Relations: Domain Decomposability, Adjoint Subspaces, and Selections, *Linear Algebra Appl.*, 153(1991), 135-159.

LEE, S.J. and WALTERS, T.

[1] Boundary conditions for linear manifolds III: infinite dimensional adjoints via solution spaces, *J. Math. Anal. Appl.*, 82(1981), 89-111.

LIN, C.S.

[1] Regularizers of closed operators, *Canad. Math. Bull.*, 17(1974), 67-71.

LINDENSTRAUSS, J. and TZAFRIRI, L.

[1] On the complemented subspaces problem, *Israel J. Math.*, 9(1971), 263-269.

[2] *Classical Banach spaces I*, Springer-Verlag, Berlin, 1977.

[3] *Classical Banach spaces II*, Springer-Verlag, Berlin, 1979.

LONGSTAFF, W.E., and ROSENTHAL, P.

[1] On Operator Algebras and Operator Ranges, *Integral Equations and Operator Theory*, 9(1986), 820-830.

LUENBERGER, D.G.

[1] *Optimization by Vector Methods*, Wiley, New York, 1969.

MABIZELA, S.

[1] Upper Semicontinuity of Parametric Projections, *Set-Valued Analysis*, 4(1996)m 315-325.

MABIZELA, S., and ZHONG, J.

[1] On the best uniform ϵ-interpolation operators, *Numer. Funct. Anal. Optimiz.*, 14(1993), 117-124.

MAC LANE, S.

[1] An algebra of additive relations, *Proc. Nat. Acad. Sci., U.S.A.*, 47(1961), 1043-1051.

MARTIN, D.H., and SWART, J.

[1] A characterisation of semi-Fredholm operators defined on almost reflexive Banach spaces, *Proc. R. Ir. Acad.*, 86A(1986), 91-93.

MARTINEZ ABEJON, A.

[1] *Semigroups of operators and ultrapowers* (Spanish), Doctoral dissertation, Univ. Cantabria, 1994.

MARTINEZ MAURICA, J., and PELLON, T.

[1] Non-Archimeadean Tauberian operators, *Proceedings of the conference on p-adic analysis, Hengelhoef 1986*, 101-111.

MARTINON, A.

[1] *Operational quantities in Fredholm theory* (Spanish), Doctoral dissertation, Univ. La Laguna, 1989.

MC ALLISTER, B.L.

[1] Hyperspaces and multifunctions (The first half century 1900- 1950), *Nieuch Arch. Wisk.*, 26(1978), 309-329.

MCGEHEE, R.

[1] Attractors for closed relations on compact Hausdorff spaces, *Indiana University Mathematics Journal*, 4(1992), 1165-1209.

MCGEHEE, R., and SANDER, E.

[1] A New Proof of the Stable Manifold Theorem, Geometry Center GCG 88, October 1995 (to appear, *ZAMP*, 1996).

MENNICKEN, R.

[1] Perturbations of semi-Fredholm operators in locally convex spaces, in *Advances in Functional Analysis, Holomorphy and Approximation Theory*, Ed.: G. Zapata, Marcel-Dekker, New York, Lecture Notes in Pure and Appl. Math. 80(1983), 232-301.

MENNICKEN, R., and SAGRALOFF, B.

[1] Eine Verallgemeinerung des Satzes von abgeschlossenen Werterbereich in lokalkonvexen Räumen, *Manuscripta Math.*, 18(1976), 109-146.

[2] Störungstheoretische Untersuchungen über Semi-Fredholmpaare und -operatoren in lokalkonvexen Vektorräumen. I. *J. reine angew. Math.*, 280(1976), 1-36.

[3] Störungstheoretische Untersuchungen über Semi-Fredholmpaare und -operatoren in lokalkonvexen Vektorräumen. II. *J. reine angew. Math.*, 303/304(1978), 389-436.

[4] On Banach's closed range theorem, *Archiv. Math.*, 33(1979), 461-465.

MILMAN, V.D.

[1] Some properties of strictly singular operators, *Funct. Anal. Appl.*, 3(1969), 77-78.

MICHAEL, E.

[1] Continuous selections I, II, III, *Annals of Math.*, 63, 361-381; 64, 562-580; 65, 375-390.

MÖLLER, M.

[1] *Zustandsdiagramme linearer Relationen und ihrer Adjungierten sowie von paaren von unterräumen in Lokalkonvexen Räumen*, Master's Thesis, Braunschweig, 1976.

[2] *Normale Auflösbarkeit linearer Operatoren in Dualräumen von (LF)-Räumen*, Doctoral dissertation, Regensburg, 1980.

[3] Necessary conditions for normal solvability in duals of LF-spaces, *Manuscripta Math.*, 58(1987), 283-294.

MÜNNICH, A., and SZAZ, A.

[1] An alternative theorem for continuous relations and its applications, *Publ. Inst. Math. (Beograd)* (N.S.), 33(1983), 163-168.

NASHED, M.Z.

[1] Inner, outer and generalised inverses in Banach and Hilbert spaces, *Numer. Funct. Annal. Optim.*, 9(1987), 261-325.

[2] Operator parts and generalized inverses of linear manifolds with applications, in: *Trends in Theory and Practice of Nonlinear Differential Equations* (V. Lakshmikantham, Ed.), Marcel Dekker, New York, 1984, pp.395-412.

NEIDINGER, R.

[1] *Properties of Tauberian operators in Banach spaces*, Ph.D. Thesis, Univ. Texas, 1984.

NEIDINGER, R., and ROSENTHAL, H.P.

[1] Norm-attainment of linear functionals on subspaces and characterisations of Tauberian operators, *Pacific J. Math.*, 118(1985), 215-228.

NEUBAUER, G.

[1] Über den Index abgeschlossener Operatoren in Banachräumen I., *Math. Annalen*, 160(1965), 93-130.

[2] Über den Index abgeschlossener Operatoren in Banachräumen II., *Math. Annalen*, 162(1965), 42-119.

[3] Homotopy Properties of Semi-Fredholm Operators in Banach Spaces, *Math. Annalen*, 176(1968), 273-301.

NEUMANN, J. VON

[1] Über adjungierte Funktional-operatoren, *Ann. Math.*, 33(1932), 294-310.

[2] *Functional Operators, Vol.2: The Geometry of Orthogonal Spaces*, *Ann. Math. Stud.*, 22, Princeton U.P., 1950.

NEWBURGH, J.D.

[1] A topology for closed operators, *Ann. of Math.*, (2) 53(1951), 250-255.

NITECKI, Z., and PRZYTYCKI, F.

[1] The entropy of the relation inverse to a map, preprint.

ONIEVA, V.M.

[1] *Adjoint operators and state diagrams in functional analysis* (Spanish), Univ. Santander, Santander, 1980.

OSTROVSKII, M.I.

[1] A note on continuous restrictions of linear maps between Banach spaces, *Acta Math. Univ. Comenianae*, 41, No.2(1992), 185-188.

[2] Topologies on the set of all subspaces of a Banach space and related questions of Banach space geometry, *Quaestiones Math.*, 17, No.3(1994), 259-319.

OSTROVSKII, M.I., and FONF, V.P.

[1] Operators with Dense Range and Extensions of Minimal Systems, *Soviet Math.*, (Iz. VUZ) 34(1990), 55-56.

PAPAGEORGIOU, N.S.

[1] On the theory of Banach valued multifunctions. I. Integration and conditional expectation, *J. Multivar. Anal.*, 16(1983), 185-206.

PARTHASARATHY, T.

[1] *Selection Theorems and their Applications*, Lecture Notes in Mathematics 263, Springer-Verlag, Berlin, 1972.

PELCZYNSKI, A.

[1] On strictly singular and strictly cosingular operators in $L(\nu)$ spaces, *Bull. Acad. Polon. Sci. Ser. Math.*, Astronom. Phy. (1) 13(1965), 37-41.

PFEIFER, P.

[1] Adjoints of nondensely defined Hilbert space operators, *J. Math. Phys.*, 22(1981), 1619-1622.

PIETSCH, A.

[1] *Operator Ideals*, North Holland, Amsterdam, 1980.

[2] *Eigenvalues and s-Numbers*, Cambridge University Press, Cambridge, 1987.

PLEIGEL, A.

[1] A survey of spectral theory for pairs of ordinary differential operators. In: *Spectral Theory and Differential Equations, Proceedings of the Symposium held at Dundee, Scotland, 1974*, Lecture Notes in Mathematics 448, Springer-Verlag, Berlin, 1975, 256-272.

PLICHKO, A.N.

[1] Selection of subspaces with special properties in Banach space and some properties of quasicomplement, *Functional Anal. Appl.*, 15(1981), 67-68.

[2] Some Remarks on Operator Ranges (Russian), Teor. Funktsii, *Funktsional Anal. i Prilozhen*, 53(1990), 69-70.

RAKOCEVIC, V.

[1] Measures of non-strict singularity of operators, *Mat. Vesnik*, 35(1983), 79-82.

RITSNER, V.S.

[1] Theory of linear relations (Russian), VINITI, 1982, no.846-82.

RITSNER, V.S. and SENDEROV, V.A.

[1] On the spectral theory of linear relations in a Pontryagin space, *Soviet Math. (Iz.VUZ)*, 28(1984) no.8, 34-37.

ROBINSON, S.M.

[1] Normed convex processes, *Trans. Amer. Math. Soc.*, 174(1972), 124-140.

ROBINSON, S.

[1] Regularity and stability for convex multivalued functions, *Math. Oper. Res.*, (1976), 130-145.

ROCKAFELLAR, R.T.

[1] Monotone processes of convex and concave type, *Mem. Amer. Math. Soc.*, 77 (1967).

[2] *Convex Analysis*, Princeton U.P., Princeton, N.J., 1970.

ROSENTHAL, H.P.

[1] On totally incomparable Banach spaces, *J. Funct. Anal.*, 4(1969), 167-175.

SADOVSKII, B.N.

[1] Limit-compact and condensing operators, *Usp. Mat. Nauk.*, 27(1972), 81-146.

SAGRALOFF, B.

[1] Eine Ergänzung zum Satze vom abgeschlossenen Werterbereich in lokalkonvexen Räumen, *Manuscr. Math.*, 22(1977), 213-224.

SAMOILENKO, Yu.S., and TUROVSKAYA, L.B.

[1] On representation of semi-linear relations, in: *Methods of functional analysis of mathematical physics*, Yu. Berzanskii (Ed.), Acad. Sci. Ukraine, Inst. Math., Kiev, 1992, pp.97-108.

[2] On $*$-representation of semi-linear relations (Russian). *Methods of functional analysis in problems in Mathematical physics*, 97-108, Acad. Sci. Ukraine, Inst. Math., Kiev, 1992.

SANDER, E.

[1] *Hyperbolic Sets of Noninvertible Maps and Relations*, PhD thesis, University of Minnesota, June, 1996.

SARASON, D.

[1] The multiplication theorem for Fredholm operators, *Amer. Math. Monthly*, 94(1987), 68-70.

SCHACHERMAYER, W.

[1] For a Banach space isomorphic to its square the Radon-Nikodym property and the Krein-Milman property are equivalent, *Studia Math.*, 81(1985), 329-338.

SCHECHTER, M.

[1] Quantities related to strictly singular operators, *Indiana Univ. Math. J.*, 21(1972), 1061-1071.

SCHECHTER, M., and WHITLEY, R.J.

[1] Best Fredholm perturbation theorems, *Studia Math.*, 90(1988), 175-190.

SEDAEV, A.A.

[1] The structure of certain linear operators, *Mat. Issled*, 5(1970), 166-175. (Russian).

SHEVCHIK, V.V.

[1] On subspaces of a Banach space that coincide with the ranges of continuous linear operators, *Revue Roum. Math. Pures Appl.*, 31(1986), 65-71.

[2] On non-strictly singular operators with non-closed ranges, *Ukr. Math. J.*, 41(1989), 1403-1407.

SHMULYAN, Yu.L.

[1] Transformers of linear relations in J-spaces, *Funct. Anal. Appl.*, 14(1980), no.2, 110-113.

SHNEPERMAN, L.B.

[1] Characterization of the dimension of linear spaces by the additive semigroup of linear relations of rank zero, in: *Semigroup studies of mappings*, Edited by E.S. Lyapin, Leningrad Gos. Ped. Inst., Leningrad, 1989, pp.129-137 (Russian).

[3] Characterization of linear spaces by their additive semigroup of linear relations (Russian), *Studies in semigroups* (Russian), 145-152, Leningrad Gos. Ped. Inst., Leningrad, 1990.

[3] Isomorphism of additive semigroups of linear relations, in: *Semigroups and their homomorphisms*, Edited by E.S. Lyapin Ross. Gos. Ped. Univ., Leningrad, 1991, pp.118-122 (Russian).

SHTRAUS, A.V.

[1] Pseudoresolvents and representations of linear operators (Russian). *Functional analysis*, 29 (Russian), 141-152, Ulyanovsk. Gos. Ped. Inst., Ulyanovsk, 1989.

SINGER, I.

[1] *Best Approximation in Normed Linear Spaces by Elements of Linear Subspaces*, Springer-Verlag, New York, 1970.

[2] *Bases in Banach Spaces, I*, Springer-Verlag, Berlin, 1970

[3] *Bases in Banach Spaces, II*, Springer-Verlag, New York, 1981.

SMITHSON, R.E.

[1] Multifunctions, *Nieuch Arch. Wisk.*, 20(1972), 31-53.

SORJONEN, P.

[1] On linear relations in an indefinite inner product space, *Ann. Acad. Fennicae Ser.A1*, (1978/1979), 169-192.

STRÖH, A.

[1] *Closed two sided ideals in von Neumann algebra and applications*, Thesis, Univ. Pretoria, 1989.

STRÖH, A. and SWART, J.

[1] Measures of noncompactness of operators in a von Neumann algebra, *Indiana U. Math. J.*, 38(1989), 365-375.

SWARTZ, C.

[1] *An Introduction to Functional Analysis*, Dekker, 1992.

SZAZ, A.

[1] Generalized preseminormed spaces of linear manifolds and bounded linear relations, *Univ. u Novom Sadu Zb. Rad. Prirod. - Mat. Fak. Ser. Mat.*, 14(1984), 49-78.

[2] Projective generation of preseminormed spaces by linear relations, *Studia Sci. Math. Hungar.*, 23(1988), 297-313.

SZAZ, A., and SZAZ, G.

[1] Additive relations, *Publ. Math. Debrecen*, 20(1973), 259-272.

[2] Linear relations, *Publ. Math. Debrecen*, 27(1980), 219-227.

[3] Absolutely linear relations, *AequationesMath.*, 21(1980), 8-15.

[4] Quotient spaces defined by linear relations, *Czechoslovak Math. J.*, 32(1982), 227-232.

[5] Multilinear relations, *Publ. Math. Debrecen*, 31(1984), 163-164.

SZ-NAGY, B.

[1] On the stability of the index of unbounded linear transformations, *Acta Math. Sci. Hungar.*, 3(1952), 49-51.

TACON, D.G.

[1] Generalised semi-Fredholm transformations, *J. Austral. Math. Soc.*, A34(1983), 60-70.

TARAFDAR, E.

[1] A note on the bounded linear operators on the spaces ℓ^p and L_p, *J. Math. Anal. Appl.*, 40(1972), 683-686.

TAYLOR, A.E.

[1] *Introduction to Functional Analysis*, Wiley, New York, 1958.

TAYLOR, A.E., and HALBERG, C.J.A.

[1] General theorems about a bounded linear operator and its conjugate, *J. Reine. Angew. Math.*, 198(1957), 93-111.

TOWBER, J.

[1] Linear relations, *J. Algebra*, 19(1971), 1-20.

TYLLI, H.-O.

[1] On the asymptotic behaviour of some quantities related to semi-Fredholm operators, *J. London Math. Soc.*, 31(1985), 340-348.

[2] Approximation conditions for closed ideals and semi-Fredholm operators, in preparation.

URSESCU, C.

[1] Multifunctions with closed convex graph, *Czech. Math. J.*, 25(1976), 438-441.

VLADIMIRSKII, J.N.

[1] Strictly cosingular operators, *Sov. Math. Dokl.*, 8(1967), 739-740.

[2] Observations on Calkin operators, *Sibirskii Mat. Zhurnal* (Russian), 17(1976), no.5, 963-966. English translation: *Siberian Math. J.*, 17(1977), no.5, 715-717.

WAGNER, D.H.

[1] Survey of measurable selection theorems, *SIAM J. Contr. Opt.*, 15(1977), 859-903.

WEIS, L.

[1] *Über strikt singuläre und strikt cosinguläre Operatoren in Banachräumen*, Dissertation, Bonn 1974.

[2] On perturbations of Fredholm operators in $L_p(\mu)$-spaces, *Proc. Amer. Math. Soc.*, (1977), 287-292.

[3] Perturbation Classes of semi-Fredholm Operators, *Math. Z.*, 178(1981), 429-442.

[4] On the computation of some quantitites in the theory of Fredholm operators, *Supplemento Rend. Circ. Math. Palermo*, Series 2, 5(1984), 109-133.

WEST, G.

[1] Generalised measures of Non-precompactness Relative to von Neumann Algebras, *J. Math. Anal. Appl.*, 194(1995), 845-857.

WEYL, H.

[1] Über beschrankte quadrätische Formen, deren Differenz volstetig ist, *Rend. Circ. Mat. Palermo*, 27(1909), 572-580.

WHITLEY, R.J.

[1] Strictly singular operators and their conjugates, *Trans. Amer. Math. Soc.*, 113(1964), 252-261.

[2] Conull and other matrices which sum, a bounded divergent sequence, *Amer. Math. Monthly*, 74(1967), 798-801,

WILANSKY, A.

[1] *Functional Analysis*, Blaisdell, New York, 1964

[2] *Summability through Functional Analysis*, North Holland, Amsterdam, 1984.

WILHELM, M.

[1] Criteria of openness for linear relations, *Fund. Math.* , 114(1981), 219-228.

WOLF, F.

[1] On the essential spectrum of partial differential boundary problems, *Comm. Pure Appl. Math.*, 12(1959), 211-228.

WROBEL, V.

[1] Strictly cosingular operators in DF-spaces, *Math. Z.*, 174(1980), 281-287.

YAGI, A.

[1] Generation theorem of semigroup for multivalued linear operators, *Osaka J. Math.*, 28(1991), 385-410.

YANG, K.W.

[1] The generalized Fredholm operators, *Trans. Amer. Math. Soc.*, 219(1976), 313-326.

YOOD, B.

[1] Properties of linear transformation preserved under addition of a completely continuous transformation, *Duke Math. J.*, 18(1951), 599- 612.

ZEMANEK, J.

[1] Geometric interpretation of the essential minimum modulus, "Invariant Subspaces and Other Topics" (Timisoara/Heruland, 1981) pp.225-227, in: *Operator Theory: Adv. Appl.*, Vol.6, Birkhäuser, Basel, 1982.

[2] On the Δ-characteristic of M. Schechter. In: *Proceedings of the Internaltional Conference, Leipzig, 1983*. Ed.: Baumgartel et al, Teubner.

[3] Geometric characteristics of semi-Fredholm operators and their asymptotic behaviour, *Studia Math.*, 80(1984), 219-234.

Index of Operational Quantities

Notation Index

Index